Lecture Notes in Computer Science 4725

Commenced Publication in 1973
Founding and Former Series Editors:
Gerhard Goos, Juris Hartmanis, and Jan van Leeuwen

David Hutchison   Randy H. Katz (Eds.)

Self-Organizing Systems

Second International Workshop, IWSOS 2007
The Lake District, UK, September 11-13, 2007
Proceedings

 Springer

Volume Editors

David Hutchison
Lancaster University, Computing Department
InfoLab21, Lancaster LA1 4WA, UK
E-mail: d.hutchison@lancaster.ac.uk

Randy H. Katz
University of California, EECS Department
RADLab, Soda Hall #1776, Berkeley, CA 94720-1776, USA
E-mail: randy@eecs.berkeley.edu

Library of Congress Control Number: 2007934515

CR Subject Classification (1998): C.2.4, C.2, D.4.4, D.2, I.2.11, H.3

LNCS Sublibrary: SL 5 – Computer Communication Networks and
Telecommunications

ISSN   0302-9743
ISBN-10   3-540-74916-0 Springer Berlin Heidelberg New York
ISBN-13   978-3-540-74916-5 Springer Berlin Heidelberg New York

Springer is a part of Springer Science+Business Media

springer.com

© Springer-Verlag Berlin Heidelberg 2007
Printed in Germany

Typesetting: Camera-ready by author, data conversion by Scientific Publishing Services, Chennai, India
Printed on acid-free paper   SPIN: 12122264   06/3180   5 4 3 2 1 0

Preface

The 2nd International Workshop on Self-Organizing Systems (IWSOS 2007) was hosted by Lancaster University and held in the beautiful English Lake District. Lancaster University is fortunate to have on its doorstep some of the finest scenery in the United Kingdom, and research groups often take the opportunity to benefit from this natural advantage. For IWSOS 2007 we chose the quiet North Lakes and specifically the Lodore Falls Hotel, which is situated in the Borrowdale valley, just south of Derwent Water and the town of Keswick.

It was a fitting location for the second workshop in the series on self-organizing systems that began last autumn in Passau, Germany.

Future networked systems will, to some degree, need to be self-organizing. For example, they will be deployed in remote and hostile environments, where manual setup and configuration may be undesirable or impossible. Some networks, such as mobile ad-hoc networks, will be spontaneously deployed, have a dynamic population, and may be short-lived. The time it takes traditional management activities to converge, where people are in the control loop, is unsuitably long for these kinds of network. Furthermore, the potential scale and complexity of future networked systems, including the future Internet, will make some form of self-organization highly desirable and perhaps a necessity. The complexity of these networked systems comes from the heterogeneity of devices, communication technologies and protocols, and the stringent user requirements (e.g., resilience) that they will need to support.

Research into self-organizing networked systems is in its infancy, and there is a multitude of open issues to be addressed. Building on the success of its predecessor, this workshop brought together leading international researchers in a visionary forum for investigating the potential of self-organization and the means to achieve it.

These proceedings, which we present to you here, constitute the technical content of IWSOS 2007. There are 17 full papers, 5 short papers and 2 invited keynote talks.

This year's submission was unusual. We received only 36 papers, but the standard was uniformly high – there was no long tail of unsatisfactory or out-of-scope papers that is sometimes experienced by workshop organizers (whose only benefit is to be able to declare a low acceptance ratio). Each paper submitted to IWSOS 2007 was reviewed by four people drawn from an outstanding Technical Program Committee. We are extremely grateful to the TPC for providing such thorough and thoughtful reviews, which made the task of selecting the successful papers so much easier.

Our technical program consisted of sessions on Ad Hoc Routing for Wireless/Sensor Networks (4 papers), Peer-to-Peer Networking (2 papers), Network Topology and Architecture (3 papers), Adaptive and Self-Organizing Networks (3 papers), Multicast and Mobility Protocols (2 papers), and a Miscellaneous session (3 papers) on other important but difficult-to-classify topics. Finally, we have a short-papers session containing 5 papers on very interesting and highly promising work in progress.

The opening keynote talk will be given by Joseph Hellerstein, formerly of IBM Research, who is now a Principal Architect with Microsoft Corporation at Redmond in the USA. Joe will apply his considerable experience to speaking about the challenges of engineering self-organizing systems.

The second keynote speaker is Ken Calvert, who is Professor of Computer Science at the University of Kentucky and a senior figure in computer networking research. Ken will speak about a current collaborative project that is investigating the infrastructure and self-organizational aspects of a possible postmodern Internet architecture.

As Technical Program Committee chairs, we wish to thank a number of key people whose contributions have made our task an easy one: first, all the authors, whether successful or not, who chose to send us such high quality papers; second, again, the TPC who did such splendid reviewing; and finally the local organizing committee who as with all events take the brunt of the hard work behind the scenes. To all the above we dedicate these proceedings of IWSOS 2007.

July 2007   David Hutchison
                                                                              Randy H. Katz

Organization

Steering Committee

Hermann de Meer	University of Passau, Germany
David Hutchison	Lancaster University, UK
Bernhard Plattner	ETH Zurich, Switzerland
James P.G. Sterbenz	University of Kansas, USA

Technical Program Chairs

David Hutchison	Lancaster University, UK
Randy H. Katz	UC Berkeley, USA

Technical Program Committee

Karl Aberer	EPFL, Lausanne, Switzerland
Ozalp Babaoglu	University of Bologna, Italy
Ernst Biersack	Institute Eurecom, France
Andrew Campbell	Dartmouth College, USA
Georg Carle	University of Tuebingen, Germany
Augusto Casaca	INESC-ID, Lisbon, Portugal
Claudio Casetti	Polytechnic of Turin, Italy
Tarik Cicic	Simula Research Laboratory, Norway
Costas Courcoubetis	AUEB, Athens, Greece
Simon Dobson	University College Dublin, Ireland
Markus Fiedler	Blekinge Institute of Technology, Sweden
Stefan Fischer	University of Luebeck, Germany
Michael Fry	University of Sydney, Australia
Christos Gkantsidis	Microsoft Research, UK
Martin Greiner	Siemens AG at Munich, Germany
Indranil Gupta	University of Illinois at Urbana Champaign, USA
Guenter Haring	University of Vienna, Austria
Gisli Hjalmtysson	University of Reyjkavik, Iceland
Amine Houyou	University of Passau, Germany
Karin A. Hummel	University of Vienna, Austria
Wolfgang Kellerer	DoCoMo Lab Europe, Germany
Anne-Marie Kermarrec	INRIA/IRISA, France
Emre Kiciman	Microsoft Research, USA
Rajesh Krishnan	BBN Technologies, USA
Guy Leduc	University of Liege, Belgium

Baochun Li   University of Toronto, Canada
J.P. Martin-Flatin   NetExpert, Switzerland
Paul Mueller   Kaiserslautern University, Germany
Manish Parashar   Rutgers University, USA
Christian Prehofer   Nokia Research, Finland
Danny Raz   Technion, Israel
Lukas Ruf   In&Out AG, Switzerland
Rolf Stadler   KTH, Stockholm, Sweden
Ralf Steinmetz   Technical University of Darmstadt, Germany
Burkhard Stiller   University of Zurich, Switzerland
John Strassner   Motorola Labs, USA
Zhili Sun   University of Surrey, UK
Kurt Tutschku   Wuerzburg University, Germany
Amin Vahdat   University of California at San Diego, USA
Maarten van Steen   Vrije University Amsterdam, Netherlands
Patrick Wuechner   University of Passau, Germany

Local Organizing Committee

Carol Airey   Lancaster University
Joe Finney   Lancaster University
Johnathan Ishmael   Lancaster University
Andreas Mauthe   Lancaster University
Nicholas Race   Lancaster University
Andrew Scott   Lancaster University
Paul Smith   Lancaster University

Supporting and Sponsoring Organizations

ACM
IEEE
IFIP

Agilent Technologies
Content NoE
EuroFGI NoE
InfoLab21 KBC
Microsoft Research
Telekom Austria
UK EPSRC

Table of Contents

Adaptive and Self-organizing Networks

Multicast and Mobility Protocols

Miscellaneous Topics

Short Papers

Engineering Self-Organizing Systems

Joseph L. Hellerstein

Microsoft Corporation
Redmond, Washington USA

Self-organizing systems (SOS) hold the promise of addressing many challenges in large scale distributed systems, especially in reducing the need for human intervention for configuration, recovery from failures, and performance optimization. While there are many principles for creating SOS such as minimizing dependencies between components and avoiding the use of globally shared state, we lack a systematic methodology. This talk explores how techniques from control theory and game theory might be used in combination to engineer SOS.

D. Hutchison and R.H. Katz (Eds.): IWSOS 2007, LNCS 4725, p. 1, 2007.
© Springer-Verlag Berlin Heidelberg 2007

Infrastructure and Self-organization in Postmodern Internet Architecture

Ken L. Calvert

Laboratory for Advanced Networking
University of Kentucky
Lexington, KY, USA

The Postmodern Internetwork Architecture project, a collaboration among researchers at the Universities of Kentucky, Kansas and Maryland, aims to produce a clean-slate design for a thin (inter)network layer. Design goals for this Internet layer include complete separation of routing from forwarding, avoidance of hierarchical or centrally-assigned identifiers in the forwarding plane, and provision of explicit mechanisms to support policies of the various stakeholders (specifically users and service providers).

The talk will begin with a high-level overview of the PoMo architecture, followed by a description of our routing/forwarding approach, which is based on loose source routing. The focus will be on design considerations that suggest which functions of the system should be assigned to the infrastructure, and which should/can be self-organized.

D. Hutchison and R.H. Katz (Eds.): IWSOS 2007, LNCS 4725, p. 2, 2007.
© Springer-Verlag Berlin Heidelberg 2007

Mercator: Self-organizing Geographic Connectivity Maps for Scalable Ad-Hoc Routing

Luis A. Hernando and Unai Arronategui

I3A, University of Zaragoza
C/María de Luna 1, Ed. Ada Byron, 50018. Zaragoza, Spain
lahernan@unizar.es, unai@unizar.es

Abstract. A fundamental problem of future networks is to get fully self-organized routing protocols with good scalability properties that produce good paths in a wide range of network densities. Current approaches, geographic routing and table based routing, fail to provide very good scalability with good paths in sparse networks. We propose a method based on the discovery of connectivity between geographic regions that are self-organized in a multilevel hierarchy. The Mercator protocol builds lightweight connectivity maps in a fully decentralized manner and shows a scalable and resilient behaviour. Each node builds and maintains its own hierarchical map that summarizes connectivity information of all the network around itself using geographic regions. Link state routing is used over the multilevel connectivity graph of the map to obtain global paths. The analysis and simulation of our approach show that routing state and communication overhead grows logarithmically with network size while producing good paths.

Keywords: Self-Organizing Hierarchical Protocol, Geographic Connectivity Maps, Link State, Scalable Ad-hoc Routing.

1 Introduction

A fundamental problem of future ad-hoc networks is to get self-organized routing protocols with good scalability properties and good enough paths. On the one hand, best scalability properties are obtained with geographic routing [9,3,10] where in the best case, there is no communication overhead, location service aside, and routing state is only related to one hop nodes. But, this kind of protocols gives efficient paths only with regular dense networks. On the other hand, table-based routing achieves best paths in all kind of networks, but some state and communication overhead is always needed, even with best ad-hoc hierarchical protocols [14]. Also, scalability with sufficient resiliency seems another challenge.

We propose a self-organizing method that achieves less state than ad-hoc routing protocols based on tables, and better paths than geographic routing protocols. Also, its assures resiliency. In our method, routing is done through multilevel geographic regions. All nodes have an identifier and an address based on GPS

D. Hutchison and R.H. Katz (Eds.): IWSOS 2007, LNCS 4725, pp. 3–17, 2007.

like absolute geographic coordinates or derived from GPS enabled neighbours [6]. These coordinates are used like node addresses and region identifiers. The Mercator protocol builds connectivity maps, C-Maps, based on geographic regions in a fully decentralized way. In this protocol, each node builds and maintains its own C-Map containing a summary of connectivity information from hierarchical regions around the node. A fisheye approach on the maps gives more detailed information in the proximity of the node. These regions are square tiles obtained from simple geographic operations. Link level connectivity information is processed and folded in region connectivity information with a recursive technique. Link state routing is applied to the graphs that represent connectivity between the regions, at different levels, from the C-Maps. With Mercator, we achieve routing state that only depends on the number of levels in the map hierarchy. Routing communication overhead is severely reduced because low level updates can remain inside a region scope, if enough density is provided. The number of levels of the hierarchy arises logarithmically from node density in regions.

The proposal in this paper takes a conservative approach with size of C-Maps to minimize stored state and traffic loads. Actually, our results in routing state and communication overhead show that even sensor devices, as the MICAz [5] with 128k of memory for programs and 256kb/s of bandwidth, can be candidates to adopt our approach. Of course, energy consumption has to be considered in sensor networks, which is out of the scope of this paper.

However, in networks with more resources, like mesh networks or vehicular networks, the detail of a C-Map can be increased to obtain better routes. A trade-off between route lengths and overhead needs to be evaluated in each network class to define the size of C-Maps.

Also, location services and node mobility can be easily included, integrating scalable proposals like [12]and [13] with our techniques.

The rest of the paper is organized as follows. Section 2 shows a review of existing and related work. In section 3, the connectivity maps are described. Section 4 presents Mercator, the maps construction protocol. Section 5 presents the routing protocol. Section 6 offers theoretical information and experimental results about costs induced by the protocol. And, finally, section 7 offers the conclusions of this work.

2   Related Work

Two main approaches have been adopted in order to achieve good scalability properties in ad-hoc networks. On the one hand, hierarchical routing protocols divide the routing problem in different spatial or temporal levels. On the other hand, geographic routing protocols use the physical network coordinates of nodes to perform position based forwarding.

Implicit hierarchy is observed in flat protocols like Fisheye State Routing (FSR) [15] and Hazy Sighted Link State (HSLS) [16]. FSR provides accurate path information from the immediate neighbourhood of a node and imprecise knowledge of paths to distant destinations. Although, this imprecision is solved with the route being more accurate as the packet draws closer to the destination. We have been inspired by this approach to limit the communication overhead while maintaining robustness. Both protocols achieve low communication overheads by selectively adjusting frequencies of routing updates. However, classical heavy O(N) routing state is needed in both protocols which could be a fundamental limitation in scalability.

Hierarchical State Routing (HSR) [14] is a multilevel clustering-based link state routing protocol. A clustering scheme recursively defines a logical hierarchical topology. Communication and storage complexity are the best, H*K, growing linearly with hierarchy levels (H) and average one hop neighbours of a node (K). But, resiliency problems can arise as there is only one clusterhead per cluster to link the hierarchy. Also, cluster needs to be assigned a hierarchical address to aggregate in the node address format. ARCH [1] decrease the cost of the organization of clusterheads avoiding the rippling effect. But, one clusterhead per cluster is still a bottleneck.

In Zone-Based Hierarchical Link State (ZHLS) [8], a GPS-based method is used to define scope of the regions without clusterheads. A peer-to-peer technique reduces traffic bottleneck, avoids single point failures, and makes mobility management easier. There are only two levels in the hierarchy. We have generalized this approach with a multilevel hierarchy and we have reduced the routing state applying a fisheye technique. The GAF [17] protocol also uses a GPS partitioning scheme of the network in square grids but it's focused exclusively on energy conservation and can run over any ad-hoc routing protocol.

Although geographic routing is simple and light, it has important routing problems with sparse networks where voids can block position based forwarding. This dead ends are avoided with recovery algorithms. GPSR [9] uses a right hand rule algorithm over a planarized graph to get around the void. However, voids with complex and big shapes can lead to very inefficient paths. So far, GDSTR [11] is the best recovery solution that improves path lengths. It is based on a spanning tree built on convex hulls. But, it also applies the recovery algorithm after getting in a dead end. Thus, it is sensitive to voids and concavities in the network in order to get shorter routes.

Terminode routing [2] and GeoLANMAR [18] are hybrid protocols with two levels that use geographic routing for distant destinations and link state in local scope. GeoLANMAR uses classical LANMAR methods to manage logical groups in group mobility applications. Terminode routing has a method called Geographical Map-Based Path Discovery for remote routing. This method only operates to get anchored paths at the same level and it is assumed that a density map is already available somewhere outside the network. Besides, no distribution technique is presented for the maps. This trend is developed with the use of geographic maps [4], from vehicular navigation systems, in geographic routing. However, good node density is assumed to assure connectivity.

Our goal has been to obtain a routing method with better routes than light state geographic routing protocols and smaller routing state than hierarchical routing protocols with managed communication overhead. All this is done assuring resiliency.

3   Connectivity Maps

Connectivity information between different areas in the network is stored into a map, the C-Map. In this approach, the network is modelled as a hierarchy of square areas defined on a 2D euclidean space where nodes are located. Each node is able to obtain its localization coordinates by means of an external positioning system like GPS or other localization techniques.

A C-Map stores information about a hierarchy of nested square regions surrounding the owner node's location. The highest level of the map, with the biggest squares, determines the area covered by the whole map. Lower levels keep information about connectivity of small areas near the node, while higher levels keep *summarized* connectivity information about large and faraway areas.

Hierarchy of Network Regions. Information stored in a C-Map is organized into M hierarchical *Map Levels*, starting from 0 (the link level), to $M - 1$. Each map level is composed of Q non-overlapping squares, known as *tiles*, properly sized to fit four tiles of level m into one tile of level $m + 1$. In addition, C-Maps are centred around their owner node, which means that all tiles of each level are placed around the node. The purposes of centring maps are:

- To balance the amount information stored in all directions around the node.
- To enable the propagation of information, by means of the addition of C-Maps, as explained later.
- To make possible the summarization of information from lower levels into higher ones, using a composition operator. The final goal is to ensure the scalability of the protocol by cutting down the amount of information stored into each C-Map.

C-Maps are constructed and centred following the next two **rules**:

- Map level m must be completely nested into map level $m+1$, so $Q = 4^n, n \in \mathbb{N}$.
- Into each Map level, the tile containing the owner node, known as *central tile*, must be completely surrounded by other tiles.

In the end, the node owner of the C-Map will be placed into one of the 4 tiles in the middle of each level. Whenever the node moves from one tile to another, its map has to be re-centred by repositioning the tiles at all levels as needed to meet the previous rules.

Connectivity Between Tiles. Connectivity is defined as the possibility for a message coming from a tile to reach a contiguous one across their common border. For each tile in the C-Map, the connectivity information with its surrounding tiles is stored, and as a result, any node having the connectivity information about all the tiles of each map level is able to quickly find out if a region of the network is reachable across a path of connected tiles.

Connectivity Inside Tiles: Fragments. At first, a tile identified as reachable through all of its borders, would seem like an ideal tile, enabling communications to traverse the tile completely, but it is not always true: nodes inside a tile might be divided into different unconnected groups known as *fragments*. This important issue implies that a fragmented tile is not always traversable from one border to other.

A fragment is defined as a connected group of nodes inside a tile and is always traversable.

4   The Mercator Protocol: C-Map Construction

In this paper we propose **Mercator**, a fully distributed protocol that builds, distributes and maintains the C-Maps for every node in the network. All the information the protocol produces is inferred from basic status of connectivity between nodes at link level. First, connectivity between the smallest tiles is discovered with an interchange of HELLO messages. Then, that information is distributed to the immediate neighbour areas, giving nodes some knowledge about its vicinity and the ability to summarize the information they know. In subsequent iterations of this process, basic and summarized connectivity information will be shared using MAP messages, extending the ability of nodes to build information at higher levels.

After some time, every node will be provided with a map of its neighbourhood composed of levels at different scales. Later updates of the map will be required to address the topology changes produced by joining or exiting nodes or even entire network areas.

4.1   Previous Considerations

A square of level m, S_m, is identified by a tuple $<m, (i,j)>$ where (i,j) are S_m's grid coordinates using columns and rows. S_m's side length, L_m, is calculated as $L_m = L_0 * 2^m$, being L_0 a parameter of the network. Given a point P placed inside S_m, (i,j) can be calculated as the integer division of P's coordinates by L_m.

4.2   Mercator Protocol's Information: The C-Map

Information produced by **Mercator** is stored into a C-Map at each node. C-Maps are composed of a fixed number of Map levels, M, each one filled with Q tiles arranged in a square matrix layout following the basic centring rules described in Sec. 3. The minimum Q value that meets the requirements explained in Sec. 3 is $Q = 16$ which is adopted by default. Q values lower than 16, do not guarantee that the central tile at any level is completely surrounded by other tiles (i.e $Q = 4$) or do not guarantee the statement $Q = 4^n, n \in \mathbb{N}$, which is a requirement for the addition and the composition operators.

C-Maps are sized in order to cover the whole space occupied by the network, thus, the number of levels stored into the Map, M, is a well known parameter dependent of the maximum diameter of the network. The 4 tiles in the middle of the highest Map Level ($M-1$) should cover the entire network. The coverage area of a C-Map, A_{Cmap}, can be calculated as follows:

$$A_{Cmap} = D_{Cmap}^2 = 2 * L_{M-1}^2 = (2 * L_0 * 2^{M-1})^2 \qquad (1)$$

Where $D_{Cmap} =$ is the guaranteed maximum diameter of the network covered by the highest Map Level.

Table 1 shows the coverage areas and maximum network diameters achieved by a C-Map with M levels in a default setup, being $Q = 16$ and $L_0 = 500$ metres. Both A_{Cmap} and D_{Cmap} increase exponentially with M.

Table 1. Maximum network diameter and coverage area achieved by a C-Map with M map levels in a worst case scenario

M :	4	5	6	7	8	9	10	11
$D_{Cmap}(Km)$:	8	16	32	64	128	256	512	1,024
$A_{Cmap}(Km^2)$:	64	256	1,024	4,096	16,384	65,536	262,144	1,048,576

As explained before, nodes inside a tile can be divided into fragments. Information stored into the C-Map about each tile comprehends the identifiers of the different fragments that the tile is divided into as well as their connectivity status with neighbour fragments and some expiration timers needed for the management of the map.

4.3   Calculation of Fragment's Identifier

Fragments are separated groups of interconnected nodes within the limits of a tile, and thus, each fragment can only be connected with other ones inside adjacent tiles. The identifier of a fragment reflects its connectivity properties: borders of every tile are associated with *cardinal directions* and, in addition, identifiers of fragments are the aggregation of those cardinal directions of the borders traversed by links with neighbour fragments. If two fragments inside a tile have similar connectivity properties their identifiers will be similar, but they will occupy different tiles at lower levels which will differentiate them. If two fragments have similar identifier at level 0, the geographic coordinates of any node which belongs to them will make the difference, since a node only belongs to one fragment per level.

4.4   Discovery of Level 0 Information

Once the C-Map is centred, HELLO messages are broadcasted[1] periodically by **Mercator** protocol to gather information about the fragment the node belongs

---

[1] Local Broadcast. Broadcasted messages are only received by nodes in range and not forwarded again by any of them.

to and its surroundings at the lowest level (level 0). HELLO messages are used also to maintain a fresh list of neighbour nodes.

Each HELLO message carries information about the sender, its coordinates and level 0 fragment identifier, in addition, it also contains the identifiers of the tile's borders through which link to neighbour fragments are established.

Upon the reception of a HELLO message, each node will update its state: if HELLO messages are received from a node in a neighbour tile, the shared border among both tiles is considered traversed by a link and the sender's fragment will be considered connected to the receiver's fragment. By contrast, if a HELLO message is received from a node which is located in the same tile, then sender and receiver nodes belong to the same fragment and the receiver node will check the HELLO message looking for information about the links from its own fragment to other ones.

The updated information will be broadcasted in subsequent HELLO messages and after some stabilization time, each node will be able to compute its fragment's identifier based on the tile's borders that seem active.

4.5   C-Map Addition Operation

Connectivity information is shared between nodes when they exchange their C-Maps. C-Map construction and centring rules guarantee that the C-Maps of two nodes placed into two adjacent tiles, must be partially or totally overlapped depending on the tile they are placed at each level (See Fig. 1). A node acquires information from its neighbours' C-Maps by applying the addition operator defined as follows:

Let C_1, C_2, C_r be C-Maps,

$$C_1 + C_2 \rightarrow C_r$$

where C_r is the result of the addition $C_1 + C_2$. C-Map addition operator works as follows: first, C_r is centred at the same point as C_1, then addition operator will select tiles from C_1 not present in C_2 and will copy its information (fragments) into C_r. Then it will select those tiles present in both maps and will add to C_r all fragments from C_1 as well as those from C_2. Finally, if central tiles of corresponding levels in C_1 and C_2 are neighbours and thus share a border, C_2 owner's fragment is marked as connected with C_1 owner's fragment into C_r.

4.6   C-Map Information Exchange

After some stabilization time, each node is able to provide information about its fragment of level 0 and its connections to other neighbour fragments and will start broadcasting[2] MAP messages to its immediate neighbours, sharing information about its local and surrounding tiles. Initially, all 16 tiles of each level will be empty except the central tile of level 0, which will include the owner's node fragment.

---

[2] Local Broadcast, as in HELLO messages broadcasting.

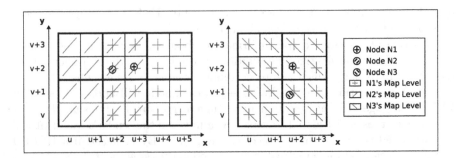

Fig. 1. Partially overlapped Map Levels in the picture on the left: the two central columns are overlapped. Fully overlapped Map Levels in the picture on the right. Note that a node is always placed inside one of the four tiles in the middle of its map level.

Each MAP message carries the C-Map stored by the node at the sending time and its owner node's coordinates.

Upon the reception of a MAP message, the receiver node will add the received C-Map to its own, acquiring the new information (as explained in 4.5). In subsequent sendings, the new information will be spread producing a propagation effect.

4.7   Higher Level Fragment Information Composition

Each node is in charge of the fragments it belongs to at each level and is able to calculate their identifiers and keep them stored into its C-Map. The calculation of identifiers of fragments with information about lower levels is known as composition. As result, information about fragments of 4 tiles at level $m - 1$ is summarized into one tile at level m. Internal connectivity details are hidden and external ones summarized.

Each node is able to compute the identifier of the fragment it belongs to at level m because it keeps stored into its C-Map the fragments of level $m - 1$ connected to its own. They all conform the fragment of level m. This is possible because levels in a C-Map are centred at the owner node's location (see 4.2). The composition algorithm works as follows: first, it looks for all fragments in the 4 tiles of level $m - 1$ which are connected (directly or by intermediate fragments) to the node's fragment at level $m - 1$ inside the tile of level m. Then, all those connected fragments will conform the node's fragment at level m and its identifier is inferred from their connectivity properties. See Fig. 2.

4.8   A Global View of Built C-Maps

The behaviour of the **Mercator** protocol on a network of about 500 nodes is shown in Fig. 3. It can be observed how connectivity information reflects the shape of the network including void areas and concavities, and how a summarized network topology is constructed for each level.

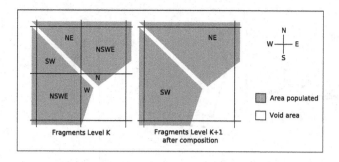

Fig. 2. Example of fragment composition

5   Routing with C-Maps

Having a C-Map, traditional ad-hoc routing algorithms can take advantage of the available connectivity information at different levels. Implicit avoidance of big concavities and empty areas, which is a big trouble in geographic routing, can be easily achieved by routing with the aid of high level information.

In this section we propose a routing technique inspired in traditional routing algorithms over a network graph. Information from the C-Map is used to construct a graph representing the fragments at all levels in the map and their interconnections.

Two differentiated protocols are involved in the routing process. The first one, known as *Fragment Routing*, acts as a high level path planner, which analyses the previously constructed network graph, looking for the best path through connected fragments of any level toward a target fragment of level 0. The second one, is a classical ad-hoc routing algorithm deployed inside every fragment of level 0 in the network, which is able to deliver a message to a particular node within the same fragment as well as to forward the packet to a node connected with a neighbour fragment.

In Fig. 3, a message is sent from *source node* toward *destination node*. Different snapshots have been taken at different steps: just before the message is sent by source node and by the time it passed through two intermediate nodes. It can be observed how the selected best path is progressively more accurate as the message gets closer to destination.

Mercator Node Addressing. Node addressing in the ad-hoc network is designed to provide fragment information available at all levels. Mercator node addresses contain the addressee's geographic coordinates and the identifier of all the fragments it belongs to (one per level). Since coordinates of tiles can be calculated from the coordinates of a point inside it, (see Sec. 4.1), localisations of fragments referred in the address are well known.

A *Mercator address* is in the form: $<(x, y), f_0, f_1, \ldots, f_{M-1}>$, where x, y are the addressee's coordinates and f_k are the identifiers of the fragments it belongs

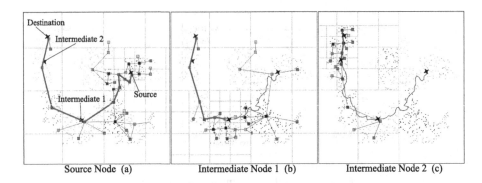

Source Node (a)   Intermediate Node 1 (b)   Intermediate Node 2 (c)

Fig. 3. A message is sent from *source node* to *destination node*. Snapshots at different moments are taken showing the network connectivity graphs up to level 2, which are known by Source Node in Fig. 3(a), by Intermediate Node 1 in Fig. 3(b), and by Intermediate Node 2 in Fig. 3(c). The planned best path in each snapshot, from current node toward destination, is displayed using the thickest grey lines. The winding line represents the exact link level path already followed by the message.

to at each level k. The size of a Mercator address, $W_{address}$, can be calculated as follows:

$$W_{address} = 2 * \left\lceil \log_2(L_0 * \sqrt{Q} * 2^{M-1}) \right\rceil + M * W_{id} \qquad (2)$$

Using 1 metre precision in 2D integer coordinates (x, y), and where W_{id} is the size of a fragment's identifier (4 bits). Table 2 shows the address length required for different network diameters in a default setup, with $Q = 16$ and $L_0 = 500$ metres.

Table 2. Network diameter covered by networks with M levels and the corresponding Mercator address length

M :	6	9	12	15	17
$D_{Cmap}(Km)$:	32	256	2048	16384	65536
$W_{address}(bits)$:	54	72	90	108	120

The Network Graph. Information stored into a node's C-Map is enough to calculate a network graph with the summarized network topology information. Each fragment is represented by a vertex. A link between two fragments is represented by a link between their corresponding vertexes while the weight of each link represents the approximated cost of routing a message from a fragment to the other and is set to the distance between the central points of the two tiles where the fragments are placed.

It is remarkable that the graph represent fragments (not nodes) which belong to different levels. The topology reflected in the graph is more detailed in areas closer to the node due to the particular construction of the C-Map. Since the number of tiles that a map stores is fixed and the number of fragments stored per tile is bounded, the cost of processing a graph is upper bounded.

Fragment Routing. This routing algorithm, which is run on every node, computes the best path over the network graph, that starting from the node's fragment at level 0, reaches the destination node's fragment at the lowest possible level. The next hop in the path is then selected and the message is forwarded toward it. A routing table at each node is periodically constructed, with entries for all available fragments from node's C-Map. The routing table is calculated applying Dijkstra's Algorithm to the network graph. Every possible destination fragment in the graph is associated with a next hop, the beginning of the shortest path to that destination, which is, a fragment of level 0 and a neighbor of the node's fragment.

At the time of routing a message, the protocol will use the Mercator address of the message's destination, to choose the best next hop fragment. A search is performed in the routing table, looking for a fragment that matches in coordinates and identifier to any of the included in the address. If more than one match is found (at different levels), the one with lowest cost is chosen.

Ad-Hoc Routing Inside Fragments of Level 0. A traditional ad-hoc routing protocol is deployed inside each fragment of level 0 with two objectives. First, once *Fragment Routing* has selected the next fragment-hop toward destination, this ad-hoc routing protocol takes charge of the message and forwards it to the nearest node within the fragment which is directly connected to the next hop fragment. And second, when a message finally arrives to the fragment of level 0 which contains the destination node, the protocol is in charge of the delivery of the message to its final destination.

Understanding How Routing with C-Maps Works. C-Maps are centred, which means that only a part of the network is mapped at one level, so, if a level does not cover a destination target, a higher map level has to be used because it contains information less accurate but broader.

As a message is being routed and approaches its destination, C-Map information into intermediate routers gets progressively more accurate toward the destination target and the path can be slightly modified and optimized.

6   Costs Analysis and Experimental Results

The **Mercator** protocol has been designed to achieve high scalability properties with target networks from a few to million nodes. The design has been focused on keeping conservative storage requirements per node and shared channel bandwidth usage.

C-Map Storage Complexity. Each node in the network stores only its own C-Map which is composed of M Map Levels. Each Map Level stores $Q = 16$

tiles which might be divided into different fragments. Storage cost of a C-Map, S_{Cmap} can be calculated as follows:

$$S_{Cmap} = M * Q * F * f \qquad (3)$$

Where F is the average number of fragments per tile and f is the storage cost of a fragment (fixed). Although F is not a fixed value, our simulations show that only in very specific cases it grows over 2.0. Thus, storage complexity is $O(M)$ and grows logarithmically with network size.

Mercator Bandwidth Usage. All **Mercator** information exchange is performed by means of local broadcast operations. Once the sender node transmits a message, it is not forwarded again by any of the receiver nodes.

Connectivity information discovery and distribution is done by means of periodic HELLO and MAP messages broadcastings. HELLO and MAP messages sending rate is calculated in order to use a low percent of the available channel bandwidth. In the simplest scenario, without using any kind of optimization like data compression or variable sending rates, bandwidth required by *Mercator* per node, M_{Bw}, can be calculated as follows:

$$M_{Bw} = R_{HELLO} * S_{HELLO} + R_{MAP} * S_{MAP} \qquad (4)$$

Where R_{HELLO} and R_{MAP} are the broadcast rates for HELLO and MAP messages and S_{HELLO} and S_{MAP} are the average size of HELLO and MAP messages. All parameters but S_{MAP} are constant, S_{MAP} complexity is $O(M)$ because each MAP message contains a C-Map. Thus M_{Bw} complexity is $O(M)$.

6.1 Experimental Results

We have implemented an ad-hoc network simulator to test the **Mercator** protocol under different scenarios. The following settings have been used in our tests: radio range is 200 metres. All nodes receive broadcasted messages within their radio range. R_{HELLO} is set to 1 message per second. R_{MAP} is set to 1 message each 3 seconds. Expiration ages are set to 4 *ticks* for each item. Tick duration is $1/R_{MAP} * 2^n$ seconds where n is the level number of the item monitored by the timer. Side's length of a tile at level 0, L_0, is closely related to expiration ages and radio range. L_0 value has been set to 500 metres, which allows information to traverse an entire tile of level 0 in less than 4 hops without expiring.

Storage and Bandwidth Requirements. Using the above parameters, a C-Map with 9 levels (area of target networks up to 65,000 square kilometres and average network population of 1300 million nodes using a density of 1 node per 200 m^2) requires 5.3KBytes of storage (according to Equation (3)). In this scenario, the bandwidth usage per node is 14kbps (according to Equation (4)).

Stabilization Time. In a first test we have measured the map's stabilization time, which provides an idea of the response speed that the protocol offers against network topology changes. Stabilization time is measured by the number of MAP messages that each node sends until the entire network map is completely built. We have simulated a cold startup in a network with 7000 nodes under different population density conditions (from 4 to 9 average neighbour count). As shown in Fig. 4, stabilization time increases linearly with the extension of the mapped area. High density conditions help to increase information propagation speed and to reduce stabilization time.

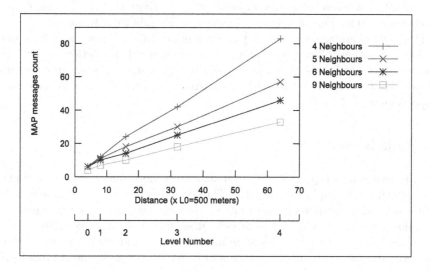

Fig. 4. Stabilization time after a cold startup

We have also simulated dynamic scenarios with nodes joining and going away from the network. If a new node does not produce a substantial change into the network topology, stabilization time is about 1 message, by contrast, if a new link is created between two fragments, the stabilization time is similar to the case of a cold startup of the level of those fragments.

6.2   Observations

Storage required by Mercator protocol grows logarithmically with the network area, maintaining a very reduced map with the network connectivity status. A C-map fits easily into today's sensors memory [5] even for very large target networks with thousand kilometres of diameter.

The storage cost reduction is made possible by the hierarchical approach of the map. Our proposal gives a generalized algorithm for M levels so it can be applied to target networks of very different natures.

A consequence of the low storage cost is that complete replication is feasible. There are no central nodes in the network: every node in the network keeps a map, which improves network resilience, avoids network congestion around critical points and helps with adaptive deployment: a node joining to a previously established network needs only one map from its immediate neighbour to get all the information it needs to become communicated.

At the highest levels, the fact that connectivity status information is summarized and the big size of network regions compared to link level radio range, turn the probabilities that a change at link level produce a change in the connectivity information of high levels very low. Information summarization helps to keep hidden into lower levels the vast majority of connectivity changes, leaving higher levels very stable. This fact greatly simplify node mobility management.

Destination addresses of moving nodes must change as they move from one region to another. Since nodes move with a low speed in relation to network diameter, in most cases, movements will produce a minor change into the address. This fact can be exploited by a distributed location service for efficiency improvement.

7   Conclusions

In this paper we propose a new approach in ad-hoc routing based on nodes with absolute geographic coordinates. The Mercator protocol uses these coordinates to build a multilevel connectivity C-Map summarizing network connectivity information with a fisheye approach. Resiliency is guaranteed with C-Map computation and distribution done in all nodes. It can nicely marry with existing mobility solutions. Routing state and communication overhead grow logarithmically with network size. Better paths than light state geographic routing can be obtained and smaller routing state than hierarchical routing protocols is provided.

Future work will explore the extension of C-Maps with information about mobility and congestion.

Acknowledgments. This work has been supported by the Spanish CICYT DPI2006-15390 project and the GISED, group of excellence recognised by the Diputación General de Aragón.

References

1. Belding-Royer, E.M.: Multi-level Hierarchies for Scalable Ad hoc Routing. Wireless Networks (WINET) 9(5), 461–478 (2003)
2. Blazevic, L., Le Boudec, J., Giordano, S.: A Location Based Routing Method for Mobile Ad Hoc Networks. IEEE Transactions on Mobile Computing 4(2), 97–110 (2005)
3. Bose, P., Morin, P., Stojmenovic, I., Urrutia, J.: Routing with guaranteed delivery in ad hoc wireless networks. Wireless Networks 7(6), 609–616 (2001)

4. Cheng, A.M.K., Rajan, K.: A digital map/GPS based routing and addressing scheme for wireless ad-hoc networks. In: Proceedings of IEEE Intelligent Vehicles Symposium, pp. 17–20. IEEE Computer Society Press, Los Alamitos (2003)
5. Crossbow Technology, Inc.: MICAz wireless modules (2007),
   http://www.xbow.com/Products/productdetails.aspx?sid=164
6. He, T., Huang, C., Blum, B.M., Stankovic, J.A., Abdelzaher, T.F.: Range-Free Localization Schemes in Large-Scale Sensor Networks. In: Proc. of the 9th Intl. Conference on Mobile Computing and Networking (MOBICOM), pp. 81–95 (2003)
7. Hong, X., Xu, K., Gerla, M.: Scalable Routing Protocols for Mobile Ad Hoc Networks. IEEE Network 16(4), 11–21 (2002)
8. Joa-Ng, M., Lu, I.-T.: A Peer-to-Peer zone-based two-level link state routing for mobile Ad Hoc Networks. IEEE Journal on Selected Areas in Communication 17(8), 1415–1425 (1999)
9. Karp, B., Kung, H.T.: GPSR: Greedy Perimeter Stateless Routing for Wireless Networks. In: Proceedings of the 6th ACM International on Mobile Computing and Networking (MobiCom 00), pp. 243–254. ACM Press, New York (2000)
10. Kuhn, F., Wattenhofer, R., Zhang, Y., Zollinger, A.: Geometric ad-hoc routing: Of theory and practice. In: Proceedings of PODC, pp. 63–72 (2003)
11. Leong, B., Liskov, B., Morris, R.: Geographic Routing without Planarization. In: Proceedings of the 3rd Symposium on Network Systems Design and Implementation (NSDI 2006) (2006)
12. Li, J., Jannotti, J., De Couto, D.S.J., Karger, D.R., Morris, R.: A Scalable Location Service for Geographic Ad Hoc Routing. In: Proceedings of the 6th Int. Conf. on Mobile Computing and Networking (MobiCom'00), pp. 120–130 (2000)
13. Li, M., Lee, W.-C., Sivasubramaniam, A.: Efficient peer-to-peer information sharing over mobile ad hoc networks. In: Proceedings of the 2nd Workshop on Emerging Applications for Wireless and Mobile Access (MobEA 2004) (2004)
14. Pei, G., Gerla, M., Hong, X., Chiang, C.C.: A Wireless Hierarchical Routing Protocol with Group Mobility. In: Proceedings of IEEE Wireless Communications and Networking Conference (WCNC'99), pp. 1538–1542. IEEE Computer Society Press, Los Alamitos (1999)
15. Pei, G., Gerla, M., Chen, T.-W.: Fisheye State Routing: A Routing Scheme for Ad Hoc Wireless Networks. Proceedings of IEEE International Conference on Communications 1, 70–74 (2000)
16. Santivanez, C., Ramanathan, S., Stavrakakis, I.: Making Link State Routing Scale for Ad Hoc Networks. In: Proceedings of MobiHOC'2001, pp. 22–32 (2001)
17. Xu, Y., Heidemann, J., Estrin, D.: Geography-informed Energy Conservation for Ad Hoc Routing. In: Proceedings of Int. Conf. on Mobile Computing and Networking (MobiCom'2001), pp. 70–84 (2001)
18. Zhou, B., de Rango, F., Gerla, M., Marano, S.: GeoLANMAR: Geo Assisted Landmark Routing for Scalable, Group Motion Wireless Ad Hoc Networks. Proceeding of the IEEE 61st Semiannual Vehicular Technology Conference (VTC2005-Spring) 4, 2420–2424 (2005)

A New Approach to Adaptive Multi-routing Protocol for Mobile Ad Hoc Network

Ung Heo, Deepak G.C., and Jaeho Choi

Division of Electronic & Information Engineering
Chonbuk National University
Jeonju, Jeonbuk, Republic of Korea
wave@chonbuk.ac.kr

Abstract. A routing protocol is designed considering a particular environment all the time which is not possible in the case of practical ad hoc networks. Because of the uncertainty in topological rate of change, mobility model, and terrain condition, the performance is severely degraded. So the concept of assigning single routing protocol does not address the problem of most modern day mobile networks. Instead, the feedback-based routing protocol which is highly adaptable in changing environment is more suitable and this concept has been proposed in this paper. The mathematical modeling of feedback parameters have been designed and analyzed for the highly unstable networks. Some of the parameters we measure here are network connection ratio, end-to-end connectivity, packet delivery ratio, and number of nodes. Those are functionally related with the routing parameter.

Keywords: MANET, Autonomy, Adaptability, Auto-configuration, Routing.

1 Introduction

Mobile Ad Hoc networks are such networks in which autonomous sets of mobile nodes are dynamically connected via wireless links without using an infrastructure network. Due to the dynamic nature of Ad Hoc networks, the allocated resources in priori are not matched with the requirement and the method of communication between them cannot be fixed in priori. Because the mobility model of nodes is random way point, the mobility function defined in terms of time cannot be formulated exactly. In such a flexible network, there must be some flexibility to choose a suitable routing protocol from a *protocol box* which is defined later. Therefore, an adaptive multi-protocol system should be developed, which transits into appropriate mode of protocol. However, the term multi-mode is different from the multi-protocol. In the former, the network is segmented and each segment has different routing policy and duty; the second one is what we have proposed here. We want to make clearer that multi-protocol does not like "Once Protocol 1, always Protocol 1" policy in a sub-network. Instead, in our proposed method, the protocol is a variable factor throughout the network.

It is clear from a circuit theory that the feedback system is more stable because it is able to adapt in any changing environment. Our proposed scheme will use the same

D. Hutchison and R.H. Katz (Eds.): IWSOS 2007, LNCS 4725, pp. 18–29, 2007.

concept on protocol design so that it can perform well in various conditions without performance degradation. Since all routing protocol can function well only under particular environment, they are categorized into different classes and sub-classes. One can refer [9] for a brief classification of routing protocols. We will discuss some of them here in short which we are going to implement. In [1], Multi-Mode routing protocol has been described considering two algorithms namely Limited Link States (LLS) and Self Organized (SO). This paper has proposed a reference area concept as the closer to the destination a node is, the more information related to that destination it will have. But, it does not consider what happens when the destinations move out of the reference area. The packet move here and there if we do not change the routing protocol, instantly.

The collaborative management of MANET, which calculates a capability function [3] as an optimizing factor, is a self-configuring strategy. The service discovery as a multi-protocol framework [2] has designed a common architecture for an individual discovery protocol to enhance configurability and re-configurability of the network, which is just a core framework and does not deal with the decision support parameters. The ref. [5] has described adaptable ad hoc routing experience as LID and Pattern Extraction mechanism, which is merely an extension of [1], but has described views of an adaptable architecture as structure learning/engaging modules. The Terminode Project [6] basically operates on a self-organized mobile Ad Hoc Network and even explores interlayer interactions. Here, the Self-Organized networks are defined as a network run solely by the end users. The AutoCom principle [7] addresses some requirements of self-organized networks as well as interoperability problems due to merge and split problems. It has proposed a heterogeneous routing protocol as a solution. The combination of Stateless Configuration Initialization (SCI) and Configuration Conflict Detection and Resolution (CCDR) is proposed in [10]. Here, it is assumed that some nodes may work on two routing protocols if both routing protocols are operated on neighbors, simultaneously. It is not significant way of auto-configuration because of the high cost of hardware and software to maintain a duality.

The remainder of the paper is organized as follows. Section 2 presents routing scenarios for Ad Hoc Networks. Section 3 discusses our proposed scheme which is further divided into three sub-sections: namely, adaptable module; configuration parameters; and configuration beacon. Section 4 describes an analysis of the overall modules. Section 5 discusses some of implementation issues. The paper draws conclusions in Section 6 with discussions on some future works.

2 Routing Scenario

For a pair of ad hoc network nodes, the communication will occur between them over a period of time until the session is finished; or one of the nodes moves away; or the battery backup power diminishes. An efficient routing protocol must support load balancing of traffic. So, each node needs some knowledge of network topology beyond the local neighbors. This concept attempts to collect and process that knowledge efficiently. Most of routing protocols assume that the nodes have homogenous resources and capabilities. The bidirectional links are often assumed.

No single protocol works well in all environments. Some attempts are made to develop adaptive/hybrid protocols. The proactive protocols are based on periodic updates, which involve a high routing overhead. In the reactive protocols, the source initiates route discovery and determines route time-to-time basis. The hybrid is a combination of reactive and proactive. Even though it is adaptive, it does not change protocol when the network is working. The curious readers can refer [9] and [11] for further details and comparison. Algorithms that are computationally complex however require significant processing cycles. So, we have chosen well studied and easy to implement routing protocols in our adaptive modules. These protocols are *DSDV, ZRP, TORA,* and *AODV,* and they are numbered *Protocols 1, 2, 3,* and *4,* respectively. All routing protocols are invalid when the network connection ratio (NCR) is smaller than unity. Because of this condition, no protocol is implemented for NCR < 1. Some properties concerning each routing protocol are given below:

- **Destination Sequence Distance Vector (DSDV):** This routing is needs periodic update transmissions and guarantees loop free paths. The latency of route discovery is very low because the source uses a ready-made route to a valid destination. For the network with highly dynamic nodes, this protocol should be avoided; otherwise, the bandwidth will be wasted due to excessive control overheads. It is suitable for a network with high bandwidth, low mobility, and lower number of nodes.
- **Ad Hoc On-Demand Distance Vector (AODV):** This routing protocol searches routes to destination when source needs to communicate. It is less secure because of its distributed nature. Though its performance is not satisfactory when nodes are highly mobile, however, it is better than DSDV. Moreover, it is more scalable because the control overhead is lower by just keeping information about the destination.
- **Zone Routing Protocol (ZRP):** This routing is a scalable and highly efficient method. Its performance comes between reactive and proactive protocols.
- **Temporally Ordered Routing Protocol (TORA):** TORA has advantage to support multiple routes. It performs in a dynamic mobile networking environment. Its NCR shown in Fig. 3 comes near to that of the AODV. It has loop free, distributed, and on-demand properties. It allows a route to be created and maintained proactively for some destinations while reactively for others. This protocol needs synchronized clock and requires an extra device such as the GPS.

3 Proposed Method

The auto-configuration strategy allows the infrastructureless networks to react when there is a change in network parameters and conditions by appropriately selecting and replacing the current running protocols. All nodes should have equal innate authority to make a decision in a fully distributed manner. The self stability algorithm should start with the power on to all the nodes and end with the power off.

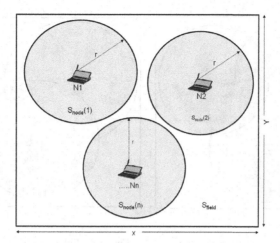

Fig. 1(a). No connectivity topology

Therefore, the protocols should be configured on the fly as it can be dynamically started up and closed down at runtime. The auto-configured routing protocol should adapt itself to the present network conditions considering the traffic level and patterns as well as the mobility patterns of whole network. They are necessary but not sufficient conditions because the larger the number of decision parameters we measure, the better will be the performance. The major question that concerns us the most is what parameters we have to measure. The paper points to that question in the following sub-sections.

3.1 Configuration Parameters

Consider the network connection ratio defined in terms of the total area covered by nodes. Assume that the coverage area is squarely proportional to the transmission range, which is equal to a circular area of a size πr^2. Suppose there are N nodes in a network field A_{field} the network coverage ratio is defined as follows:

$$\text{NCR} = R = \frac{\sum_{i=1}^{N} S_{node}(i)}{A_{field}} \tag{1}$$

where $S_{node}(i)$ is the coverage area of node i, and A_{field} is the total area of the network.

The value of R is maintained by varying two areas mentioned in Eq. (1). The transmission range of a station can be increased or decreased to maintain these areas. If so, it becomes a power control routing mechanism, which has been described in large volume of previous research works. However, we follow a different way since adaptive transitions between protocols may not be easy, if not impossible, just by considering the transmission range. Instead, this range is used to develop a new parameter for making a protocol selection criterion. Figures 1(a), 1(b), and 1(c) show

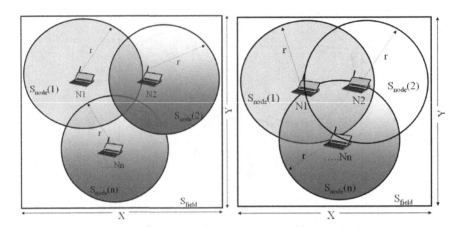

Fig. 1(b). Low connectivity topology   **Fig. 1(c).** High connectivity topology

the conditions for different value of R. In Figure 1(a), the communication is not possible; in Figure 1(b), a node must come inside overlap area for proper communication; and in Figure 1(c), the proper communication can be done between the nodes.

Define the connectivity of the network in terms of number of nodes and network coverage ratio. More specifically, the connectivity is defined in terms of reachability i.e., the number of neighbor nodes with good connection [4]. The reason we define connectivity is that "A good coverage in MANET network means a good connectivity." The connectivity is defined as follows:

$$C = \frac{R}{1 - \gamma e^{-R}} \quad (2)$$

where, γ is a protocol decision constant. In fact, this parameter is a function of R and its relation to R is shown later in this paper. This parameter indicates whether the communication is possible among the nodes and helps to determine a particular instant of choosing a better-fitting routing protocol.

We like to maximize the value of C because the higher connectivity is always desirable in networking. To achieve such higher connectivity, Eq. (2) is differentiated with respect to R. The first derivative of C with respect to R is as follows:

$$\frac{dC}{dR} = \frac{1 - \gamma e^{-R} - R\gamma e^{-R}}{\left(1 - \gamma e^{-R}\right)^2} \quad (3)$$

Similarly, the second derivative of C with respect to R is as follows:

$$\frac{dC^2}{d^2R} = \frac{\gamma e^{-R}(-2+R+2\gamma e^{-R}+R\gamma e^{-R})}{(1-\gamma e^{-R})^3} \tag{4}$$

We calculate the minimum required connectivity to maintain the communication between the nodes by evaluating the function $\dfrac{dC}{dR}=0$. The result becomes as follows:

$$(1-\gamma e^{-R}-R\gamma e^{-R})=0$$

$$\gamma = \left(\frac{1}{R+1}\right)e^{R} \tag{5}$$

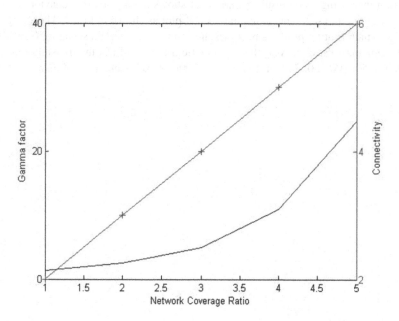

Fig. 2. The curves for connectivity C and parameter γ with respect to the coverage ratio R using γ in Eq. (2)

Fig. 2 graphically shows changes in γ-parameter and the connectivity C with respect to R, respectively. For all R, γ-parameter and C monotonically increases. For a large R, γ shows somewhat divergence behavior. Also, the flexibility increases such that the system has more options to select routing protocols. The routing protocol is chosen only on this case. Now, we implement the definition of network coverage ratio so that we are able to calculate its nominal value. Given A as the physical area of network field, the total coverage area A_N covered by N nodes can be determined as follows:

$$A_N = \sum_{i=0}^{N-1} A_{node}(i) = \pi r^2 N \qquad (6)$$

It is assumed that all the nodes have an equal transmission range r and isotropic pattern of antenna such that the coverage area becomes circular. From the definition of the NCR, the total coverage area is $\pi r^2 N$, and the threshold or minimum value for the transmission range making the communication possible is as follows:

$$r_{min} = \sqrt{\frac{A_T.R}{N.\pi}} = \sqrt{\frac{A.R}{N.\pi}} \qquad (7)$$

The scalability is taken into consideration by increasing the value of N, and then NCR goes on increasing. As a result, γ-parameter shows an asymptotic behavior where we can determine the segments along the γ vs. R plot as shown in Fig. 3. Those segments decide which routing protocol becomes appropriate at a certain scenario. Considering N the number of nodes, A_{field} the network field area, and r_{min} the transmission range, with N = 50, 100,150, 200, or 250, A_{field} = 700m × 700m, and r_{min} =80m, respectively.

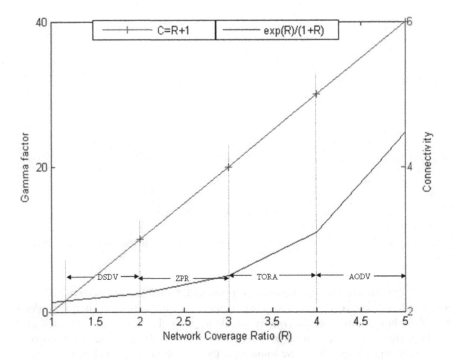

Fig. 3. Multi-protocol operation ranges and decision values for protocol transitions

For the different values of N= 50, 100, 150, 200, or 250 with the minimum connectivity γ valued defined in Eq. (5), the values of C becomes R+1. As shown in Figure 3 we can differentiate four distinct areas which have different properties. When R is high, the coverage increases as well as the connectivity. In this situation, the AODV routing protocol is suitable. As shown in that figure, the multi protocol operation ranges are divided into four segments with respect to C and R, and we can assign a specific routing protocol for each one of those segmentations. The lower the value of R, the lower the number of nodes N. DSDV is used for the lower valued segment R, which is less scalable protocol. Also, TORA and ZRP are assigned according to their properties. TORA is nearer to AODV because it is on-demand routing protocol. When nodes are partly clustered and slowly moving, ZRP could be the better choice.

In addition, the same figure contains two dotted line indications drawn for mobility and bandwidth requirements. They are also helpful in decision making process. DSDV needs more bandwidth because it needs to maintain routes every time, and a higher volume of control signals need to be sent. On the contrary, AODV does require less bandwidth. Hence, DSDV is more suitable for networks having less speed in comparison to AODV. Depending on networking conditions and scenarios, the adaptive transition may occur from one protocol to another.

3.2 Adaptable Module

The architecture of auto-configuration routing module is shown in Figure 4. The Multi-Routing Protocol Decision Support Module is provided by data packets and *CONF_Beacon* packet. The module gets either data or beacon at a time by some switching mechanism between them. Generally, *CONF_Beacon* is transmitted first

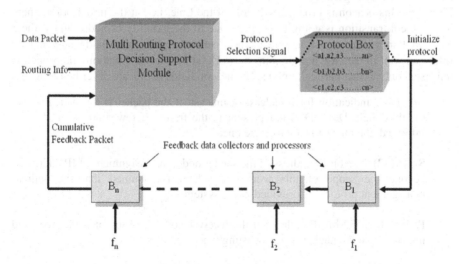

Fig. 4. Architectural modules for auto-configuration routing

throughout the network and this is called *configuration session*. Once the system is stable, the data packet *transmission session* starts. The time it takes to enter from first session (configuration) to second session (data transmission) depends on the network size, *CONF_Beacon* length, and bandwidth of the network.

As shown in Figure 4, there are **n** feedback modules referred to a Decision Support Function. They are denoted by *{f1,f2,f3,.......,fn}* and the corresponding processing modules are *{B1,B2,......,Bn}*. Here, the functional parameters f_i, $i=1,2,...,n$, can be NCR, number of nodes, transmission range, or mobility patterns, etc. All the calculations are performed in *Bi*, $i=1,2,...,n$, which are shown in Eqs. (1) - (5). All these parameters go into the multi-routing protocol decision support module as cumulative feedback packets. This module decides more suitable routing protocol on the present scenario. It then designs *CONF_Beacon* control packet and data packet, appropriately.

The Protocol Box is a fairly simple module which keeps track on routing protocols. It stores the control frame format of each protocol to be used in future. When it gets hints about the very next protocol to be used, it generates *CONF_Beacon* and broadcasts it. After some specified time period, it sends control packets. The dotted line in the module means that it keeps on working on the same protocol unless it gets another control packet to change the protocol. The same algorithm runs on each node so that there will be a same decision throughout the network.

3.3 Configuration Beacon

When there is transition from one routing protocol to another, all neighbor nodes are informed first. Because of this signaling, a network layer beacon has to be defined, which carries information regarding the appropriate protocol changes. This beacon is named as the configuration beacon (*CONF_Beacon*) and is designed as shown in Figure 5. This beacon is broadcasted most of the time, i.e., at the initial point where the protocol transition is going to happen. Due to the mobility of nodes and without any fixed infrastructure of mobile ad hoc network, a neighbor node may not be informed well in the case a node joins network lately. At this situation, the beacon is unicasted only to that particular node. The beacon fields in bits are described follows:

B/U (4) – indication for broadcast or unicast: if the packet is a broadcast type, the third field Dest_ID is not present in the frame. If fewer nodes need to be informed, the unicast is appropriate one:

S_ID (32) – 32-bit IP address of the sender node: the assignment of IP address is beyond the scope of this work. Instead, it is assumed that the address management protocol is provided in the system.

R_ID (32) – 32-bit IP address of the receiver node: It is absent if the first field indicates that the packet is broadcasting type.

Old_P1 (8) – the current routing protocol.

New_P2 (8) – the selected adaptable routing protocol.

M (4) – the indication for protocol transition method.

BC_Id (8) – broadcast identity field: it avoids loop formation. When a node generates this *CONF Beacon*, this field is incremented by one. If a node receives a beacon having the broadcast id, which it has already broadcasted, is simply discarded. It is reinitialized when it reaches to maximum value.

TTL (8) – *time to live* field: it is decremented each time a node relay this packet to its neighbors. The beacon having TTL field zero is discarded. The value of this field is determined according to the size of network.

1	32	32	8	8	4	8	8	Bits
B/U	S_ID	R_ID	Old_P	New_P	M	Seq_Num	TTL	

Fig. 5. Frame format of *AUTO_Conf* beacon control packet

4 Analysis

One of the major issues is to minimize transition time from one protocol to another. There are a lot of factors which plays vital role to maintain transition time. Those include processing capacity, memory unit, information acquisition of feedback module, however, but we do not mention them here. In this analysis, we are focused to the issue regarding the configuration beacon.

Field R_Id is not needed for broadcasting and it is known by the first field B/U. in this case the frame size is 72 bits. For the unicast with $N_b=1$, the frame size is $72+32$ bits, where N_b is the number of neighbors nodes to be broadcasted. Sometimes, a node may have to inform to some of the neighbors in the case a few nodes join the network a few moments later. It takes more time to make them informed about the routing protocol changes by sending the unicast packets. The limited-broadcasting may be a better idea for information dissemination to limited number of users. For example, with $N_b=5$, the frame size increases and becomes $72+32(5)$ bits. Considering a mobile device with a data rate 2Mbps, then T_c the minimum ideal time needed to change the routing protocol from one to another is as follows assuming δ represents the transmission and processing delay:

For Broadcasting, $T_c = (36+ \delta)$ µsec
For Unicasting, $T_c = (51+ \delta)$ µsec and
For limited broadcasting (for 5 nodes), $T_c = (116+ \delta)$ µsec

5 Implementation Issues

The investigated self-organizing network and its auto-configuration routing protocol design are drawing immense attention because it is new and emerging field in ubiquitous and mobile computing. When considering implementation of architectural modules and their procedure, there are two important issues to be taken account.

5.1 Internal Issues

The design of module is important because it determines the overall cost of the mobile device. It deals with memory requirement, processing capacity, and/or inter-module communications. The important one is the decision module itself and its algorithm. The description of design of these modules is beyond the scope of this paper.

5.2 External Issues

The external issues deals with the parameters required for internodes communication. The important one is "data gathering module." It may be a sensor or actuator, which feeds bandwidth requirements, speed of nodes, coverage area of nodes, N, and/or NCR as inputs to B_i, $i=1,2,...,n$. Another issue is to design the frame size. There must be a compromise between a larger frame size and the amount of bandwidth for broadcasting. If the packet type is unicast, there must be a field indicating the IP address of destination. For the existing IPv4 addressing scheme, the beacon size is increased by at least 32 bits plus other additional information. This dilemma makes system design and implementation more complex. Both issues in our work are taken into consideration.

6 Conclusions and Future Works

The recent development in ad-hoc networking is fast growing and a large amount of literatures are accumulating. In key point of the proposed scheme is not to design new one but to use existing protocols in efficient way. With that strategic point in mind, our investigation has been focused on how to collect parametric information and use them to adaptively control the routing protocol from one to another on a fly depending on various network conditions and scenarios. We have measured the protocol convergence factor γ which has a distinct relation to the network coverage ratio R. These two parameters provide the basic building block of our multi-routing protocol method along with other important network parameter such as number of nodes, speed of nodes, transmission range, and connectivity. We could differentiate each protocol operation range that is defined by γ gamma and R. From the basic properties of MANET routing protocols, we have assigned them to a particular and appropriate protocol operation ranges. In addition, we have designed configuration beacon frame format and also presented the length of the control and data packets under the consideration of protocol change decision time and the network bandwidth. From the analytic work, we can conclude that the performance degradation on the transition period is very low and yields a negligible effect. The importance of this novel approach will play more significant role for the future multi-dimensional networks.

In our best understanding, a scheme to change the routing protocol according to feedback parameters is first time proposed here. By fully extending the investigation, we expect more new ideas can be derived and real-world like results can be produced. Specifically, we need to obtain a better design on multi-protocol decision module that can comprehensively model the dynamics of ad hoc network. Some exotic techniques such as neural networks and/or fuzzy logic theory can also be adopted for such purposes

Acknowledgements

This research work has been supported by the second phase of **Brain Korea 21 Projects**.

References

1. Santivanez, C., Stavrakasis, I.: A framework for a Multi-mode Routing Protocol for (MANET) Networks. In: Proceedings of IEEE WCNC '99, New Orleans, LO, pp. 515–519 (September 1999)
2. Flores-Carets, C.A., Blair, G.S.: A Multi-Protocol Framework for Ad Hoc Service Discovery. In: IEEE-MPAC'06, Melbourne, Australia (Nov-Dec 2006)
3. Malarias, A., Palou, G., Gounares, S.: Self-Configuring and Optimizing Mobile Ad Hoc Networks. In: Proc. of Second IEEE Conference on Automatic Computing (ICAC) (2005)
4. Badonnel, R., state, R., Festor, O.: Monitoring End-to-End Connectivity in Mobile Ad Hoc Networks. In: Lorenz, P., Dini, P. (eds.) ICN 2005. LNCS, vol. 3421, pp. 83–90. Springer, Heidelberg (2005)
5. Santivanez, C., Stavrakasis, I.: Towards Adaptable Ad Hoc Networks: The Routing Experience. In: Smirnov, M. (ed.) WAC 2004. LNCS, vol. 3457, pp. 229–244. Springer, Heidelberg (2005)
6. Hubaux, J.P., Gross, T., Boudec, J.Y.L, Vetterli, M.: Towards Self-Organized Mobile Ad Hoc Networks: The Terminode Project. IEEE Communications Magazine 2001, 118–124 (2001)
7. Legendre, F., de Amorim, M.D., Fdida, S.: Some Requirements for Autonomic Routing in Self-organized Networks, In: Smirnov, M. (ed.) WAC 2004. LNCS, vol. 3457, pp. 13–24. Springer, Heidelberg (2005)
8. Vieu, V.B., Mikou, N.: Distributed Mobility Prediction-Based Weighted Clustering Algorithm for MANETs, In: Kim, C. (ed.) ICOIN 2005. LNCS, vol. 3391, pp. 717–724. Springer, Heidelberg (2005)
9. Royer, E., Toh, C.T.: A review and Current Routing Protocols for Ad Hoc Mobile Wireless Networks. IEEE Personal Comm., 46–55 (April 1999)
10. Forde, T.K., Doyle, L.E., O'Mahony, D.: Self-stabilizing Network-Layer Auto-Configuration for Mobile Ad Hoc Network Nodes. In: Proceedings of the IEEE International Conference on Wireless and Mobile Computing, Networking and Communications (Wimob 2005), Montreal, Canada, 22 - 24 August 2005, pp. 22–24. IEEE Computer Society Press, Los Alamitos (2005)
11. Broch, J., Maltz, D.A., Johnson, D.B., Hu, Y.-C., Jetcheva, J.: A Performance Comparison of Multi-Hop Wireless Ad Hoc Network Routing Protocols. In: Proceedings of the Fourth nnual ACM/IEEE InternationalConference on Mobile Computing and Networking (MobiCom'98), Dallas, Texas, USA, October 25-30, 1998, IEEE Computer Society Pres, Los Alamitos (1998)

The Development of a Wireless Sensor Network Sensing Node Utilising Adaptive Self-diagnostics

Hai Li[1], Mark C. Price[1,2], Jonathan Stott[1], and Ian W. Marshall[1]

[1] Computing Laboratory, University of Kent, Canterbury, UK, CT2 7NF
[2] School of Physical Sciences, University of Kent, Canterbury, UK, CT2 7NH
mcp2@star.kent.ac.uk

Abstract. In Wireless Sensor Network (WSN) applications, sensor nodes are often deployed in harsh environments. Routine maintenance, fault detection and correction is difficult, infrequent and expensive. Furthermore, for long-term deployments in excess of a year, a node's limited power supply tightly constrains the amount of processing power and long-range communication available.

In order to support the long-term autonomous behaviour of a WSN system, a self-diagnostic algorithm implemented on the sensor nodes is needed for sensor fault detection. This algorithm has to be robust, so that sensors are not misdiagnosed as faulty to ensure that data loss is kept to a minimum, and it has to be light-weight, so that it can run continuously on a low power microprocessor for the full deployment period. Additionally, it has to be self-adapative so that any long-term degradation of sensors is monitored and the self-diagnostic algorithm can continuously revise its own rules to accomodate for this degradation. This paper describes the development, testing and implementation of a heuristically determined, robust, self-diagnostic algorithm that achieves these goals.

1 Introduction

1.1 Background: The PROSEN Project

PROSEN (PROactive condition monitoring of SEnsor Networks) is an EPSRC funded, multi-university project [1] which is investigating techniques to enable automated control and proactive management of sensor arrays. The project aims to develop a proactive Wireless Sensor Network (WSN) to enable condition monitoring of a wind farm in an uncontrolled, unsupervised, outdoor environment that will be deployed for a minimum of one year.

Each sensor node will measure temperature, wind speed, humidity, rainfall and cloud cover and store the raw data on-board. Preliminary data checking, analysis and sensor diagnosis will also be performed on-board. As long-range wireless communication is power intensive, in order to prolong their life, each node must pass only "events" (not raw data) to the management system which will be located at one of the investigating universities. What is deemed an event is determined from an overall system policy which is made up of a set of adapatable policy rules which can be modified on individual nodes. For example, an event

D. Hutchison and R.H. Katz (Eds.): IWSOS 2007, LNCS 4725, pp. 30–43, 2007.
© Springer-Verlag Berlin Heidelberg 2007

could be generated when a sensor records a measurement above (or below) a certain (policy determined) threshold, when a possible sensor fault is detected, or when the battery voltage of the node reaches a certain critical level.

Figure 1 is a schematic showing the information flow within the PROSEN WSN. Each node will also have a short range (174 MHz) radio in order to communicate with its nearest neighbour. This enables a second level of data verification if, for example, one node is measuring an abnormally high temperature, it can query its neighbour to verify the validity of the reading. If the sensor reading is invalid then a possible sensor fault condition is flagged, and reported to the management system.

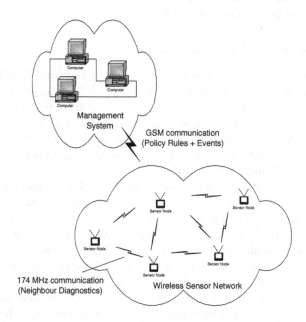

Fig. 1. Schematic showing information flow within the PROSEN WSN

In traditional condition monitoring systems, the sensor nodes acquire data under the control of a local microcontroller located on the node, and then raw data is transmitted to a central base-station (e.g. PC). This central base-station then performs high level, CPU intensive, functions such as data analysis and decision making (e.g. [2]).

This type of approach is purely reactive, and prone to catastrophic failure in reponse to unanticipated failure modes, degradation, changing operating conditions or adverse environmental conditions. Moreover, a sole controlling central station consitutes a single point failure, and should it fail the whole network could be rendered ineffective.

To tackle the drawbacks of such a system, we are investigating and demonstrating techniques that enable the automated control and management of sensor arrays to be proactive. In order to achieve this goal, we need to give the sensor nodes much more on-board 'intelligence' such as self-diagnosis, data analysis, asessment of data quality and decision making routines.

The obvious challenge with such an approach, is that it requires a sophisticated processor on each node to handle the data analysis, self-diagnosis and decision making processes. Such processing power comes at the expensive of increased power useage, thus further constraining the frequency and duration of power hungry, long-range communications. There is therefore a requirement to develop a self-diagnostic algorithm that is not only robust and adaptable, but will run on a low power microprocessor.

1.2   Approaches to Sensor Self-diagnosis

Automated fault detection techniques have been widely studied and developed during the last few years (for example, Angeli *et al.* [3]) and the most popular methods include model-based methods [4,5] and artificial intelligence methods [6]. These methods are highly reliable and are robust, but all are based on highly complicated computation, thus requiring a high speed processor, large amounts of memory, and therefore have a high power consumption. Two further examples are Nithys *et al.* [7] and Farinaz *et al.* [8].

Nithys *et al.* [7] developed a cross-validation based technique for on-line detection of sensor faults. Their idea is to compare the results of multisensor fusion with, and without, each of the sensors involved using non-linear function minimization and then identify the faulty sensor using non-parametric statistical techniques. Their simulation results indicate the high accuracy of the approach, but the implementation complexity of non-linear function minimization is too high for a low power microprocessor with limited memory and processing speed.

Farinaz *et al.* [8] propose a distributed, localized, sensor fault detection algorithm for WSNs. In their algorithm, each node monitors its health status and that of its nearest neighbours. This data is correlated and exchanged between the nodes. Each node therefore has knowledge of its own status and all its neighbours. The drawback of this algorithm is that there is a large amount of information transferred between nodes, resulting in a high power overhead due to the wireless communications required.

A further avenue is to use a rule-based approach. Betrand-Krajewski *et al.* [9] present a formal approach to the establishment of a such a rule-based system. The major advantage which such a system is the low processing power required, and the rapidity in which a working rule-set can be tested, evaluated, modified and retested.

Jinran *et al.* [10] explore rule-based fault detection techniques for helping improve the quality of the data collected by their WSN in Bangladesh. Their research is based on the idea that fault diagnosis and repair are knowledge-intensive and experiential tasks. After analysing a dataset, some rules have been established to suggest actions a user can take to remedy, or validate data. For example, such a directive could be: *"If measurements from a sensor are identified as noisy, either check the battery or the connectors on the sensor and to the sensor-board"*. Their approach is a high level fault detection technique running on the base station side, but requires a large amount of node-human interaction to quickly identify, then remedy problems.

In this paper, we describe and evaluate a light weight heuristically determined rule-based algorithm to identify possible sensor faults for each of our sensor nodes. It has a low computing complexity and (so far) has achieved a 100% sucess rate in detecting faults, with no false-positives reported.

We describe the architecture of our prototype platform in Section 2. Section 3 describes our low level self-diagnosis routines, and presents some practical results, and Section 4 gives conclusions and future work.

1.3   Hardware Development Strategy

From the outset, our design strategy has been to minimise the duration and frequency of long range communications, and limit such communications to the transmission of events (ie. alarms, alerts, node health status etc.), and the reception of policy rules which determine the conditions under which these events are generated.

We therefore adopted the following methodology:

1. Deploy a "first generation" prototype node in a controlled external environment to measure base-line operational parameters of the sensors and communications components. This has a simple self-diagnostic rule set, and event generating capability.
2. Develop a "second generation" node that will have full system functionality. This will have an adaptive self-diagnostic rule-set, full event generating capability and be able to receive updates from the management system. It will also incorporate two processors, a low power micro-processor that performs low-level tasks, and a higher power processor which is powered up intermittently to perform more CPU intensive tasks.
3. Using data from the first and second generation nodes, fully optimise the hardware architecture and system parameters for the finalised "third generation" node to maximise the node lifetime.

2   A Prototype Sensor Node

In order to minimise to development time (and cost), we built the first generation prototype sensor node using readily available commercial products in order to quickly obtain experimental baseline data to establish an intital rule-set. The station selected is called the Davis Vantage Pro2 [11] and consists of two major components: the Integrated Sensor Suite (ISS), which houses and manages the external sensor array, and the console (connected to a PC) which provides the user interface and data display. The ISS and console communicate via a 868 MHz RF transmitter and receiver. We also integrated a Campbell Scientific

CR216 wireless datalogger [12], which has five 12-bit analogue inputs, two pulse inputs, two digitial I/O lines, a RS-232 port and a RF416 spread spectrum radio (operating at 2.4 GHz) so that we can monitor the sensors in parallel with the Davis ISS (via the custom built sensor interface) [13]. This logger has a user-programmable 8-bit microprocessor with 6.5 KBytes of program space, and 250 KBytes of data storage, and it is on this platform we have implemented our data acquisition and self-diagnosis algorithm for our first generation node.

Figure 2 shows the various components of the node. The "Base Station" side is located within the School of Physical Sciences building at the University of Kent, and the "Sensor Node" is deployed on the first floor roof of the same building (Figure 3). Initial deployment was carried out in July 2006.

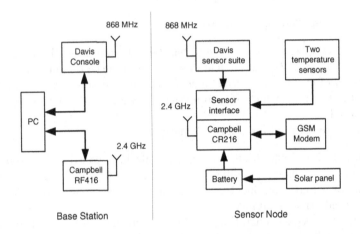

Fig. 2. Components of our first generation prototype node

In addition we have also connected the Campbell data logger to a GSM modem via its RS-232 interface. This allows us to send events and alarms to a remote system via SMS. This has proved to be very effective, felixble and reliable and will be developed into a two-way process in our second generation node [14].

As we also anticipate that the time between battery replacements in the field could be anywhere between one and two years, we have added a 0.18 m^2, (6 watts maximum output) solar panel to keep the 12 Volt (7 Ah) lead-acid battery topped-up. This has provided sufficient power to keep the battery fully charged, even over the winter.

3 Self-diagnostic Methods

3.1 Establishing the Initial Rule-Set

To establish an effective rule set, we used heuristic, phenomenological and statistical methods to establish:

Fig. 3. A photograph of the deployed first generation sensor node

1. Sanity levels. This is simply a set of values based upon possible non-physical readings, ie. a humidity reading greater than 100% (or less than 0%), temperatures less that -40° centigrade, or greater than +40° centigrade etc. Any reading outside of these values is a probable sensor malfunction.
2. Maximum and minimum environmental parameters (ie temperature, humidity, wind speed) over a long period. This was achieved by analysing a data set from a nominally identical weather station that has been deployed for two years within a mile of our prototype node [15], plus additional data obtained from the met office [16]. Any deviation of the measured values outside of these values could be indicative of a sensor fault.
3. Noise parameters. Specifically the standard deviation of the noise of a sensor over a long period. Again, any increase (or decrease) in these values may suggest a sensor problem.
4. The correlation between different, but complementary sensors. For example, the solar radiation sensor and solar panel both output a voltage proportional to the intensity of the solar radiation incident upon them. Thus they should be strongly correlated, and any deviation from this correlation could be characteristic of a malfunction in either sensor.

In order to illustrate our methodology, we now discuss three examples of how we obtained our base-line performance parameters for the temperature sensors, the solar radiation sensor and the anemometer.

We have installed two Campbell Scientific (Model 109) temperature sensors on our prototype node. They are housed within their own radiation shields and are approximately 1 metre from the floor with a horizontal separation of 25 cm.

In Figure 4 (top) we have plotted the temperature as measured by our two temperature sensors for a three day period, and (bottom) the residuals between the two readings.

In order to calculate a base-line noise value, we calculated the standard deviation, σ, of the residuals for 50,000 readings (equivalent to 33 days of data), also plotted on the bottom graph of Figure 4 are the ± 3, 7 and 11σ levels.

As these temperature sensors are nominally identical, in the absence of systematic effects, the error between the two readings should be $\pm 1\%$ [17] and the residual values should be normally distributed.

In Table 1, we show the number of records that should deviate more than 3, 5, 7, 9 and 11σ assuming a normal distribution, and the *actual* number of records from our 50,000 data points sample that do deviate.

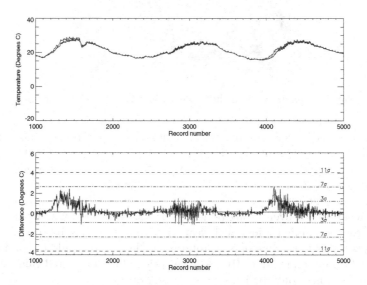

Fig. 4. Plot of the temperature recorded from the two temperature sensors (top) and the difference (residuals) between the two readings (bottom)

As can be seen from Table 1 and Figure 4, the noise is clearly not normally distributed and has a bias consistent with a regular systematic effect. Closer investigation revealed that this effect was caused by the physical location of the two temperature sensors. One of the sensors is on the east side of the node, and the other on the west side of the node. As the sun rises in the morning, the east sensor warms more quickly than the west sensor causing a large (~ 2 degree) temperature differential. However, during the course of the day, this difference reduces and becomes unnoticeable. However, as this is a regular, systematic effect, it does not change our methodology for detecting sensor faults.

Table 1. Analysed results from 50,000 readings

Noise deviation	Number of records (Normally distributed)	Number of records (Measured)
3σ	67	1132
5σ	< 1	130
7σ	0	19
9σ	0	8
11σ	0	0

3.2 Solar Radiation Sensor

In order to do self-diagnosis on the solar radiation sensor, we have adopted a different approach to that used for the temperature sensors. As we only have one solar radiation sensor, we cannot use the same statistical method described above.

However, we do have access to the output voltage measured where the solar panel connects to the battery and this give us a direct reading of the output voltage of the solar panel (plus the battery voltage). Therefore, we should see a strong correlation between the voltage from the solar radiation sensor and the solar panel voltage.

Figure 5 shows the correlation for a period of three days, and Figure 6 shows the solar radiation sensor output voltage plotted against the solar panel voltage for 72,000 readings. As can be seen, there is a clear envelope that all the data lie within, showing a strong correlation between the two readings.

The observed hysterisis type appearance is due to the charging cycle of the battery. In the early morning (before dawn), the battery level is low (typically ~13 volts) and during the course of the day the battery charges up, so that after dusk its voltage level is ~14 volts. The battery then discharges back to 13 volts during the course of the night, and the cycle repeats.

In order to quantify this correlation we calculate the linear correlation coefficient, r, via:

$$r = \frac{\sum_i (x_i - \overline{x})(y_i - \overline{y})}{\sqrt{\sum_i (x_i - \overline{x})^2}\sqrt{\sum_i (y_i - \overline{y})^2}} \qquad (1)$$

where $(x_i, y_i), i = 1, \ldots, N$ represent the measured values of the battery + solar panel voltage and the solar radiation sensor respectively, and $\overline{x}$ is the mean of x and $\overline{y}$ is the mean of y [18].

By using Equation 1, we calculated the daily correlation coefficient, r, between solar radiation and battery level for a 24 hour period. Figure 7 is a the plot of r for fifty days for the data set shown in Figure 6. These data illustrate that the daily correlation coefficient between solar radiation and battery + solar panel voltage is always greater than 0.6, even on very overcast days. Therefore, based on this

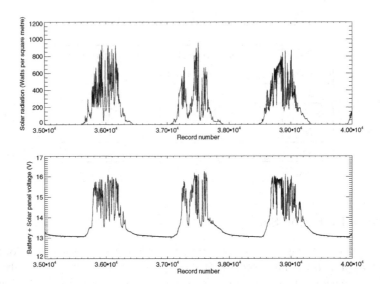

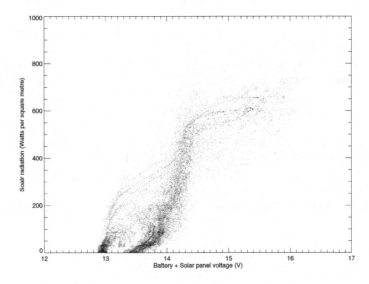

Fig. 5. Three days of data illustrating the correlation between the solar radiation intensity, and the measured solar panel + battery voltage

Fig. 6. Correlation between solar radiation intensity and solar panel + battery output voltage for 72,000 records

analysis, we set a threshold value, r_{th}, of 0.6 in our self-diagnostic algorithm. A calculated value of $r < r_{th}$ will flag an alert of a possible degradation in the performance of either the solar panel, or the solar radiation sensor.

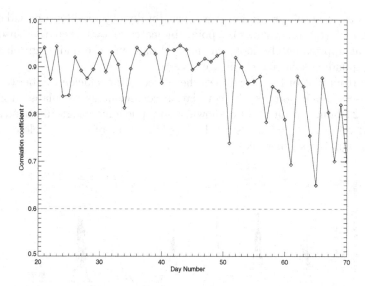

Fig. 7. Correlation coefficient, r, plotted for a 50 day period

3.3   Anemometer Diagnosis

In order to check the operation of the anemometer, we again used the fact that it consists of two different, but complementary sensors; a wind direction sensor, and a wind speed indicator. Due to the mechanical nature of these sensors, the most probable failure mode is a "sticking" of the sensor in a fixed position. However, a wind speed of zero, and/or an unvarying wind direction could just be indicative of a very still day and not necessarily a failed sensor. We therefore analysed our weather data for the last two years [15] in order to establish what were the longest periods of exceptional stillness, ie. where the wind speed indicator was zero, and when the wind direction was unvarying.

This established the following rule.

$$\text{IF } (W_d \text{ is changing}) \text{ and } (W_s \text{ is unchanged for 30 mins})$$

$$\text{THEN ReportFault} \tag{2}$$

and

$$\text{IF } (W_s > 2 \text{ mph}) \text{ and } (W_d \text{ is unchanged for 30 mins})$$

$$\text{THEN ReportFault} \tag{3}$$

where W_d is the measured wind direction and W_s is the measured wind speed.

3.4   Algorithm Testing

In order to test out self-diagnosis routines we forced a failure condition upon several of the sensors to ascertain the robustness of the self-diagnosis routines, and their ability to generate the appropriate alarm event.

In one test, the solar radiation sensor was totally obscured for a period of one hour. Figure 8 (top plot) shows the point (indicated arrow) where the sensor was covered, at approximately noon, on the 78th day, and the middle graph shows the corresponding solar panel voltage.

The bottom plot of Figure 8 shows the correlation coefficient, r, between the two datasets. Each point is the correlation coefficient as calculated from the previous 24 hours of data. Also shown is our phenomenologically determined threshold value of $r_{th} = 0.6$. As can be seen, the value of r drops below r_{th} and an alarm signal was generated.

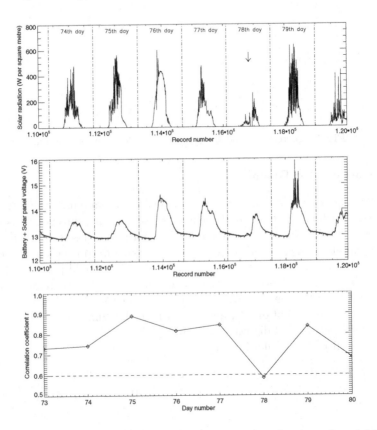

Fig. 8. Solar radiation and solar panel + battery voltage, indicating a forced failure of the solar radiation and detection of the failure

3.5   Detection of a Real Sensor Failure

During the latter part of the tests conducted above, we frequently received alerts indicating a failure of one of the temperature sensors. In Figure 9 (top), we have plotted the residuals for the two temperature sensors over the period in question, and our 11σ threshold level. As can be seen, there are many points where the

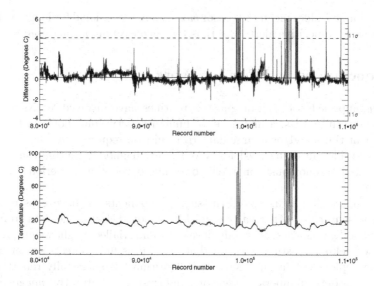

Fig. 9. Detecting a real failure of one of the temperature sensors

data exceeded this threshold. The bottom graph of Figure 8 shows the raw data for the sensor. Clearly one of the sensors is faulty as it is intermittently recording temperatures in excess of 100° centigrade!

In this section, we introduced our base-line self-diagnosis routines, explained some technical methods for low level sensor fault detection and showed some experimental results. Our self-diagnosis algorithm is a rule-based system, where knowledge obtained from analysis of a large dataset has helped determine these rules. Practical results have shown that these rules sucessfully report sensor failure and it can be easily implemented on a low power microcontroller.

3.6   Towards Self-adaptability

In the proceeding analysis and examples, our node reacted solely on a fixed set of conditions imposed upon it; ie., "if X > Y generate event". However, we have anticipated the need for these set of conditions to be modifiable, either by the management system or the node itself, as the base-line performance of the node changes during its deployment. As a simple example, we consider the possible long term degradation of the solar radiation sensor caused by buildup of deposits on the transparent external casing of the sensor. This would manifest itself as a weakening of the correlation between its output and that of the solar panel. In order to compensate for any such degradation, the node can actively update the value of r_{th} required to generate an alert by performing a running average over the last 50 days worth of data. Any sensor degradation would lead to a gradual decrease in the value required to generate an alert. Such a method does not preclude the self-diagnosis system failing the sensor in the case of a catastrophic malfunction, but does mean the sensor can remain operational for longer without generating false-positive alerts and thus (erroneously) discarding useful data.

4   Conclusions and Further Work

Previous approaches to self-diagnostics routines have involved WSNs with access to powerful CPUs, a high level of human supervision, short (in the field) deployment times and/or a large data transmission requirement.

We have identified the need for a WSN self-diagnostic routine that can be implemented autonomously on a low power microprocessor for periods in excess of a year.

By using readily available off-the-shelf components we have constructed a prototype sensor node that can be quickly deployed. Using the data from this deployed node, we have successfully developed and trialled a light weight, robust, rule-based self-diagnostic algorithm that very sucessfully detects sensor faults. Since its deployment in July 2006, the algorithm has sucessfully reported the failure of one of the temperature sensors, and (just as importantly) *not* generated any false-positive alarm events.

Our experimental results shows that this approach has a low computing complexity and achieves a high probability of correct diagnosis. It can be implemented on a broad set of low power microprocessors that have limited memory and processing speed.

We thus intend to migrate our current Campbell Scientific datalogger based system to our second generation node within the next two months. This node will be a hybrid node, incorporating a low-power microcontroller (Texas Instruments' MSP430F1611) to acquire data and run the low-level algorithm discussed here, and an Intel PXA-255 embedded Linux machine (such as a "Gumstix" [19]) which is swtiched on intermittently to do more CPU intensive tasks, such as double-checking the low-level diagnostic routine to validate alarm events.

Acknowledgments

The authors would like to thank EPSRC (Engineering and Physical Sciences Research Council) for funding this project and the mechanical workshop of the Electronics Department of the University of Kent for their invaluable technical assistance.

Addendum

Since this work was initially carried out and submitted, the authors, with the exception of Dr. Jonathan Stott, have relocated to Infolab21, Department of Computing, University of Lancaster, Lancaster, UK, LA1 4WA.

References

1. PROSEN. Prosen project homepage (2007), [Online]. Available:
   http://www.prosen.org.uk
2. Caselitz, P., Giebhardt, J., Mevenkamp, M.: Application of condition monitoring systems in wind energy converters. In: European Wind Energy Conference (EWEC'97), Dublin, October 1997 (1997)
3. Angeli, C., Chatzinikolaou, A.: On-line fault detection techniques for technical system: A survey. International Journal of Computer Science & Applications I(1), 12–30 (2004)
4. Roumeliotis, S.I., Sukhatme, G.S., Bekey, G.A.: Sensor fault detection and identification in a mobile robot. In: 1998 IEEE International Conference on Robotics and Automation, May 1998, pp. 2223–2228. IEEE Computer Society Press, Los Alamitos (1998)
5. de Freitas, N., Dearden, R., Hutter, F., Morales-Menendez, R., Mutch, J., Poole, D.: Diagnosis by a waiter and a mars explorer. Proceedings of IEEE 92(3), 455–468 (2004)
6. Goel, P., Dedeoglu, G., Roumeliotis, S.I., Sukhatme, G.S.: Fault detection and identification in a mobile robot using multiple model estimation and neural network. In: IEEE International Conference on Robotics and Automation, Leuven, Belgium, May 1998, IEEE Computer Society Press, Los Alamitos (1998)
7. Ramanathan, N., Balzano, L., Burt, M., Estrin, D., Harmon, T., Harvey, C., Jay, J., Kohler, E., Rothenberg, S., Srivastava, M.: Rapid deployment with confidence: Calibration and fault detection in environmental sensor networks, Center for Embedded Networked Sensing, UCLA and Department of Civil and Environmental Engineering, MIT, Tech (April 2006)
8. Koushanfar, F., Potkonjak, M., Sangiovanni-Vincentelli, A.: On-line fault detection of sensor measurements. In: Sensors, 2003. Proceedings of IEEE, October 2003, pp. 974–979. IEEE Computer Society Press, Los Alamitos (2003)
9. Bertrand-Krajewski, J., Bardin, J., Mourad, M., Beranger, Y.: Accounting for sensor calibration, data validation, measurement and sampling uncertainies in monitoring urban drainage systems. Water Science and Technology 47(2), 95–102 (2003)
10. Chen, J., Kher, S., Somani, A.: Distributed fault detection of wireless sensor networks. In: DIWANS '06: Proceedings of the 2006 workshop on Dependability issues in wireless ad hoc networks and sensor networks, Los Angeles, CA, USA, pp. 65–72 (2006)
11. Davis instruments, wireless vantage pro2 specifications (2006), [Online]. Available:
    http://www.davisnet.com/support/weather/
12. CR200 Series Datalogger with Spread Spectrum Radio, Campbell Scientific, Inc. (2005)
13. RF401/RF411/RF416 Spread Spectrum Data Radio/Modem, Campbell Scientific, Inc. (2005)
14. Li, H., et al.: (In prep. 2007)
15. Stott, J.: Canterbury weather website (2006), [Online]. Available:
    http://www.canterburyweather.co.uk/
16. Uk weather extremes (2007), [Online]. Available:
    http://www.metoffice.gov.uk/climate/uk/extremes/index.html
17. Model 109 Temperature Probe User guide. Campbell Scientific Inc. (2002)
18. Press, W.H., Teukolsky, S.A., Vetterling, W.T., Flannery, B.P.: Numerical recipes in C. Cambridge university press, New York, USA (1992)
19. Gumstix. Gumstix homepage (2007), [Online]. Available:
    http://www.gumstix.com

Efficient and Resilient Overlay Topologies over Ad Hoc Networks

Sandrine Calomme and Guy Leduc

Research Unit in Networking
Electrical Engineering and Computer Science Department
University of Liège, Belgium

Abstract. We discuss what kind of overlay topology should be pro-actively built before an overlay routing protocol enters a route search process on top of it.

The basic overlay structures we study are the *K-Nearest Neighbours* overlay topologies, connecting every overlay node to its K nearest peers.

We introduce a family of optimizations, based on a pruning rule. As flooding is a key component of many route discovery mechanisms in MANETs, our performance study focusses on the delivery percentage, bandwidth consumption and time duration of flooding on the overlay. We also consider the overlay path stretch and the overlay nodes degree as respective indicators for the data transfer transmission time and overlay resilience.

We finally recommend to optimize the K-Nearest Neighbours overlay topologies with the most selective pruning rule and, if necessary, to set a minimal bound on the overlay node degree for improving resilience.

1 Introduction

Ad hoc networks are formed without the use of any existing network infrastructure nor centralized administration. The devices in contact can have different hardware capabilities, software, application needs, and mobility pattern. Plenty of multi-hop routing protocols have been proposed for MANETs and, in such heterogeneous networks, the best one may be different for each set of communicating nodes. The preferred routing solution could also change along time, because of mobility and varying network conditions. Consequently, the requirement of choosing a routing protocol and imposing it to all ad hoc devices in order to form a MANET is a limitation. To overcome this restriction, we propose to copy the layered approach of Internet [1]: agree only on a few unspecialized protocols at the physical, data link and routing layers, imposed by their proved qualities or de facto, and over this basic architecture, develop plenty of more specialized solutions, from routing to application, thanks to the overlay technique. Overlay routing could promote the deployment of ad hoc networks, offering a very flexible ground for a variety of applications using the overlay routing protocol that best fits their specific needs.

In this paper, we discuss what kind of overlay topology should be pro-actively built before an overlay routing protocol enters a route search process on top

D. Hutchison and R.H. Katz (Eds.): IWSOS 2007, LNCS 4725, pp. 44–58, 2007.

of it. As flooding is a key component of many route discovery mechanisms in MANETs, our study focusses on the bandwidth consumption and time duration of flooding on the overlay. We also consider the overlay path stretch and the overlay nodes degree as respective indicators for the data transfer transmission time and overlay resilience.

The interference level is not directly addressed. We let the task of reducing interferences to the underlay topology control algorithm and assume that reducing the number of packets emitted per flood is an efficient way to pace collisions due to the overlay use. We also do not present how the studied topologies can be built or maintained. For example, the reader will not find any test on mobile networks, which does not mean that we assume a static network. The overlay topology control protocol itself will be presented in a follow-up paper. Its design guideline will be to maintain, in a mobile context, the overlay topology as close as possible to a target overlay topology, chosen in accordance with the conclusions drawn in Section 6.

2   Related Work

A major part of the current litterature about overlays addresses peer-to-peer applications. Although this work was not conducted for P2P networks, it could probably be exploited in unstructured peer-to-peer middleware. We did not explore this open issue but can compare in some points our work to what have been done for ad hoc P2P networking. In several works, for example ORION [2] and [3], it is assumed that all nodes run the proposed protocol. The use of an overlay allows to get rid of this restriction. In [4], the Gnutella protocol is optimized for ad hoc networks. XL-Gnutella proactively builds an overlay on top of which queries can be efficiently disseminated by the underlying routing protocol. This is similar to our objective of building an overlay for the propagation of overlay routing requests. In Section 5.3, we compare the topologies we recommend for overlay routing to the XL-Gnutella ones.

Topology control (TC) consists of selecting a subset of edges in a graph representing the communication links between network nodes, with the purpose of maintaining some global graph property (e.g., connectivity), while reducing energy consumption and/or interference [5]. Similarly, our problem requires to select a subset of paths between overlay nodes, with the purpose of maintaining their connectivity, while reducing the number of messages they induce in the whole network when they flood an overlay message. Hence, the roots of this work can be found in the TC literature. Mechanisms for building the presented overlay topologies are inspired from two topology control protocols : k-Neigh [6] and XTC [7]. We first present overlay topologies where each overlay node must be aware of a minimal number of the closest other overlay nodes, a process identical to the one used by k-Neigh, but without discarding asymmetric neighbours. We then propose optimizations of the obtained topologies based on an XTC-like criterion.

3   Study Overview

In the sequel, all concepts related to the whole ad hoc network, and not only to overlay nodes, will be identified by the term "underlay".

We consider a connected underlay and assume that a routing protocol that builds the shortest symmetric paths is available to all nodes. Overlay nodes are randomly and uniformly distributed on the set of ad hoc nodes. The proportion of overlay nodes is called the overlay density.

3.1   Fundamental Properties

The overlay topologies we discuss are strongly connected, i.e. there exists a path on the overlay graph between any pair of overlay nodes, at least with a high probability.

They can be built by a fully distributed algorithm. We also take care of locality: The topology can be built even if each overlay node is allowed to exchange only a few messages with a limited number of nearest overlay nodes. As we do not make any assumption about the underlay routing protocol type, locality is an important feature. With reactive on-demand protocols, like AODV, the control traffic necessary for building a data path between overlay neighbours increases exponentially with the number of hops that separates them.

3.2   Performance Criteria

The objective of the overlay creation and maintenance is to offer a logical communication structure between the overlay nodes which allows the deployment of efficient overlay routing protocols. From this angle of view, the quality of an overlay is strongly linked to desired properties of overlay routing protocols. We translate this in terms of the following objectives.

1. Bandwidth: as routing control traffic is often generated by flooding, the bandwidth necessary to send a message from one overlay nodes to all other ones by using a simple flooding procedure must be as low as possible.
2. Diffusion time: in order to quickly compute valid routes, the overlay control traffic must be flooded rapidly.
3. Delivery: in order to find routes, the overlay control traffic must be received by all overlay nodes.
4. Stretch: the average cost of the shortest overlay path between any pair of overlay nodes must be as close as possible to its value in the underlay. Its maximal cost must also be kept reasonable. In this paper, we use the hop metric. Other metrics, as for example the path delay, could be considered.

3.3   Flooding Technique

As flooding is a key component of many route discovery mechanisms in MANETs, the above performance criteria mainly focus on the flooding of a message on the overlay. We assume that, once the overlay is built, each overlay node knows the hop distance to every neighbour it has selected. In order to spare bandwidth, an overlay node employs the following flooding technique:

1. For all overlay neighbours located only one hop away, it emits a single overlay message, which is actually broadcast in the underlay with a *Time To Live* (TTL) field set to one.
2. For every overlay neighbour located further away, an individual overlay message is created, which will be unicast to it by the underlay routing protocol.

3.4 Simulations Description

All simulations in this paper, except for Section 4.4, were realized with ns-2.29.

The ad hoc nodes are randomly and uniformly distributed on a square field. We vary their number from 50 to 250. Overlay nodes are randomly chosen in the set of ad hoc nodes. All experiments were made for overlay densities ranging from 10 to 90%. For the sake of brievity, we only present graphics for the 50% overlay density. Analysis is identical for all densities.

For a given set of ad hoc nodes, the more efficient underlay topology control (TC) algorithm is used, the more traffic is needed for the construction, use and maintenance of overlays built on top of its resulting logical topology [8]. Hence, in order to test the overlay topologies in a stringent environment, we employ the logical topologies obtained after the use of an ideal homogeneous underlay TC technique which assigns the same value r to each node's radio transmission range, r being the minimal value that makes the underlay connected (i.e. the critical radio transmission range).

The underlay routing protocol used is AODV [9]. The performance criteria are only evaluated on strongly connected overlay topologies. The overlay topologies are calculated offline and provided as input to the ns simulator. A source node emits 23 overlay messages of 64 bytes, at the rate of one message per second. The performance study ignores the period elapsed during the transmission of the first 3 messages. Their flooding necessitates the building of AODV paths between the overlay neighbour pairs. Consequently, the AODV traffic is heavier at the beginning of the simulations and the diffusion time of the first overlay messages is higher than for the following messages. When there is no congestion, the latter must be forwarded on AODV paths that are already up. Each point on the graphics is a mean calculated on 20 trials.

4 Building Topologies That Fullfill the Locality and Connectivity Properties

4.1 Ropt: The Critical Neighbourhood Range

One simple way to give preference to nearest neighbours, and thus to respect the locality principle, is to fix the maximal hop distance between overlay neighbours, the *neighbourhood range*. For any underlay and subset of overlay nodes, one can compute the critical neighbourhood range, that is the minimal neighbourhood

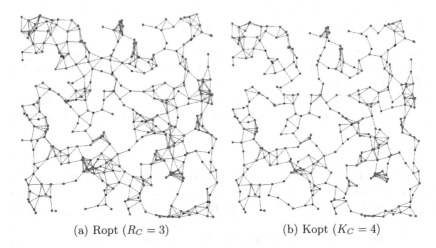

(a) Ropt ($R_C = 3$)   (b) Kopt ($K_C = 4$)

Fig. 1. Example of the Ropt and Kopt overlay topologies for the same underlay topology

range R_C such that the overlay is connected [8]. We denote *Ropt* (R optimal) a topology obtained when each overlay node considers as a neighbour any overlay node that is located at a distance less than or equal to R_C.

4.2 Kopt: The Critical Neighbourhbood Cardinality

Another simple way to respect locality is to fix the maximal number of overlay neighbours. For any underlay and subset of overlay nodes, one can compute the critical number of overlay neighbours, that is the minimal neighbourhood cardinality K_C such that the overlay is connected. We denote *Kopt* (K optimal) a topology obtained when each overlay node considers as a neighbour its K_C nearest neighbours, the distance metric being the number of hops. Let k_i be the number of overlay nodes located at i hops from a given overlay node U. If there exists an integer j such that $\sum_{i=1}^{i=j} k_i < K$ and $\sum_{i=1}^{i=j+1} k_i > K$, the required number of overlay neighbours is randomly picked in the set of overlay nodes located at distance $j + 1$ from U[1].

4.3 Ropt and Kopt Delivery Percentage

Figure 1 shows an example of the Ropt and Kopt overlay topologies for the same underlay. There are 500 nodes and the overlay density equals 50 %. The 250 overlay nodes are represented with (red) squares. The remaining nodes, represented with (blue) circles, are drawed if and only if they are on the shortest path between a pair of overlay neighbours. For this particular underlay and assignment of overlay nodes, the critical neighbourhood range equals 3 and the

---

[1] We evaluated some more sophisticated policies, but none provided significantly better performance.

critical neighbourhood cardinality equals 4. This figure also illustrates that the Ropt overlay topologies are much denser than the Kopt ones. This is confirmed on Figure 2(a) wich shows their average overlay nodes degree. The high overlay nodes degree of Ropt topologies explains their very weak delivery percentage for flooded messages. Severe congestion problems arise for a moderate amount of overlay nodes. Figure 2(b) illustrates this problem for an overlay density equal to 0.5.

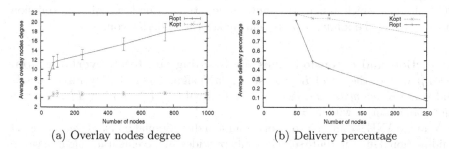

(a) Overlay nodes degree   (b) Delivery percentage

Fig. 2. Average overlay nodes degree and overlay flooding delivery percentage for Ropt and Kopt topologies

4.4 KNN: The Minimal Number of Overlay Neighbours Needed for Connectivity

Problem statement. *Kopt* topologies provide better delivery percentages but are difficult to build in practice. There is no analytical function that gives the optimal number of overlay neighbours needed for connectivity. One could imagine a distributed algorithm that determines K_C. For example, the algorithm employed in [10] for electing the best radio transmission range could be adapted. However, this would require the exchange of a lot of information in the whole network. We reject this solution because of its high bandwidth demand.

It the TC field, it has been demonstrated that for any protocol which preserves worst-case connectivity of the ad hoc network, there exists a placement of n nodes such that the maximum physical node degree in the controlled topology equals $n-1$ [5] . In other words, there is no given number of physical neighbours k, with $k < n - 1$, that implies the connectivity of every network composed of n nodes. However, it has also been shown that setting the minimum number of physical neighbours to 9 is sufficient to obtain connected networks with high probability for ad hoc networks with the number of nodes ranging from 50 to 500 [6].

Similarly, an extensive set of simulations allowed us to determine empirically a parameter K that assures with a high probability the overlay connectivity for a wide range of ad hoc network sizes and overlay densities. We denote KNN (K-Nearest Neighbours) a topology obtained when each overlay node considers as neighbours its K nearest overlay nodes. We now describe the experiment we conducted in order to obtain the K value necessary for the simulations described on Section 3.4.

Testbed. We model the ad hoc network by a graph. Vertices represent the ad hoc nodes that we randomly and uniformly distribute on a unitary square field. We vary their number from 50 to 1000 and the overlay density from 10 to 90%.

For a given set of ad hoc nodes and communication links, the traffic needed for the construction, use and maintenance of overlays is higher on top of sparse logical underlay topologies, that is when an efficient underlay TC algorithm is used [8]. Hence, in order to test the overlay topologies in a stringent environment, we employ the logical topologies obtained after the use of an ideal homogeneous TC technique which assigns the same value r to each node's radio transmission range, r being the minimal value that makes the underlay connected.

Reduction and extension rules for building the KNN overlay graph. Let L_U^K denote the set of K nearest overlay neigbhours of U. Overlay nodes U and V are *K-symmetric neighbours* if and only if $U \in L_V^K$ and $V \in L_U^K$. Figure 3 shows an example with $K = 1$.

Many MANET routing protocols assume bidirectionnal links. Moreover, using unidirectional links in route searches only provides an incremental benefit because of the high overhead needed to handle them [11]. We thus fix as an objective to build overlay topologies where the neighbourhood relation is symmetric.

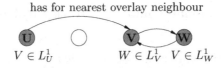

Fig. 3. With $K = 1$, V is a K-asymmetric neighbour of U. V and W are K-symmetric neighbours.

Let L_U denote the set of overlay neighbours selected by overlay node U. For each pair of overlay nodes U and V, there could be two rules to ensure symmetry of the overlay topology:

1. Reduction rule: $V \in L_U$ iff. $U \in L_V^K$ AND $V \in L_U^K$,
2. Extension rule: $V \in L_U$ iff. $U \in L_V^K$ OR $V \in L_U^K$ (graph symmetric closure)

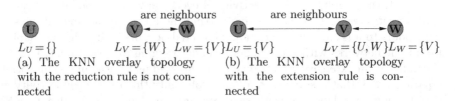

Fig. 4. Result of the reduction and extension rules on the same example topology, with asymmetric neighbours

With the reduction rule, only the symmetric K-neighbours of a node are included in its neighbourhood. With the extension rule, asymmetric K-nearest neighbours are also considered. For a given value K, the topology obtained with the extension rule is a super-graph of the topology obtained with the reduction rule. Its connectivity probability is thus higher. An example is given on Figure 4.

Results. Figure 5 shows the evolution of the percentage of overlays that are connected, for 200 tests with 500 nodes, as a function of the number of nearest overlay nodes (K) for both rules.

The same experiment has been conducted for nodes ranging from 50 to 1000. For a given overlay density, the curves obtained for the different underlay sizes were very close from each other. In other words, we observed that the percentage of connected overlays was more influenced by the overlay density than by the number of nodes.

The three lowest curves are obtained with the reduction rule and the three highest with the extension rule. With both rules, there is a phase where the connectivity probability is very low and a phase where it is very high. The transition from the low-probability phase to the high-probability one arrives sooner and is sharper with the extension rule.

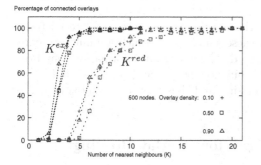

Ov. density	K_{95}^{ext}	R_{95}^{ext}	K_{95}^{red}	R_{95}^{red}
0.1	5	5	11	8
0.3	7	3	17	5
0.5	8	3	19	4
0.7	8	2	25	4
0.9	6	2	15	3

Fig. 5. Percentage of connected overlays as a function of the number of neighbours, with the reduction and extension rules, and for 500 nodes

Fig. 6. Neighbourhood cardinality needed for a connectivity probability equal to 0.95 for the extension and reduction discovery rules (1000 nodes)

Table on figure 6 shows the minimum number of nearest overlay neighbours that must be considered for obtaining 190 connected overlay topologies over 200, for 1000 nodes and different overlay densities. We respectively denoted K_{95}^{red} and K_{95}^{ext} this value for overlays built with the reduction and with the extension rule. All results show that the value of K_{95}^{ext} is far less than K_{95}^{red}. The maximal value of K_{95}^{ext} on our whole set of experiments equals 8, while the maximal value of K_{95}^{red} reached 30.

This table also shows the neighbourhood range (resp. R_{95}^{ext} and R_{95}^{red}) that must be admitted in order to allow the corresponding K_{95}^{red} and K_{95}^{ext} number of overlay neighbours. For each overlay density, the needed neighbourhood range is one to three hops longer, which is not negligible. Assume that the overlay nodes discover their neighbours by sending hello packets. In this case, the bandwidth consumption for building the overlay rapidly grows with the distance at which these packets must be diffused.

The number of overlay neighbours K needed to obtain a high probability of connectivity for a KNN overlay is much lower with the extension rule, that is if we do include the K-asymmetric neighbours, than with the reduction rule. Moreover the K_{95} value with the extension rule is more reliable than with the reduction one because of the sharper transition from the disconnected to the connected phase. Finally, the discovery uses less bandwidth when accepting K-asymmetric neighbours.

Our conclusion is that the extension rule should be used. It however has a side-effect: Though the nearest neighbour lists have a limited size, a given overlay node could be included in the neighbourhood of a larger number of overlay nodes, due to the symmetric closure. Hence, there is no bound on the overlay nodes degree in the KNN extended topology. Nevertheless, as will be discussed in Section 5, a sufficiently selective pruning criterion moderates a lot this drawback.

In this testbed, with the extension rule, the number of nearest overlay neighbours needed for ensuring the connectivity of 95% of the overlay graphs for any overlay density equals 8.

We would like to point out that the solution we propose is not restricted to the simple underlay model used in these simulations, which are only presented as illustrations of the principles exposed. The important information they bring is not the particular value of $K = 8$ but how it can be determined and why it is preferable to use the symmetric closure and thus let the overlay node degree unbounded.

5   Optimizing the Topologies for Overlay Routing

KNN topologies are connected with a high probability. However, as Ropt topologies, they are too dense. Their delivery percentage of flooded overlay messages is low. In this section, we explore methods for eliminating edges while preserving the connectivity property.

5.1   Shortest Path Pruning

Consider figure 7. The overlay nodes (U, V and W) are grey-shaded. Thick arrows represent the flooding of an overlay message from U and thin ones the corresponding underlay packets. In fig. 7(a), the Kopt overlay topology is used; it is composed of the three edges (U, V), (V, W) and (U, W). The flooding of the overlay message on the Kopt topology generates 6 packets on the underlay. However, as illustrated in fig. 7(b), the propagation from U to V, followed by the forwarding from V to W would have been sufficient for all overlay nodes to receive the messages and would have generated only 3 packets. The longest edge of the triangle is unnecessary.

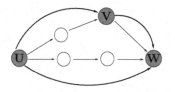 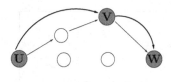

(a) Kopt topology without SPP   (b) Kopt topology with SPP

Fig. 7. Motivation for the Shortest Path Pruning. Thick arrow = overlay message, thin arrow = packet.

We thus introduce the following *Shortest Path Optimization*. Consider three overlay nodes U, V and W, and a distance metric d. The distance metric can be the hop count, the path average delay or any other real positive and symmetric function. Assume that the edge (U, W) is the longest: $d(U, V) <= d(U, W)$ and $d(V, W) <= d(U, W)$. The Shortest Path Optimization sets aside the edge (U, W) if and only if $d(U, V) + d(V, W) <= d(U, W)$. It preserves the connectivity of any overlay graph because an overlay edge is suppressed if and only if an alternative path exists on the overlay.

5.2 Maximal Pruning

Shortest Path Pruning improves the delivery percentage of flooded messages on KNN topologies. However, this pruning method is not sufficiently selective. It can be generalized by setting aside any overlay edge (U, W) such that $d(U, V) + d(V, W) <= \alpha d(U, W)$, with $\alpha >= 1$. Connectivity is still preserved.

The higher value is assigned to α, the more edges are pruned. We call this parameter the *pruning selectivity*. Maximal Pruning is reached when any edge (U, W) is suppressed as soon as there exists two shorter edges (U, V) and (V, W). This behaviour is already obtained for $\alpha = 2$: (U, W) being the longest edge, the inequality $d(U, V) + d(V, W) <= 2d(U, W)$ is always satisfied.

Let us make the distinction between the one-hop overlay neighbours, or *broadcast neighbours*, and the overlay neighbours located farther, the *unicast neighbours*. The emission of only one broadcast packet is sufficient for an overlay flooded message to reach all the broadcast neighbours. Thus, keeping all broadcast neighbours does not increase the bandwidth consumed per overlay flooding. On the other hand, it densifies the final overlay, without increasing the number of unicast neighbours of any overlay node. The consequence is a lower diffusion time and stretch. It also improves the overlay resilience. We thus modify a little the generalized rule in order to maintain as neighbours every pair of overlay nodes located at one hop from each other.

Therefore we finally define the following generic pruning rule.
Consider three edges $E_1 = (U, V)$, $E_2 = (V, W)$ and $E_3 = (U, W)$, a distance metric d, and a real number α with $1 <= \alpha <= 2$. Assume E_3 is the longest edge: $d(E_1) <= d(E_3)$ and $d(E_2) <= d(E_3)$.

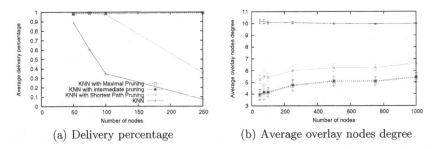

(a) Delivery percentage   (b) Average overlay nodes degree

Fig. 8. The delivery percentage and average overlay nodes degree of KNN overlay topologies pruned with various selectivity factors. A common key for both figures is given on the left one.

Edge E_3 is pruned if and only if:

1. E_3 is longer than one hop, and
2. $d(E_1) + d(E_2) \leq \alpha d(E_3)$.

Figure 8 shows the delivery percentage and the overlay node degree for various pruning selectivity (α) on KNN overlay graphs. The distance metric used is the hop count. For the intermediate pruning selectivity, parameter α is set to 1.5. The delivery percentage increases with the selectivity of the pruning method. It is correlated with the average number of overlay neighbours. Flooding an overlay message consumes much bandwidth. Congestion is avoided on sparse overlay graphs.

The average overlay nodes degree of KNN overlay topologies with Maximal Pruning is above 4, with a tight 95%-confidence interval. Maximal Pruning thus preserves some resilience on KNN overlay topologies. Note that resilience is also provided by the underlay topology and routing protocol. The underlay often offers several different paths between each pair of overlay nodes, and a new route can be built when a path between two overlay neighbours breaks.

5.3   Final Performance Study

A brief comparison with XL-Gnutella. We do not criticize the XL-Gnutella protocol, which is intended to be used for P2P data search, not for overlay routing applications. The point here is to show the utility of our own work in the context of overlay routing.

XL-Gnutella is an optimization of the Gnutella protocol for ad hoc networks. To remain fully compatible with the legacy Gnutella protocol, an overlay edge selection algorithm maintains the number of neighbouring peers between 4 and 8.

The delivery percentage of flooded messages on the XL-Gnutella and KNN topologies are compared on Figure 9(a). Recall that the underlays we use for our simulations are very sparse. In this environment, forcing every overlay node to reject neighbours once the overlay node degree has reached the highest water

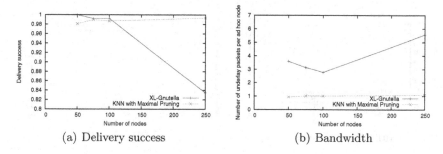

(a) Delivery success   (b) Bandwidth

Fig. 9. XL-Gnutella overlay topologies are intended to be used in a P2P networking context, not for overlay routing

mark of 8 leads a lower connectivity probability than for KNN overlay topologies, for which such restriction does not exist. For the same reason, some overlay edges are longer in XL-Gnutella than KNN topologies. This increases a lot the bandwidth required per overlay message flooding (Figure 9(b)). We also expect, when the underlying routing protocol is reactive, the discovery of XL-Gnutella topologies to consume much more bandwidth than the discovery of KNN topologies, again because some overlay neighbours are selected very far away. In the XL-Gnutella paper, authors use a proactive routing protocol, OLSR, and a cross-layer architecture that allows the P2P middleware to be aware of every overlay node identity and distance, with a low bandwidth consumption. They mention that experiences were also successful with AODV, but that results are better with OLSR.

Comparison of Kopt and KNN with Maximal Pruning. The performance of flooding a message on KNN and Kopt with Maximal Pruning topologies are compared on Figure 10. These are similar, which indicates that the use of the empirical value $K = 8$ before optimization, common for all simulations, instead of the exact minimal number of nearest neighbours needed for overlay connectivity K_C, which value must be determined for each simulation, is not a handicap.

We can also observe that the flooding of an overlay message, which can collect and propagate interesting information for the overlay routing applications, does not consume much more bandwith than the flooding of a packet on the underlay (exactly 1 packet per node). Note also the reasonable value of the overlay path stretch.

Improving resilience. One could use an intermediate value for α instead of Maximal Pruning, for the purpose of improving the overlay topology resilience. Performance obtained on the KNN topologies pruned with $\alpha = 1.5$ and $\alpha = 2$ for instance are very close (their delivery percentage is compared on fig. 8). However, the gain in resilience is difficult to quantify.

Setting a minimum overlay node degree is another way to increase the redundance of the overlay, is easier to evaluate and simple to implement. In some cases, it is even required. This is the case, for example, when one wants to deploy

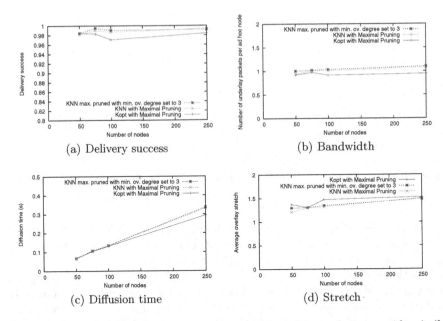

(a) Delivery success   (b) Bandwidth

(c) Diffusion time   (d) Stretch

Fig. 10. After pruning, flooding a message on KNN overlay topologies provides similar performance results than on Kopt ones. Setting the minimal overlay degree to 3 does not modify significantly the performance obtained on KNN with Maximal Pruning.

multipath routing on the overlay. A minimal number K_{min} of overlay neighbours is easily guaranteed by reading the nearest overlay nodes list in increasing order of distance and beginning to apply the pruning rule only at the $K_{min} + 1$ element. On Figure 10, we also compare the performance obtained with Maximal Pruning on KNN topologies when applying the pruning rule to the 3 nearest overlay nodes and when sytematically keeping them in the final neighbourhood.

6   Conclusions

Overlay routing is well-suited to ad hoc networks. In an ad hoc network, there is no centralized administration. Overlay routing would ease the test and the introduction of new routing protocols, without preliminary agreement between the whole nodes set. Furthermore, an ad hoc network is often composed of several groups of users with specific routing needs. The network conditions, for example the available bandwidth and the mobility level, can vary a lot. Overlay routing would allow each group of users to employ a customized routing protocol for their common application, or to adapt it to the network conditions.

In this paper, we discussed what kind of overlay topology should be proactively built before an overlay routing protocol enters a route search process on top of it.

The basic overlay structures we studied are the *K-Nearest Neighbours* overlay topologies, connecting every overlay node to its K nearest peers. These overlays can be established with respect to the locality principle, whatever the underlay routing protocol type. This feature is necessary for providing a sustainable building and maintenance cost of the overlays. Parameter K must be empirically tuned. In order to obtain an overlay where the neighbourhood relation is symmetric, the symmetric closure of the K-nearest neighbour graph is preferable to its reduction. The extension method is expected to consume less overlay topology control traffic and is also more reliable, as the corresponding K_{ext} value depends less on the number of ad hoc devices and overlay density. The extension rule lets the overlay node degree unlimited. However, an optimization of the resulting overlay topology cancels this drawback.

We introduced a family of optimization rules of the K-Nearest Neighbours topologies, based on a pruning rule. As flooding is a key component of many route discovery mechanisms in MANETs, our performance study focusses on the delivery percentage, bandwidth consumption and time duration of flooding on the overlay. Simulations illustrated the gain in performance when flooding a message on pruned topologies. The most selective rule, Maximal Pruning, suppresses any overlay edge such that there exists an alternative path in the overlay graph, while preserving from pruning any pair of overlay neighbours that are in the direct communication range of each other. We also considered the overlay path stretch and the overlay nodes degree as respective indicators for the data transfer transmission time and overlay resilience. Maximal Pruning does not increase a lot the path stretch, but can have an undesired effect on the overlay resilience. It can be easily adapted such as to impose a minimal overlay node degree K_{min}. For reasonable values of K_{min}, the performance remains very close to the one obtained with the primary Maximal Pruning rule.

The overlay topology control protocol itself will be presented in a follow-up paper. Its design guideline will be to maintain, in a mobile context, the overlay topology as close as possible to the target K-Nearest Neighbours overlay topology with Maximal Pruning and a minimal bound on the overlay node degree.

Acknowledgements

This work has been partially supported by the European Union under the ANA FET project (FP6-IST-27489).

References

1. Clark, D.: The design philosophy of the DARPA Internet protocols. Computer Communication Review 18(4), 106–114 (1988)
2. Klemm, A., Lindemann, C., Waldhorst, O.: A special-purpose peer-to-peer file sharing system for mobile ad hoc networks. In: Proc. of IEEE Semiannual Vehicular Technology Conference (VTC2003-Fall), IEEE Computer Society Press, Los Alamitos (2003)
3. Duran, A., Shen, C.C.: Mobile ad hoc p2p file sharing. In: Proc. of IEEE Wireless Communications and Networking Conference (WCNC'04), IEEE Computer Society Press, Los Alamitos (2004)

4. Conti, M., Gregori, E., Turi, G.: A cross-layer optimization of gnutella for mobile ad hoc networks. In: Proc. of ACM MobiHoc 05, pp. 343–354. ACM Press, New York (2005)
5. Santi, P.: Topology control in wireless ad hoc and sensor networks. ACM Comp. Surveys 37(2), 164–194 (2005)
6. Blough, D., Leoncini, M., Resta, G., Santi, P.: The k-neigh protocol for symmetric topology control in ad hoc networks. In: Proc. of ACM MobiHoc 03, pp. 141–152. ACM Press, New York (2003)
7. Wattenhofer, R., Zollinger, A.: XTC: A practical topology control algorithm for ad-hoc networks. In: Proc. of 4th International Workshop on Algorithms for Wireless, Mobile, Ad Hoc and Sensor Networks (WMAN) (2004)
8. Calomme, S., Leduc, G.: The critical neighbourhood range for asymptotic overlay connectivity in ad hoc networks. Ad Hoc & Sensor Wireless Networks journal 2(2) (2006)
9. Perkins, C., Royer, E.M.: Ad hoc on-demand distance vector routing. In: Proc. IEEE Workshop on Mobile Computing Systems and Applications(WMCSA'99), IEEE Computer Society Press, Los Alamitos (1999)
10. Narayanaswamy, S., Kawadia, V., Sreenivas, R., Kumar, P.R.: Power control in ad-hoc networks: Theory, architecture, algorithm and implementation of the compow protocol. In: Proc. of European Wireless 2002. Next Generation Wireless Networks: Technologies, Protocols, Services and Applications, Florence, Italy, pp. 156–162 (February 2002)
11. Marina, M.K., Das, S.R.: Routing performance in the presence of unidirectional links in multihop wireless networks. In: MobiHoc '02: Proceedings of the 3rd ACM international symposium on Mobile ad hoc networking & computing, pp. 12–23. ACM Press, New York, NY, USA (2002)

A Generic, Self-organizing, and Distributed Bootstrap Service for Peer-to-Peer Networks

Michael Conrad and Hans-Joachim Hof

Institute for Telematics, Universität Karlsruhe (TH), Germany
{conrad,hof}@tm.uka.de

Abstract. In many scenarios, self-organization is the driving force for the use of a peer-to-peer (p2p) network. However, most current p2p networks are not truly self-organizing, as little attention has been paid on how new nodes join a p2p network, the so-called *bootstrapping*. Current p2p network protocols rely on prior-knowledge of nodes like a list of IP addresses of bootstrap servers or like a list of known peers of a p2p network. However, this kind of prior knowledge conflicts with the self-organization principle and the distributed character of p2p networks. In this paper, we present the design of a *generic, self-organizing, and distributed bootstrap service* which can be used to bootstrap p2p networks of arbitrary size, even very small, private p2p networks. This bootstrap service works in today's Internet and it can be easily integrated into existing p2p applications. We present an evaluation of the proposed bootstrapping service showing the efficiency of our approach.

1 Introduction

Nowadays, peer-to-peer (p2p) networks are used in many applications, e.g. for VoIP, Instant Messaging, or filesharing. For most of these applications, decentralized control and self-organization are desired. However, most current p2p networks do not achieve true self-organization or true decentralized control because they often use well-known central servers or a list of known p2p network member nodes to bootstrap new nodes. In this paper, we propose a generic, distributed and self-organizing bootstrap service which allows nodes to join into arbitrary p2p networks. The proposed bootstrap service itself uses a p2p network, the bootstrap p2p network, for distributed storage of bootstrap information. The bootstrap information may include nodes which can be used to join the p2p network of the corresponding p2p application. For example if the user of a filesharing application starts the filesharing client for the first time, the client joins the bootstrap network and retrieves bootstrap information for the filesharing network. Then, it joins the filesharing network itself. Of course, this approach only shifts the problem of joining a p2p network to joining the bootstrap p2p network.

However, if the bootstrap service is implemented in more than one application, synergy effects may be used to join the bootstrap network. The synergy effect results from the larger number of nodes in the bootstrap network, which

D. Hutchison and R.H. Katz (Eds.): IWSOS 2007, LNCS 4725, pp. 59–72, 2007.

occurs because all nodes join the same bootstrap p2p network. One method for bootstrapping is Random Address Probing, which probes randomly chosen IP addresses to find active nodes. These nodes are used to establish initial contact with the p2p network. We use Local Random Address Probing for the bootstrapping of the bootstrap p2p network. While Random Address Probing may be successfully used by very large p2p networks, it is not efficient for small to medium networks, as many probes are necessary. To evaluate the performance of our bootstrap p2p network, we collected real world data about the distribution and number of peers of a deployed, large p2p filesharing network (eDonkey). Assuming that all eDonkey clients would use our bootstrap service, we evaluate the performance of our protocol.

This paper is structured as follows: in section 2, requirements for a bootstrap service are defined. Section 3 reviews related work. In section 4, the design of our generic, distributed, and self-organizing bootstrap service is presented. The bootstrap service is evaluated in section 5. Section 6 concludes this paper.

2   Requirements for a Generic Bootstrap Service

A generic bootstrap service for p2p applications and p2p networks must fulfill the following requirements:

- *Self-Organization and Distributed Control (R1):*
  A p2p network can only be self-organizing if every network protocol step, including the bootstrapping, is self-organizing. To support the distributed character of most p2p networks, a bootstrap service may not rely on prior knowledge (e.g. a list of IP addresses of bootstrap servers).
- *Heterogeneity (R2):*
  A decentralized bootstrap service should support p2p networks of arbitrary size and function. The service should provide bootstrap support for very small (private) p2p networks, only consisting of a few nodes, as well as for huge p2p networks with up to millions of active nodes.
- *Scalability and Robustness (R3):*
  The decentralized bootstrap service itself should scale well with an increasing number of nodes. The bootstrap service should also scale with the number of participating networks which is especially important if the bootstrap service is used to bootstrap a huge number of small, private p2p networks. The bootstrap service should work as expected, even in case of malfunction of several nodes.
- *Practicability (R4):*
  The bootstrap service should be designed for today's Internet. Hence, the bootstrap service may not rely on currently undeployed protocols like multicast etc.
- *Seamless Integration (R5):*
  It should be easy to integrate the bootstrap service into an existing p2p application.

- *Modularity and Extensibility (R6):*
    A bootstrap service should be extendable to be able to react on changes of the network environment and to react on innovations. If, for example, multicast gets widely deployed in the Internet one day, it should be easy to extend the bootstrap service.

In this paper, we do not consider other requirements like privacy issues of the bootstrapping service etc.

3   Related Work

Cramer et al. [1] compare different bootstrapping techniques for p2p networks, including static bootstrap servers, out-of-band node caches, random address probing, and network layer mechanisms using any- or multicast. The results of the Random Address Probing method are of interest for our work. We enhance this mechanism for the proposed bootstrap service. The focus of [1] is on locality aware bootstrapping to offer an optimal topology for the join of new peers. However, no detail is given about the design of a generic bootstrap service. We fill this void with the proposed bootstrap service.

The idea of a generic bootstrap service for p2p networks was already discussed in [2] and [3]. The first paper is proposing an approach relying on a distributed hash table running on top of a structured overlay network, whereas the second paper uses a prefix based routing on top of a structured overlay network. The universal ring, which was proposed in [2], relies on a distributed hash table to store and query informations about services and to provide bootstrap information to use these services. While this approach can be used to provide bootstrap support for p2p networks, it does not meet some of the requirements of section 2: At first, the requirements R1 (Distributed Control) and R4 (Practicability) can not be met. To join the universal ring, a server or a globally known multicast address are used. A central server is prohibitive and multicast is a mostly undeployed technology in today's Internet. In addition, the scalability of the proposed for huge peer-to-peer networks seems unclear. Providing bootstrap support for large p2p networks using a distributed hash table results in storage of a huge amount of data on a very small set of nodes, hence overloading these nodes. In our opinion, a special distribution of bootstrapping information for huge peer-to-peer network is necessary to avoid overloading nodes of the bootstrapping nodes. The paper [3] lacks information on how nodes join the bootstrap service. The authors propose to use the NEWSCAST protocol [4] for data storage instead of a distributed hash table. As NEWSCAST relies on multicast, it does not meet requirement R4 (Practicability).

Even existing public peer-to-peer frameworks like JXTA [5] use static bootstrapping nodes (so-called seeds) to integrate new nodes into the peer-to-peer network. The JXTA framework also includes decentralized bootstrapping support using multicast, however due the missing deployment of multicast in the Internet infrastructure, this method is inapplicable for the use in the public Internet.

4 Design

This section presents the design of the proposed generic, distributed, self-organizing bootstrap service. The design meets all requirements of section 2.

4.1 Overview

Instead of creating a stand alone bootstrap mechanism for each existing peer-to-peer network (p2p network) we propose a single dedicated p2p network for the bootstrapping of all peer of arbitrary p2p networks. This so called *bootstrap p2p network* provides bootstrap information for peers which want to join a p2p network. Our bootstrap p2p network provides a service similar to the domain name system (DNS): while DNS resolves domain names to IP addresses, our bootstrap p2p network resolves p2p network names to bootstrap information (which is in most cases an IP address of a peer or a list of peers already connected to the p2p network). These bootstrap information can be used by a node to join the p2p network. After successfully joining a p2p network, the peer publishes bootstrap information to the bootstrap p2p network to supporting queries for bootstrap information for future peers. The peer also joins the bootstrap p2p network itself.

A dedicated bootstrap p2p network shifts the problem of joining an arbitrary p2p network to joining the bootstrap p2p network. However, as only one bootstrap p2p network exists for all p2p networks and all peers join the bootstrap p2p network, the bootstrap p2p network is larger than any other p2p network, allowing bootstrapping techniques, which are prohibitive for smaller networks, for example Random Address Probing. Especially very small p2p networks consisting of only tens or hundreds of peers can profit from this bootstrap p2p network, but it is also of benefit for large p2p networks because it simplifies the initial deployment of any p2p network.

4.2 Components of the Bootstrap Service

To achieve a flexible and extensible design our *bootstrap service* consists of two separate modules: The first module implements the initial bootstrapping of the bootstrap p2p network. The second module is responsible for providing bootstrap information (like a list of IP addresses of active peers) to nodes which want to join a distinct p2p network. Figure 1 shows the public interface and the schematic composition of the bootstrap service.

The public interface of the bootstrap service offers two methods. The method `lookup(name)` is used by a new node to search for bootstrap information of a p2p network whose identifier is `name` (e.g. *ed2k* for eDonkey peers). The method returns a list of `BootstrapData` objects. Each of these object contains a bootstrap information record (e.g. the IP address of one active node of the p2p network). After a node successfully joined a p2p network, it uses the method `publish(name,info)` to publish bootstrap information (`info`) about the p2p network `name`.

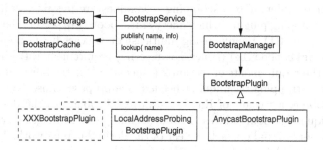

Fig. 1. Schematic composition of the bootstrap service

Every bootstrap service instance runs the *BootstrapManager*, which is responsible for the initial bootstrapping of the bootstrap p2p network. The BootstrapManager offers the interface *BootstrapPlugin* to support plugins to meet requirement R6 (Modularity and Extensibility). Each plugin implements one distinct bootstrap mechanism, e.g. Random Address Probing. Plugins are not limited to pure self-organizing bootstrapping methods. For example, there may be bootstrap plugins which use a node cache, or there may even be server-based plugins. However, it is recommended to offer at least one self-organizing bootstrap plugin as fallback to maintain the self-organizing character of the boostrap service.

The component *BootstrapStorage* is responsible for storing bootstrap information. The proposed bootstrap service uses a soft sate approach, hence old bootstrap information will be deleted after a given time. The component *BootstrapCache* is used by the bootstrap service to cache information about peers of bootstrap p2p network to simplify the bootstrapping in case of a reconnect to bootstrap p2p network.

The design of the proposed bootstrap service does not contain any constraints about programming languages or other programming paradigms. Therefore, it can be integrated into arbitrary p2p applications. Hence, seamless integration (requirement R5) can be achieved. We expect that the easy and seamless integration will lead to a rapid deployment of our bootstrap service, especially if the open source community can be convinced to use the bootstrap service. Hence, the critical mass for the bootstrap service can be easily achieved.

The following sections give a detailed overview of the two modules Bootstrap-Manager and BootstrapStorage.

4.3   BootstrapManager: Bootstrapping of the Bootstrap p2p Network

In our proposed bootstrap service, the *BootstrapManager* is responsible for the initial bootstrapping of a new node. To join the bootstrap p2p network, the BootstrapManager instance of the bootstrap service running on the node may uses one or more of its bootstrap plugins to join the bootstrap p2p network.

A number of different bootstrapping techniques are possible. Starting from simple bootstrapping plugins, which query static bootstrap servers (hence do not meet requirement R1), plugins using multicast or anycast based bootstrapping (`AnycastBootstrapModule`) (hence not meeting requirement R4) are possible. To meet all the requirements of section 2 (especially Practicability (R4)) a more advanced bootstrapping technique is needed. We propose to use *Local Random Address Probing*, a variant of Random Address Probing, as the default bootstrap technique. Random Address Probing sends probe messages to random IP addresses hoping to find one active peer of the desired p2p network. The performance of Random Address Probing [1] strongly depends on the number of peers and the distribution of these peers in the IP space. For most p2p networks, Random Address Probing is prohibitive because these networks do not have enough users, resulting in a very poor performance of Random Address Probing.

However, the proposed bootstrap service uses the bootstrap p2p network, which is potentially very large. Hence, Random Address Probing may be used. *Local Random Address Probing* is similar to the classical Random Address Probing, but instead of probing IP addresses uniformly distributed over the complete IP address space, the Local Random Address Probing limits the probed IP range to the local range around the current IP address of the user. This behavior of Local Random Address Probing improves the performance of Random Address Probing because local communication may be faster than remote communication and because the distribution of p2p nodes may not be uniform. Hence, the improved performance of Local Random Address Probing is partly based on the assumption that most users of p2p applications are private users which get Internet access via dialup networks (DSL, cable). Hence, we expect a higher locality of p2p users in dialup networks than in other network ranges. We verify this assumption in section 5.1. We exploit the higher density of p2p network peers in dialup networks to further improve the performance of the initial bootstrapping. Using Local Random Address Probing for our bootstrap service, requirements R1 and R4 can be met as there are no more dependencies to a central infrastructure. In contrast to other bootstrap services (see section 3), our proposed bootstrap service can be deployed in today's Internet.

4.4   BootstrapStorage: Efficient Distribution of Bootstrap Information

Regarding the potentially large number of nodes in the bootstrap p2p network, the distributed management must be very efficient to avoid overloading of peers. Management duties includes storage of potentially many published bootstrap information and an efficient search for existing bootstrap information. These requirements can be met by using distributed hash tables based on structured overlays like Chord [6], CAN [7] or Pastry [8]. Distributed hash tables offer a distributed storage of data and an efficient search. They scale well with an increasing number of peers. Another advantage of structured peer-to-peer overlays is a common abstract interface described in [9]. Hence, the bootstrap service

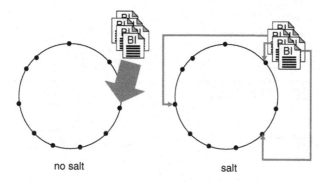

no salt   salt

Fig. 2. The use of a salt value for our bootstrap service

can be implemented independent from the used overlay network, offering the possibility to change it later.

The *BootstrapStorage* module uses the `put(key,data)` method to store bootstrap information in the bootstrap p2p network. The bootstrap information `data` is stored under the key `key`. The key is calculated from the name of the p2p network. The bootstrap information can be retrieved using the `data=get(key)` method. Both methods are provided by the common API of [9]. However, the usage of these methods implies, that all bootstrap information for one distinct p2p network is stored under the same key, as the key is derived from the name of the p2p network in which a peer wants to join (e.g. "ed2k" for eDonkey). Figure 2 ("no salt") shows this problem. Regarding very large networks like the eDonkey filesharing network, it is clear, that the bootstrap information of this network would surely overload the peer of the bootstrap p2p network which stores all bootstrap information. We propose an adaptive algorithm for a better distribution of bootstrap information over the bootstrap p2p network:

Instead of using only the name of the requested p2p network we propose to include an additional random value, the so called *salt*, in the calculation of the key. The key is calculated using the name of the p2p network concatenated with the salt as input of a hash function H:

$$key = H(name + salt)$$

Using this generation of the key results in a uniform distribution of bootstrap information of one distinct p2p network across the whole bootstrap p2p network. To search for bootstrap information, a node randomly selects a salt, calculates the key and queries the bootstrap p2p network for that key. While the improved distribution avoids overloading single peers of the bootstrap p2p network, it is not suitable for small p2p networks, as the probability of a successful search query is very small, hence resulting in many search queries with different salts. To provide a solution for this problem, we propose an *Adaptive Salt Window Algorithm*, which is able to adjust the distribution of bootstrap information automatically to the size of the particular p2p network without knowing the number of peers. The Adaptive Salt Window Algorithm uses an interval (the so

```
01    publish( name, info) {                                lookup( name) {
02
03        for( window=0, i=n; i>3 && window==0; i--) {          for( i=n; i>3; i--) {
04            salt = random( 2^i);                                  salt = random( 2^i);
05            key = hash( name + salt);                             key = hash( name + salt);
06
07            result = p2p_lookup( key);                            result = p2p_lookup( key);
08
09            if ( |result| >= 10 ) { // sufficient data available  if ( |result| > 0 ) { // data available
10                window = i + 1 // increase window                     return result
11            } else if ( |result| > 0 ) { // data available        }
12                window = i;                                   }
13            }
14        }                                                    for( i=0; i<8 && |result|==0; i++) {
15                                                                  key = hash( name + random( 8));
16        salt = random( max( 2^window, 8));                   }
17        key = hash( name + salt);                             result = dht_get( key);
18                                                          }
19        dht_put( key, info);                              return result
20    }                                                   }
```

Fig. 3. Pseudo code of `publish` and `lookup`

called Salt Window) in which all salt values are contained. The Salt Window is
adapted to the current number of peers of the p2p network. Small p2p networks
only have a small window, whereas large p2p networks have a large window.

Figure 3 shows the pseudo code of the `publish` method and of the `lookup`
method using the Adaptive Key Window Algorithm.

The `publish` method and the `lookup` method both automatically detect the
current Salt Window size of the requested p2p network. The publish method
starts with a maximum length interval of $[0, 2^n]$ in which a random salt is contained. In each step, the publish method halves this interval and tries to retrieve
data using the name of the p2p network and a random salt in $[0, 2^i]$. This step
is repeated unless bootstrap information is found. The corresponding interval
is the Salt Window used for the storage of the bootstrap information. The size
of the Salt Window is slightly adopted if more than 10 values to allow for an
increase of the Salt Window size.

When a node uses the lookup method to query for bootstrap information of
a p2p network the range of the salt value starts from the maximum value and
will be halved until one valid bootstrap information was found.

The adaptive adjustment of the key window guarantees the support for arbitrary p2p network independent of their number of peers. It prevents overloading
single peers hence results in an efficient storage of bootstrap information. The
Adaptive Salt Window Algorithm allows to react on an increasing or decreasing
number of peers. If the number of peers changes significantly, the Salt Window
will be resized with high probability. Therefore the proposed bootstrap service
meets requirements R2 (Heterogeneity) and R3 (Scalability and Robustness).
Other load balancing mechanisms will be addressed in future work.

5   Evaluation

In this section, we evaluate the performance of the proposed bootstrap service
which uses Local Random Address Probing for the initial bootstrapping of the
bootstrap p2p network. We also analyze the scalability of the bootstrap p2p
network.

5.1 Performance of Local Random Address Probing

In section 4.3 we assumed a non-uniform distribution of p2p network users in the IP address space. Furthermore we assumed, that computers connected to the Internet via dialup networks (e.g. DSL, cable) have a higher probability of running a p2p application than computers in the rest of the Internet. In this section, we provide strong evidence that these assumptions are justified. To verify these assumptions, we measured the distribution of active nodes of the eDonkey network[1], one of the biggest filesharing networks in Germany today. By inspecting a real-world p2p network, we provide a lower bound for the expected performance of the proposed bootstrap service, because we expect several p2p networks and not just one to integrate our generic, distributed and self-organizing bootstrap service. The eDonkey protocol runs as default on Port 4662. To detect eDonkey nodes, we used nmap[2] to perform a TCP SYN scan on port 4662. This scan provides a lower bound of active eDonkey nodes because it scans only for the default port. However, it is easy to change the default port and some users do this to prevent rate regulation of their provider. We also do not take the blocking of the default port into consideration, which is done sometimes by several Internet service providers.

We limit our examination of active eDonkey nodes to the German IP address space to avoid distortion of the results by different time zones and similar effects. We use the public daily database snapshot of the European Internet Registry (RIPE)[3] to get the currently allocated IP address space of Germany. In April 2007 this list contained about 160.000 IP ranges allocated with the German country code. These IP ranges are equivalent to 450.000 /24 networks, each network containing about 250 valid IP addresses. We used this list (GIPL, German IP List) as basis of our experiment. The list of German dialup networks (GDUL, German dialup network List) was created manually from a set of 38 dialup IP ranges, consisting of 29.000 /24 networks.

The experiment itself runs on a standard desktop computer connected via 100 MBit Ethernet to the campus network. Every 6 minutes a set of 5 /24 networks was extracted randomly from both lists (GIPL and GDUL). Each of these 5 networks was scanned for active eDonkey nodes as described above. Starting from the given network address, nmap was configured to scan the corresponding /20 network, which consists about 4000 valid IP addresses.

Our experiment was running from April 10th till April 18th. We scanned 21.600 /20 networks (containing 80 million ip addresses) for active eDonkey nodes, 10.800 from each of the lists (GIPL and GDUL).

Probing ranges from GIPL, 2.987 out of 10.800 probes found at least one active eDonkey node, whereas 9.193 out of 10.800 probes in the range of GDUL found active eDonkey nodes.

From over 1.100.000 online computers found in networks of GIPL only 4.39% (48.525) run the eDonkey software. The probing of the dialup networks of GDUL

---

[1] original website down, see http://en.wikipedia.org/wiki/EDonkey2000
[2] network mapper - website: http://insecure.org/nmap
[3] website: http://www.ripe.net/ripe/index.html

discovered about 880.000 online computers out of which 11,46% (100.781) were identified as active eDonkey node.

These results show that the distribution of p2p applications in dialup network is higher than in other networks. Hence, our assumptions are justified at least for the eDonkey network. Furthermore, in over 85% of probed dialup networks of GDUL at least one active eDonkey node was found whereas this is the case for only 27.65% of networks in GIPL. At the same time, the amount of active eDonkey nodes in dialup networks is twice higher than in other networks.

Figure 4 shows the distribution of eDonkey nodes across online computers for random networks (GIPL) and dialup networks (GDUL). It can be easily seen that the distribution of active eDonkey nodes is higher for dialup networks than for random networks. The maximum of active eDonkey nodes will be reached on the weekend (14-th and 15-th of April), corresponding with the Internet usage of dialup users.

As we will use Local Random Address Probing for the bootstrapping of the bootstrap p2p network, the number of probes before the first active bootstrap node stands for the overhead which our proposed bootstrap service generates and, more important, the number of probes is directly correlated to the time a user has to wait before it can join the p2p network which uses our bootstrapping service.

Figure 5 shows that in average about 600 probes are necessary to find an active node of the eDonkey p2p network in dialup networks (GDUL). Finding the first active eDonkey node requires in average less than 20 seconds. For random networks the equivalent value can not be given, because only about 30% of the network scans discover at least one active eDonkey node.

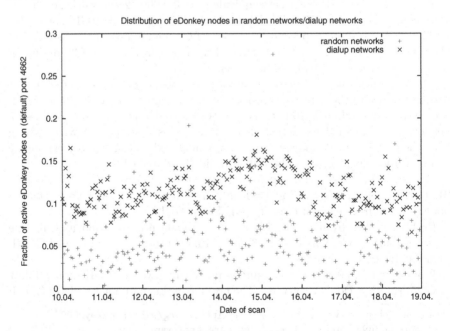

Fig. 4. Distribution of active eDonkey nodes in random or dialup networks

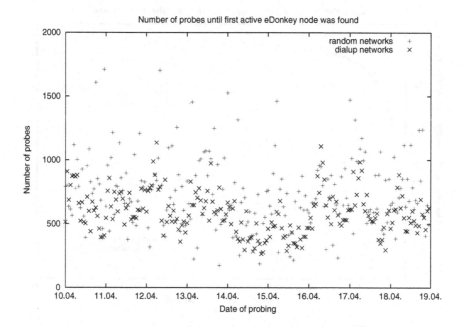

Fig. 5. Number of required probes to find first eDonkey node

In comparison with random networks (GIPL) a lower number of probes is required in dialup networks (GDUL) until the first active eDonkey nodes is found. At the same time, the deviation is significantly lower. This result support our assumption, that local random address probing is well suited for finding other peer nodes in dialup networks. The lowest number of required probes for dialup networks was reached in our scan on the weekend of 14-th/15-th April.

Figure 6 shows the cumulative distribution of time needed for probing until the first eDonkey node was found in dialup networks (GDUL) and in random networks (GIPL). Thereby the results for random networks only rely on the 30% of successful probes. It shows the efficiency of our approach. For example, after 20 seconds, over 80% of nodes found the first eDonkey node. For our bootstrap service this is equivalent to a successful bootstrapping.

Analyzing the eDonkey p2p network as a synonym for a distributed bootstrap service the initial bootstrapping into the bootstrap overlay can be satisfyingly realized by Local Random Address Probing if the bootstrap service has a large enough number of nodes. These findings show that Local Random Address Probing can be efficiently used for the bootstrapping of our bootstrap p2p network.

5.2 Performance of the Bootstrap Information Storage

This section evaluates the performance of the Adaptive Salt Window Algorithm (see section 4.4) used during storage and retrieval of bootstrap information.

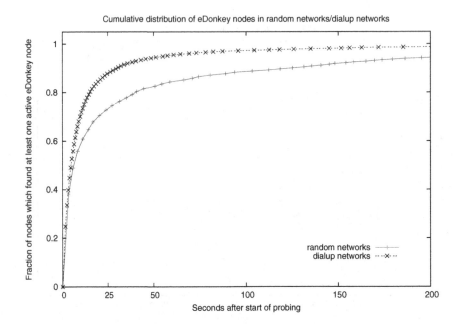

Fig. 6. Cumulative distribution of eDonkey nodes

We define 4 different scenarios of p2p networks distinguished by their number of users (10^6, 10^5, 10^4 and 10^3). Starting from a running bootstrap p2p network with 4.200.000 (2^{22}) peers, all nodes bootstrap into one network using the proposed bootstrap service. Therefore new peers use to the bootstrap p2p network to retrieve bootstrap information by generating a lookup query. In all scenarios the number of overlay queries, resulting from lookup-queries, were inspected. After a successful join into the desired p2p network each node publish bootstrap information for that p2p network.

Figures 7 shows the number of overlay queries generated by lookup requests for a large p2p network with 10^6 nodes, where all nodes joining the bootstrap p2p network subsequently. For the other scenarios, only differing in the number of joining nodes, similar results for the average number of overlay queries generated by lookup requests, will be archived.

At the beginning, when only few bootstrap information for the requested p2p network are available, up to 25 overlay queries are necessary for each lookup query. With an increasing number of nodes which join the new p2p network and publishing bootstrap information the number of overlay queries decreases significantly.

For 10^3 nodes the average number of lookup queries is about 12, for 10^4 or 10^5 peers 9 respectively 6 lookup queries are required in average. For large p2p networks with 10^6 nodes in average only 3 lookup queries are necessary. The simulation shows, that the proposed distribution of bootstrap information scales with a increasing number of peers. At the same time, small p2p networks are also supported, although a higher number of overlay queries is required.

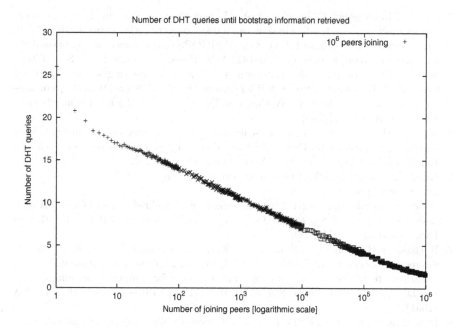

Fig. 7. Number of DHT queries to obtain bootstrap information

6   Conclusion and Future Work

We presented a generic, distributed, and self-organizing bootstrap service for arbitrary peer-to-peer networks (p2p networks) which can be used in today's Internet. Our proposed bootstrap service offers bootstrapping for small private p2p networks as well as for large p2p networks. The bootstrap service is easy to integrate into existing p2p applications and can be extended by plugins.

For storage of bootstrap information, our bootstrap service uses a distributed hash table. The Adaptive Salt Window Algorithm is used to achieve an efficient distribution of bootstrap information across the nodes which run the bootstrap service, hence preventing the overloading of single nodes.

We evaluated the proposed bootstrap service using real world data of a large p2p network, the eDonkey filesharing network. The results show that our bootstrap service can be efficiently used to bootstrap arbitrary p2p networks.

Future work will address a simulator implementation and a prototype implementation of the proposed bootstrap service.

References

1. Cramer, C., Kutzner, K., Fuhrmann, T.: Bootstrapping locality-aware p2p networks. In: Proceedings of the IEEE International Conference on Networks (ICON 2004), Singapore, November 16–19 2004, vol. 1, pp. 357–361. IEEE Computer Society Press, Los Alamitos (2004)

2. Castro, M., Druschel, P., Kermarrec, A.-M., Rowstron, A.: One ring to rule them all: service discovery and binding in structured peer-to-peer overlay networks. In: EW10: Proceedings of the 10th workshop on ACM SIGOPS European workshop: beyond the PC, Saint-Emilion, France, pp. 140–145. ACM Press, New York, NY, USA (2002)
3. Jelasity, M., Montresor, A., Babaoglu, O.: The bootstrapping service. In: ICD-CSW '06: Proceedings of the 26th IEEE International ConferenceWorkshops on Distributed Computing Systems, Washington, DC, USA, p. 11. IEEE Computer Society Press, Los Alamitos (2006)
4. Jelasity, M., van Steen, M.: Large-scale newscast computing on the Internet, Vrije Universiteit Amsterdam, Department of Computer Science, Amsterdam. Technical Report IR-503. Amsterdam, The Netherlands (October 2002)
5. Gong, L.: Project JXTA: A technology overview (August 2001), http://www.jxta.org
6. Stoica, I., Morris, R., Karger, D., Kaashoek, M.F., Balakrishnan, H.: Chord: A scalable peer-to-peer lookup service for internet applications. In: SIGCOMM'01, San Diego, California, USA (2001)
7. Ratnasamy, S., Francis, P., Handley, M., Karp, R., Schenker, S.: A scalable content-addressable network. In: SIGCOMM '01: Proceedings of the 2001 conference on Applications, technologies, architectures, and protocols for computer communications, San Diego, California, United States, pp. 161–172. ACM Press, New York, NY, USA (2001)
8. Rowstron, A., Druschel, P.: Pastry: Scalable, decentralized object location, and routing for large-scale peer-to-peer systems. In: Guerraoui, R. (ed.) Middleware 2001. LNCS, vol. 2218, pp. 329–350. Springer, Heidelberg (2001)
9. Dabek, F., Zhao, B., Druschel, P., Kubiatowicz, J., Stoica, I.: Towards a common API for structured peer-to-peer overlays. In: 2nd Int. Workshop on P2P Systems (2003)

CSP, Cooperative Service Provisioning Using Peer-to-Peer Principles

Michael Kleis[1], Kai Büttner[1], Sanaa Elmoumouhi[2], Georg Carle[3], and Mikael Salaun[2]

[1] Fraunhofer FOKUS,
Kaiserin-Augusta-Allee 31, 10589 Berlin, Germany
[2] France Télécom R&D,
avenue Pierre Marzin 2, 22307 Lannion, France
[3] University of Tübingen,
Sand 13, 72076 Tübingen, Germany

Abstract. In this paper we describe a self-organising and self-managing system for a Cooperative Service Provisioning (CSP) of media transport and processing services. The term cooperative is used since we assume that CSP providers as well as users offer resources to be utilised for media delivery and processing based on an Overlay Network principle. The core building block of the proposed system is a Distributed Hash Table extended with a CSP specific indexing principle and recursive search algorithm. The task of QoS constraint verification for a requested service is distributed between participating nodes. In this paper we describe CSP based on a Content Addressable Network (CAN) [1] DHT. The resulting system is evaluated based on a theoretical analysis as well as simulations.

1 Introduction

One of the central challenges of service provisioning in current and future networks is to incorporate an increasing number of wireless and wired network technologies, a variety of heterogeneous end user terminals and the requirements of QoS sensitive and realtime multimedia services. To optimise the perceived quality of service for its customers, a service provider has to adapt a multimedia service to the corresponding end user terminals as well as the transport and error characteristics of the used access technologies. This may include: Multimedia processing as transcoding and/or downscaling of audio/video data, QoS aware transport and routing, traffic shaping as well as the implementation of network based error correction techniques. The network and device management actions required in this context have to be based on information received through interaction with the network infrastructure as well as the source and the sink of the realtime multimedia flow. From the viewpoint of a service provider, core requirements of a platform to meet these demands are: low costs, low management and configuration complexity as well as scalability. Based on these requirements the problem addressed in this paper is the exploration of a self-*[2] system enabling a cooperative service provisioning (CSP) of media transport and processing services. In this context self-* denotes self-organising, self-managing and self-repairing. The term cooperative is used since we act on the assumption that endusers as well as network or service providers cooperate via the provisioning of processing modules (PMs) to be used for

D. Hutchison and R.H. Katz (Eds.): IWSOS 2007, LNCS 4725, pp. 73–87, 2007.

media processing as well as media routing. The main focus of this paper will be on the question, how principles from the area of Peer-to-Peer (P2P) networks can be used to establish an on demand service provision platform through coordination of a distributed set of *PM*s into a *Situated Overlay Network*. In this context we address the following three essential steps, which are *Decomposition of Services*, *Service Discovery* and *Situated Overlay setup*. In contrast to related approaches as [3], [4], [5], [6] CSP does not rely on a central entity having global knowledge during the task of Overlay setup. The central idea of our approach is to study how the DHT principle can be extended to realise a distributed control plane for the setup and maintenance of Overlays. The main incentive for using DHTs is the fact, that they are in general designed to be used in error prone P2P scenarios and can be considered as self-organizing, self-managing and self-repairing structures. As the anticipated benefit, the resulting CSP system can be realised in a decentral and distributed way while inheriting the self-* properties of DHTs.

The remainder of the paper is organised as follows: In section 2 we provide an overview of the proposed CSP principle, a formal definition of the addressed core problem as well as requirements for a SLA based service decomposition. In section 3 a DHT based approach to the addressed core problem is presented, followed by an analysis in section 4 as well as simulations in section 5. The paper is concluded with a collection of related work, conclusions, future work items and acknowledgements.

2 System Model

During this paper we consider an overlay network as an *ordered sequence (or more general Directed Acyclic Graph) of processing modules (PMs) connecting a service source and a service sink*. We call an overlay situated in case it has the capability to adapt to critical (situational) changes in network or service parameters in a self-* manner enabling an Autonomic Service Control. To approach this we adopt the DHT principle from the area of P2P networks to realise a distributed control plane for an overlay based service platform.

Standard DHTs are optimised to provide a functionality to map keys to transport addresses where data corresponding to a given key can be fetched. In contrast for the proposed CSP system, it is required to incorporate service discovery, QoS verification and control functionality into a search process for chains of keys. Therefore we need to develop a DHT+++ principle supporting: Service specific registration functions for embedding a systems *service graph* into a DHT address space, recursive *chain* queries as well as the on demand verification of QoS constraints between any arbitrary pair of nodes hosting processing modules. To describe the required DHT extensions for CSP based on a CAN [1], we introduce the following notation to formalise multimedia transport and processing services (c.f. [7]). A Processing Module PM is formalised as a triple of the form (I, P, O) where O refers to the possible input formats the PM can read, P refers to the processing function provided by the PM, and O refers to the output format of the PM.

For simplicity we select $I, O \in \{x_1, ..., x_m\} \subset \mathbb{N}$ and $P \in \{y_1, ..., y_n\} \subset \mathbb{N}$. In a real world scenario numbers referring to different I, O values could e.g. be defined

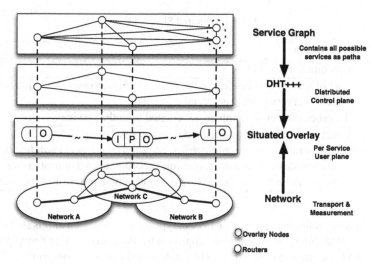

Fig. 1. CSP Approach

using a principle similar to RTP payload types[8]. Further it is assumed that neither Media Clients (MCs) nor Media Servers (MSs) do media processing. Therefore they are formalised using an (I, O) notation [1]. A MC, requesting content from a MS, can be served directly if and only if the input I of the client is compatible to the output O of the Server. In the case of non-compatibility, a PM implementing the required transforming functionality has to be inserted between the MS and the MC in order to start media delivery using a pipelining principle. To denote compatibility we use the symbol $\sim$. Based on this formalisation we define the *Service Graph* determined by a set of PMs as:

Definition 1 (*Service Graph*). *Let* $V = \{PM_1, PM_2, \cdots, PM_n\}$ *be a set of n processing modules. The Service Graph associated to V is defined as the graph $SG(V, E)$ with* $e = (PM_i, PM_j) \in E :\Leftrightarrow PM_i \sim PM_j$.

This means, that every path in a *Service Graph* corresponds to a valid processing chain for a service that can be realised by the set of PMs, and in case we need to determine the set of all processing chains to be available for instantiating a requested service it is possible to search the corresponding service graph for valid pathes connecting the media server and client.

2.1 Decomposition of Services

In this paper we focus only on tasks that are related to multimedia processing and that can be assumed to be decomposable. For instance we address services as first downscaling of content with regard to the resolution of a given end user terminal, then tagging of

---

[1] MCs and MSs are considered to be software instances running on a network node. The same node can also host independently a set of PMs performing media processing. Therefore a scenario where e.g. a Server is running a MS and several PM modules is also possible.

Table 1. Required SLA fields

SLA field	Description
SID	A unique Service ID
S_{SID}	A set of service sources (e.g. a list of transport addresses or unique names of media servers hosting a special content)
P_{SID}	The processing chain template associated with the service
C_{SID}	A vector of constraints associated with the service e.g. QoS constraints as *acceptable delay, required (bottleneck) bandwidth* and a monetary constraint i.e. maximal acceptable *cost of processing* in the form of a fee per processing demanded by a provider or enduser

the content using watermarking and finally encrypting it, where each of this steps is performed by a PM. For CSP the proposed approach for the actual service decomposition is based on SLAs, where the SLA has to be established between a content provider and a CSP enabled third party provider in advance of the first service request. In the case of a more classical P2P CSP scenario, the same principle can also be applied by replacing the service provider by a peer willing to offer a service and the third party provider by a community of peers offering processing capabilities. The specification of a concrete *SLA* principle is beyond the scope of this paper, but the required set of negotiated and/or derivable information during SLA agreement is collected in table 1.

As soon as a SLA has been established for a service, the information about the new SIDs can be made available by using e.g. a webpage or portal. To establish a mapping between the SID and the content of the SLA, a DHT principle can be used.

2.2   Service Requests

After a request for a known SID is received by any node of a CSP enabled system, the corresponding processing chain template has to be located and downloaded. In addition, information about the possible input formats acceptable by the client as well as the possible output formats of the server can be exchanged via standard session or capability description protocols as e.g. SDP[9]. Thus it can be assumed that after a request phase and analysis, the information collected in table 2 is available.

Thus, the processing chain for the requested service can completed in part as

$$PC_{SID} = (I_S, O_S) \sim (I, P_1, *) \sim \cdots \sim (*, P_i, O) \sim (I_C, O_C),$$

where an asterisk denotes an arbitrary input or output format. To be able to setup a Situated Situated Overlay for this SID it is required to find an instantiation of the corresponding processing chain PC_{SID} while fulfilling the QoS constraints C_{SID}. Addressing this problem directly results in a multiconstrained routing or path finding problem which is NP-complete[10]. In contrast, in most of the practical cases it is suitable to find a feasible solution, that can be a solution fulfilling all QoS constraints while having lowest possible *costs of processing* (c.f. Table 1). As a consequence, we address in the remainder of this paper the following Least Cost Constraint Based Routing Problem (LCBRP):

Table 2. Information available after service request

Information	Description
Service Specific	The Processing Chain template P_{SID} and a QoS constraint vector C_{SID}
Client Specific	At minimum one acceptable input format of the client I_C
Server Specific	At minimum one output format of the server O_S
Ingress PM Specific	The required input format I and processing function P_1
Egress PM Specific	The required output format O and processing function P_i

Problem 1 (LCBRP). Assume a Graph $G(V, E)$, a processing chain

$$PC = (P_1, P_2, \ldots, P_l)$$

and two vertices $u, v \in V$. Further we assume three constraints

$$C_{Max}, D_{Max}, B_{Min} \in \mathbb{R}_0^+$$

corresponding to acceptable maximal (processing) cost and delay as well as minimum required (bottleneck) bandwidth. The *Least Cost Constraint Based Routing Problem* is to find the shortest path $p = (u, \ldots, v)$ in $G(V, E)$ with regard to processing costs such that the C_{Max}, D_{Max} and B_{Min} constraints are not violated and the order of processing is retained.

3   Separable Constraint Based Routing, a DHT Centric Approach

After a service request, there are two main steps required to establish the Overlay for a service SID, which are *Service Discovery* and selection of at least one solution to the corresponding LCBRP for *Overlay Network Setup*. [2] In CSP we integrate this two steps into the DHT search process since:

- No support from the underlaying network infrastructure is required (as e.g. support for a special routing protocol, etc.)
- Integrating Service Discovery and setup of the Situated Overlay Network can save communication overhead.
- A DHT based system can be deployed fast with a low state per node (i.e. in the order of $O(\log N)$ where N is the number of nodes in a CSP system.)
- The resulting system inherits the self-* properties of it's DHT.

However, DHTs are Overlay networks optimised to accomplish a search task. To be able to adopt them it is required to reduce the addressed LCBRP problem to a search problem.

---

[2] In this paper paper we omit Overlay Maintenance strategies and leave them for future work.

3.1   Reduction of LCBRP Problem to a Distributed Search Problem

Based on the results summarised in section 2 we can assume to have after a service request a partly competed processing chain of the form

$$(I_S, O_S) \sim (I, P_1, *) \sim ... \sim (*, P_i, O) \sim (I_C, O_C) \tag{1}$$

as well a constraint vector C_{SID}. The task of a CSP control plane would be now to instantiate a valid processing chain that is a solution to the corresponding LCBRP problem. We will address this by reducing the LCBRP problem to a distributed search in a DHT.

To realise this we aim at exploiting the following facts:

1. By using a problem specific indexing scheme, the information about available valid processing chains can be stored implicitly in the DHT address space.
2. Using 1, the LCBRP problem can be addressed by utilising a distributed search principle in combination with a hop-by-hop verification of QoS constraints.

A prerequisite of 1 is, that the $\sim$ relation has to be an invariant with regard to the indexing scheme of the used DHT. I.e. it must be possible to derive compatibility of two PMs by comparing their hash values in the DHT address space. To address this we use a CAN with an address space $[0, t] \times [0, t] \times [0, t] \subset \mathbb{R}^3$, for a suitable $t \in \mathbb{R}$. After a new node n has joined successfully the CAN, the address where to store a pointer to data referencing a $PM_1 = (I_1, P_1, O_1)$ hosted by n is calculated as the coordinate

$$\text{hash}_{CAN}(I_1, P_1, O_1) := (\text{h}(I_1), \text{h}(P_1), \text{h}(O_1)) \in \mathbb{R}^3$$

where $\text{h} : \mathbb{N} \mapsto [0, t] \subset \mathbb{R}$ is a suitable hash function. In case it is required to find a PM_2 being compatible to PM_1, the CAN address where to find such information can be directly calculated as $(\text{h}(O_1), \text{h}(P_2), \text{h}(O_2))$ (where O_2 is an arbitrary output format). I.e. we have

$$PM_i \sim PM_J \Leftrightarrow \text{hash}_{CAN}(PM_i) \sim \text{hash}_{CAN}(PM_J).$$

To establish 2 we recall that the information about all possible services to be realisable is represented through the *Service Graph* of the system.

As illustrated in figure 2, each branch of the search results in case no QoS constraints are violated in a valid processing chain from the server to the client. A valid processing chain for the composed service

$$(0, 1) \sim (1, 1, 2) \sim (2, 2, 1) \sim (1, 3, 3) \sim (3, 4, 1) \sim (1, 0)$$

is marked grey.

3.2   CSP Specific Range Queries

Following expression 1 above, e.g. in case of the ingress processing module PM_1 there is not enough information available so far to formulate an exact search query.

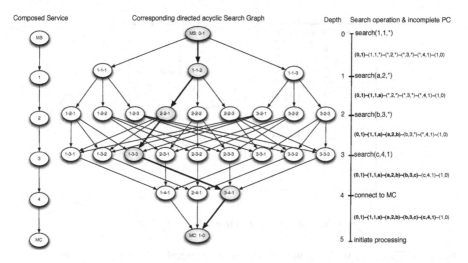

Fig. 2. Service, Search Graph and required search operations

Thus, instead searching for a concrete $PM = (I, P, O)$ we are looking for a list of available PMs capable of accepting the input format I, having a processing function P_1, and an arbitrary output O. Such a *Range Query* is defined to be of the form $\text{search}_{RQ}(I, P, *)$.

In figure 3 we illustrate how it can be realised using a two or three dimensional CAN and the before introduced indexing scheme. Since the two dimensional case, shown on the left side of figure 3 is intuitive, we describe the three dimensional case for query of the form $\text{search}_{RQ}(I_A, P_A, *)$. Because of the properties of the used indexing scheme, information about all PMs with $PM_A = (I_A, P_A, *)$ is stored along the line $g(\lambda) = A + \lambda \cdot (0, 0, 1)$. For a *Query Node* (QN) B with CAN address (I_B, P_B, O_B), the closest point on g with regard to euclidean distance $d_{b,c}$ is $C = (\text{h}(I_A), \text{h}(P_A), O_B)$. Thus we define the node owning the CAN territory including C as the *Range Query Initiator (RQI)*. As soon as this node receives the range query from B it is forwarding it to all its neighbours following the line g. For a query of the form $\text{search}_{RQ}(*, P_A, O_A)$ we have $g(\lambda) = A + \lambda \cdot (1, 0, 0)$ and $C = (I_B, P_A, O_A)$. With regard to the scope of the described *CSP range query* principle in a CAN we provide the following lemma.

Lemma 1 (Scope of a CSP range query in CANs). *Given a Content Addressable Network in a d-dimensional geometric space containing n nodes. We further assume that the CAN is partitioned into territories of equal size. The average scope of a CSP range query is $\left(\frac{d}{3} + 1\right) * n^{(1/d)} - 1$ and is therefore $O(n^{1/d})$.*

Proof. We note that for a CAN in a d-dimensional space partitioned into territories of equal size, the average routing path length is $(d/4) * n^{(1/d)}$ [1]. The scope of a CSP range query can be identified with the scope of two separate CAN routing requests. The first one, with source Query Node (QN) and destination Range Query Initiator (RQI), can be assume to be of average scope $(d/4) * n^{(1/d)}$. Now the RQI node is forwarding the range query along a line in $\mathbb{R}^d$ crossing the territory of $n^{1/d} - 1$ nodes (see also

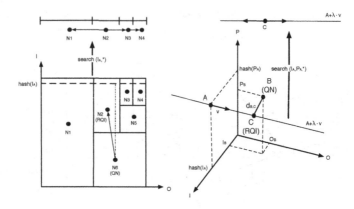

Fig. 3. Range Queries in a 2D and 3D CAN

figure 3). Therefore we have an average message complexity of $\left(\frac{d}{3}+1\right) * n^{(1/d)} - 1$ which is $O(n^{1/d})$.

3.3 Search Algorithm

Utilising the above introduced range query principle we describe now the basic CSP search algorithm. The underlying principle of the algorithm is to distribute the search for processing chains between the set of nodes hosting the required processing functionality in the form of PMs. As described above, we can start with the following partly completed processing chain

$$(I_S, O_S) \sim (I, P_1, *) \sim \cdots \sim (*, P_l, O) \sim (I_C, O_C)$$

together with IP_{MC}, the IP address of MC and three additional scalar constraints C_{Max} for costs, D_{Max} for *delay* and B_{Min} for acceptable *bottleneck bandwidth*. For the simulations performed and described in section 5, we implemented search$_{RQ}$ using a *breadth fist* as well as *depth first* principle with forward checking and backtracking. To realise the *depth first* approach, the search$_{RQ}$ principle can be modified in a way that the Range Query Initiator is collecting the results of the range query and selects the next PM based on a the used forward checking principle. However, in this section we focus on the *breadth first* like algorithm.

After receiving the request, a media server MS formulates the initial CSP search query which is of the form:

$$((I, P_1, *) \sim \cdots \sim (*, P_l, O)), IP_{MC}$$
$$\{c < C_{Max}\}, \{d < D_{Max}\}, \{b > B_{Min}\}$$
$$c = 0, d = 0, b = \infty$$

The MS as the root of the search for the service, initiates now a query$search_{RQ}(I, P_1, *)$, piggybacking:

$$((*, P_2, *) \sim \cdots \sim (*, P_l, O)), IP_{MC}$$
$$\{c < C_{Max}\}, \{d < D_{Max}\}, \{b > B_{Min}\}$$
$$c = c_s, d = 0, b = \infty$$

where c_s denotes the cost of the media provisioning. Each DHT node passed by the range query, storing information where to find a node n hosting a $PM = (I, P_1, O_{1,x})$ is forwarding the piggybacked search query and the IP address of the range query originator to n. n can now actively determine the values of d_1, b_1 e.g. using measurements or estimation techniques between itself and the range query originator. In addition n can get the value c_1 corresponding to the costs for using its P_1 related PM. If $c + c_1 \leq C_{Max}, d_1 \leq D_{Max}$ and $b_1 > B_{Min}$, n is recursively sending a range query of the form $search_{RQ}(O_{1,x}, P_2, *)$ piggybacking Q_1.

$$Q_1 := \begin{cases} ((*, P_3, *) \sim \cdots \sim (*, P_l, O)), IP_{MC} \\ \{c < C_{Max}\}, \{d < D_{Max}\}, \{b > B_{Min}\} \\ c = c_s + c_1, d = d_1, b = d_1 \end{cases}$$

This recursive process is continued and at depth $i < l - 1$ the corresponding CSP search query is

$$Q_i := \begin{cases} ((*, P_{i+1}, *) \sim \cdots \sim (*, P_l, O)), IP_{MC} \\ \{c < C_{Max}\}, \{d < D_{Max}\}, \{b > B_{Min}\} \\ c = c_s + c_1 + ... + c_i, d = d_1 + ... + d_i, b = \min\{d_{i-1}, d_i\} \end{cases}$$

The recursion stops either if the depth of the resulting search graph is equal to the number l of processing steps or the constraints are violated. In the case the search depth l is reached, the actual node is sending a report message to MC, including information about the c, d and b values corresponding to available service paths. Based on this information the MC can select the path to be instantiated backwards from MC to MS.

4   Analysis

For a complexity analysis of the described CSP search principle we will focus on the question "How many nodes are actively involved into the search for a given processing chain of length n?". A node is considered to be actively involved into a CSP search in case it has to trigger a search request. The main reason for this focus is the fact that a search request with its associated QoS verification task can be assumed to be disproportional expensive with regard to computation and communication requirements compared to standard DHT routing tasks.

In order to address this in the broadest fashion, we analyse the complexity of the underlying service composition problem while assuming that no branch of the search is stopped because the QoS constraints cannot be fulfilled. Thus we examine the question: "Given a relation between the I, P, O values of the available PMs in the system as well as a processing chain template of length n, how many possible solutions can we expect for a service request?".

To answer this, we consider I, P, O as random-variables supposing m different input and output formats i.e. $I, O \in \{x_1, ..., x_m\}$ independently. Further we assume n different processing functions $P \in \{y_1, ..., y_n\}$. In addition, for $y_1, ..., y_n$ we define the the following random-variables:

$$(I_i, O_i) := ((I, P, O)|P = y_i)$$

This means that the (I_i, O_i) are the random-variables which result from conditioning (I, O) by $P = y_i$. We further define

$$P(O = x_i | P = y_j) =: p_{j,i}(O) \text{ and } P(I = x_i | P = y_j) =: p_{j,i}(I)$$

and a processing chain as any vector

$$((I', O_1(\omega)), ..., (I_n(\omega), O')).$$

A chain is valid iff $O_i(\omega) = I_{i+1}(\omega)$ for any $1 \le i \le n - 1$, and the respective set of valid chains is denoted by C^*.

There are two possible ways of engendering a set of chains C with the cardinality $\#(C) = N_C$:

1. By simply generating N_C chains randomly.
2. By generating N_i times the random-variable (I_i, O_i) for $1 \le i \le n$ with

$$\prod_{i=1}^{n} N_i = N_C. \quad (*)$$

We are now interested in $\mathbb{E}(\#(C^*))$, the expectation of the random-variable $\#(C*)$ when the corresponding C is engendered as described in (2). Assuming that this value will heavily depend on N_C we can proceed as if C was engendered as described in (1), i.e. as if we had sampled $\#(C) = N_C$ chains independently.

Doing so the following reasoning leads to the simplified formula given below: We define

$$\frac{\#(C*)(\omega)}{\#(C)} =: p(\omega),$$

the *proportion* of correct chains in $C(\omega)$. The expectation of the proportion p corresponds to the probability to engender one single chain according to $(*)$. Therefore we have:

$$\mathbb{E}(p) = P(O_1 = I_2, ..., O_{n-1} = I_n) = \prod_{i=1}^{n-1} P(O_i = I_{i+1})$$

Because of the independence of O_i and I_{i+1} we have

$$P(O_i = I_{i+1}) = \sum_{j=1}^{m} p_{j,i}(I) p_{j,i}(O).$$

Now the expectation of the number of correct chains in a random sample can be computed as follows:

$$\mathbb{E}(\#(C^*)) = \#(C)\mathbb{E}(p)$$

$$= \prod_{k=1}^{n} N_k \mathbb{E}(p)$$

$$= \prod_{k=1}^{n} N_k \prod_{i=1}^{n-1} P(I_i = O_i)$$

$$= \prod_{k=1}^{n} N_k \prod_{i=1}^{n-1} \sum_{j=1}^{m} p_{j,i}(I) p_{j,i}(O)$$

Combining this result with the fact that for a processing chain of length n, $n - 1$ nodes[3] have to perform a QoS constraint validation step we can state that the expected number of nodes that are actively involved in the corresponding search process are

$$n \cdot \prod_{k=1}^{n} N_k \prod_{i=1}^{n-1} \sum_{j=1}^{m} p_{j,i}(I) p_{j,i}(O)$$

To illustrate the above result, note that under the simplifying assumptions

(i) For each processing function we assume the same number of PMs, which is: $N = N_i$, for any $1 \leq i \leq n$
(ii) Each processing function comes with all possible I, O combination, which is: $p_{j,i}(I) = p_{j,i}(O) = \frac{1}{m}$

we obtain:

$$\mathbb{E}(\#(C*)) = \frac{N^n}{m^{n-1}}.$$

Thus the expected number of nodes in a CSP system that are actively involved in a breadth first search process for a chain of length n is $n \cdot \frac{N^n}{m^{n-1}}$. If $N > m$ the expected number grows exponential with regard to the length n of the processing chain. As a consequence, to be able to limit the *scope* of a CSP search, it is important to use search principles where the number of concurrent branches can be controlled.

5 Simulations

In this section CSP will be evaluated using a simulation approach. The performed experiments are based on a network topology generated by the Georgia Tech GT-ITM[11] topology generator using a hierarchical transit-stub model containing 1740 nodes. For each simulation run we selected randomly a subset of 500 nodes as CSP Overlay Nodes (ONodes). Each ONode is hosting one PM with independent I, P, O values, randomly selected out of the set $\{1, ..., 5\}$. In addition a corresponding cost value was selected randomly out of the set $\{1, ..., 100\}$. Delay and bottleneck bandwidth values have been calculated based on the GT-ITM link weights using shortest path routing. For each experiment a processing chain of length $l = 1, 2, 3, 4$ was selected out of the corresponding service graph and the corresponding constraint vector with regard to *Delay*, *Bandwidth* and *Cost* values was determined. After this, the CSP approach was used to find a processing chain fulfilling the constraints while having same or lower costs. We repeated each experiment 10 times and show the averaged results. For the used range query function SEARCH$_{RQ}$ we implemented two variants:

1. *All Branches (breadth first):* Each PM found meeting the constraints continues the search starting a new branch if required.

---

[3] We assume the case of one PM per node.

Table 3. Simulation Results

Length	Active nodes	Scope	num. paths	disjoint paths	hops
1	4.6 / 2.4	6.8 / 6.8	2.1 / 1	1.5 / 1	2 / 2.4
2	35.3 / 5.3	123.7 / 27.0	36.4 / 1	13.1 / 1	3 / 5.6
3	122.7 / 8.5	2207.3 / 82.2	776.6 / 1	14.3 / 1	4 / 11.0
4	233.9 / 9.8	55792.9 / 96.2	14829.4 / 1	19.3 / 1	5 / 14.2

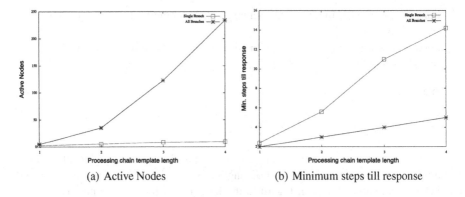

(a) Active Nodes   (b) Minimum steps till response

Fig. 4. Comparison of selected results

2. *Single Branch (depth first):* Only the PM meeting the constraints with lowest costs is used to continue the search. The same strategy was also used for backtracking in case the search could not be continued because of constraint violations.

We evaluated the experiments based on the following metrics

- **Active Nodes:** The number of CSP ONodes actively trigger search requests, as a primary complexity measure for CSP.
- **Scope of Search:** The number of CSP DHT nodes involved in a search, including active nodes and nodes involved in DHT routing. This metric covers the DHT related communication complexity of CSP.
- **Min steps till response (hops):** The minimal number of search steps performed until a solution has been found, as a rough indicator for the time required to complete a request. In case of the *all branches* approach this is always $l + 1$, where l is the length of the processing chain.

The averaged results of the performed experiments are shown in figure 4 and table 3. In each field of the table, the left value shows the result for the *all branches*, and the right value for the *single branch* based $\text{SEARCH}_{\text{RQ}}$ function. Up to a processing chain length of 2, the *all branches* approach can be considered as interesting because it has acceptable scope values while outperforming the *single branch* with regard to the hops metric. A further benefit using this method is that disjunct solutions (paths) are returned, a fact that can be exploited for resilience. In case of longer processing chains the results

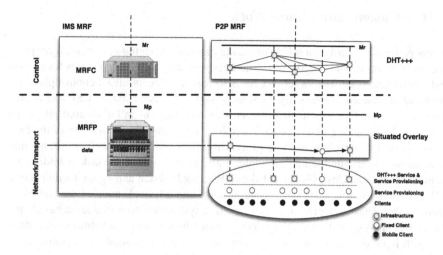

Fig. 5. CSP based MRF

of our theoretical analysis are confirmed by the simulation. The complexity of the *all branches* approach is rising close to exponential while the *single branch* approach still shows good results, even with respect to the *minimum steps till response* metric. The question how to define a hybrid search strategy with input parameters as required number of disjunct solutions, acceptable scope and response time will be one main part of future CSP work. In addition we work on a larger setup for the validation of CSP based on real network data.

6 Related Work

In contrast to other service composition proposals like [3], [4], [5], [6], CSP does not rely on a central entity having global knowledge during the task of overlay setup. Compared to [12] which is also utilising a DHT, CSP aims to include the overlay setup directly into the search process using a DHT routing integrated approach in combination with a hop-by-hop QoS constraint verification. From a provider point of view the IMS standard Rel.6, developed inside the third generation partnership project (3GPP) covers similar functionality in a broader sense. The IMS Media Resource Function (MRF), can be considered as its multimedia related core. It is responsible for resource consuming tasks as e.g. playing, transcoding and mixing of media streams. Following the standard, the MRF is a conceptional centralised entity thus scalability problems related to the ones in case of a classical client/server approach can be anticipated. To cope with such problems, a "distributed" MRF based on P2P-Principles can be an interesting option. As the most promising candidates to be used to realise a distributed MRF we suggest dedicated infrastructure nodes as well as *home gateway* devices and *set-top boxes* under partial control of the network provider. Figure 5 illustrates conceptual how a distributed MRF may be realised based on CSP using the IMS Mr interface to link the resulting situated overlays to the IMS control.

7   Conclusions and Future Work

In this paper we described a self-configuring and self-managing system for a cooperative service provisioning (CSP) of media transport and processing services for current and future Networks. The target of CSP is to realise a decentralised control plane for the setup of situated overlays. the most important difference between CSP and related approaches is the fact, that its main target is to investigate in an DHT based integrated approach: I.e. by using a DHT related search principle also for a discovery of the final service path. The algorithm used for processing chain discovery as well as instantiation is based on a distributed search principle. It allows to distribute the task of finding the solution to a (Least Cost) Constraint Based Routing Problem among the members of a P2P network. To introduce CSP, a system model and a formalisation of the addressed problem space has been provided. The resulting system has been evaluated based on a theoretical analysis and simulations. The question how to define a hybrid search strategy with input parameters as required number of disjunct solutions, acceptable scope and response time will be one main part of future CSP work.

Acknowledgments

The authors would like to thank the anonymous reviewers for their insightful comments which helped to improve the paper. This research was partly performed in the context of the Situated Autonomic Service Control (SASCO) project, funded by France Télécom R & D.

References

1. Ratnasamy, S., Francis, P., Handley, M., Karp, R., Shenker, S.: A scalable content addressable network. In: Proceedings of ACM SIGCOMM, ACM, New York (2001)
2. Babaoğlu, Ö., Jelasity, M., Montresor, A., Fetzer, C., Leonardi, S., van Moorsel, A.P.A., van Steen, M. (eds.): Self-star Properties in Complex Information Systems. LNCS, vol. 3460. Springer, Heidelberg (2005)
3. Xu, D., Nahrstedt, K.: Nahrstedt, Finding service paths in a media service proxy network. In: Proceedings of the ACM/SPIE Conference on Multimedia Computing and Networking (2002)
4. Gu, X., Nahrstedt, K., Chang, R., Ward, C.: Qos-assured service composition in managed service overlay networks. In: Proceedings of IEEE 23rd International Conference on Distributed Computing Systems, IEEE Computer Society Press, Los Alamitos (2003)
5. Jingwen Jin, K.N.: Source-based qos service routing in distributed service networks. In: Proceedings of IEEE International Conference on Communications, IEEE, Los Alamitos (2004)
6. Gu, X., Nahrstedt, K.: A scalable qos-aware service aggregation model for peer-to-peer computing grids. In: Proceedings of the IEEE HPDC-11. IEEE Computer Society Press, Los Alamitos (2002)
7. Mathieu, B., Song, M., Kleis, M.: A p2p approach for the selection of media processing modules for service specific overlay networks. In: Proceedings of International Conference on Internet and Web Applications and Services (ICIW) (2006)

8. Casner, S.: Media Type Registration of RTP Payload Formats, RFC 4855(Proposed Standard) (February 2007), [Online]. Available:
http://www.ietf.org/rfc/rfc4855.txt
9. Handley, M., Jacobson, V., Perkins, C.: SDP: Session Description Protocol, RFC 4566 (Proposed Standard) (April 2006), [Online]. Available:
http://www.ietf.org/rfc/rfc4566.txt
10. Garey, M.R., Johnson, D.S.: Computers and Intractability: A Guide to the Theory of NP-Completeness. Freeman, San Francisco (1979)
11. Zegura, E.W., Calvert, K.L., Bhattacharjee, S.: How to model an internetwork. Proceedings of IEEE Infocom. IEEE Computer Society Press, Los Alamitos (1996)
12. Gu, X., Nahrstedt, K., Yu, B. (eds.): Spidernet: An integrated peer-to-peer service composition framework. Proceedings of IEEE HPDC-13. IEEE Computer Society Press, Los Alamitos (2004)

Generic Emergent Overlays
in Arbitrary Peer Identifier Spaces

Wojciech Galuba and Karl Aberer

School of Computer and Communication Sciences,
Ecole Polytechnique Fédérale de Lausanne (EPFL), Switzerland
{wojciech.galuba, karl.aberer}@epfl.ch

Abstract. Unstructured overlay networks are driven by simple protocols that are easy to analyze and implement. The lack of structure,
however, leads to weak message delivery guarantees and poor scaling.
Structured overlays impose a global overlay topology that is then maintained by all peers in a complex protocol. In contrast to unstructured
approaches the structured overlays are efficient and scalable, but leave
little flexibility in how their topology can be adapted to the needs of the
application.

We propose a generic overlay maintenance and routing algorithm that
combines the simplicity of the unstructured overlays and the scalability
of the structured approaches, while allowing the application to define its
own peer identifier space. The overlay topology is not explicitly defined
but emerges in a self-organized way as the result of simple maintenance
rules. Independently of the identifier space used, our algorithm exhibits
logarithmic scaling of the average routing path length and the average
node degree.

The proposed maintenance and routing algorithm is simple and places
few constraints on how peers can open their connections. This together
with the ability to adjust both the identifier space and the tradeoff between the path length and the node degree makes the overlay customiz-
able in ways that are not possible in the existing approaches.

1 Introduction

Most of the state-of-the-art structured overlay networks [1, 2, 3, 4, 5, 6, 7, 8]
follow a similar design paradigm. First, a global network structure is defined
and the peers are placed in some identifier space [9]. The structure has properties desirable from the application point of view such as logarithmic routing
path length, high routing path redundancy and resilience to failures. Then, this
global network structure is expressed as invariants which are maintained by each
node locally to ensure the coherence of the global network structure. Structured
overlays despite their performance guarantees, have complex distributed protocols and the applications have little flexibility and control over how the overlay
topology is formed.

In contrast, in unstructured overlays [10, 11], the design process starts from
the local goals and rules without any particular target global structure in mind.

D. Hutchison and R.H. Katz (Eds.): IWSOS 2007, LNCS 4725, pp. 88–102, 2007.

The resulting algorithms and protocols are simple and offer much flexibility in forming the overlay topology and routing the messages, however they lack the routing efficiency and scaling guarantees of structured overlays.

In this paper we take an approach to overlay network maintenance that combines the flexibility and ease of implementation of the unstructured overlays with the efficiency and scalability of the structured overlays.

We start out by observing the common characteristics of all structured overlays. Each node is endowed with an identifier and in every structured approach there is some notion of distance defined between the identifiers. Through the distance function each node can know its position relative to other nodes in the topology. In each routing hop the message routed in the overlay is brought closer to its destination in terms of the identifier space distance. To ensure the progress of messages towards the destinations each node maintains a set of connections. The global overlay connection topology guarantees efficient and scalable delivery of messages.

We abstract out the concept of distance between overlay identifiers and allow the application to specify it. We propose a generic overlay routing and maintenance algorithm that relies solely on knowledge of the identifier distance function and does not specify any global topology that needs to be maintained. The global topology emerges in a self-organized way as a result of a simple connection opening rule.

The node identifiers are selected uniformly randomly from the identifier space. In each routing step we greedily route the message to the next hop that is closest to the destination. We compute the rate at which the message approaches its destination, i.e. how much the distance to the destination is shortened during one hop. We require that the rate for each hop be at least γ, a design parameter. If this condition is not satisfied then additional connections are opened by the maintenance algorithm to ensure the minimal rate of message progression towards the destinations.

To verify the claim that the above simple algorithm is indeed able to form an efficient overlay, we evaluate our overlay in simulation. We observe that:

- The resulting overlay has logarithmic scaling properties. Both the average routing path length and the average node degree are logarithmic in terms of the number of nodes in the network.
- The routing path length vs. node degree tradeoff can be controlled by adjusting the single design parameter γ.
- Logarithmic overlay scaling can be achieved for any identifier distance metric by adjusting γ
- Local structures emerge tightly interconnecting nodes in the identifier space on short distances. This common characteristic is shared by all state-of-the-art structured overlays and is crucial for e.g. last hop routing and key replica management in DHTs
- The overlay is robust and has low maintenance overhead even in presence of high churn

The overlay networks generated by our algorithm have characteristics comparable to many of the well-known approaches, while offering a number of additional advantages:

- Up to our knowledge it is the first overlay maintenance algorithm that is driven by the overlay traffic, i.e. connections are created only when they are needed.
- The abstract space of node identifiers and distances between them generalizes over the previous approaches.
- In contrast to other structured approaches there is no pre-defined rigid global structure that has to be maintained, in our case the topology emerges in a self-organized way as a function of the underlying identifier space. The lack of pre-defined topology greatly simplifies the algorithm and minimizes the implementation effort.
- The proposed algorithm leaves plenty of room for adjusting the routing efficiency and the number of connections the nodes need to maintain, this combined with the generalized identifier space gives a level of customizability not available in other overlays.

2   The Model

In this section we present the basic assumptions followed by the formulation of the proposed overlay routing and maintenance algorithm.

2.1   Graph Embedded in a Metric Space

Let the network be represented by a graph $G(V, E)$, where V is the set of overlay nodes and E is the set of overlay connections between them. Let $id : V \rightarrow I$ be a function that assigns an identifier from the set I to each of the nodes in V. Let $d : I \times I \rightarrow \mathbb{R}$ be the distance function. The id function embeds the nodes in the metric space defined by d, hence d must satisfy the four properties of the distance function in a metric space: non-negativity, symmetry, identity and the triangle inequality. The pair (I,d) is the *identifier space* and the *mapping function id* maps the overlay nodes into that space.

The communication in the network proceeds by sending messages. The messages can only be sent along connections. Once a connection is established between two nodes it can be used to send messages in both directions.

2.2   Routing

Assume a source node $m.src$ wants to send a message m to a destination node $m.dest$ ($m.f$ denotes the field f in message m). The routing proceeds in the standard hop-by-hop way. The next hop is selected by greedily minimizing the identifier space distance to $m.dest$ and at the same time avoiding previously visited nodes.

Let $m.visited$ be the set of nodes through which m has already been routed. Let v_c be the node that currently holds m and needs to forward it. Let $neigh(v_c)$

be the set of neighbors of v_c. The node v_c selects the next hop v_{nh} based on the following rule: take the set $neigh(v_c) \setminus m.visited$, and from it select node v_{nh} for which the value $d(id(v_{nh}), id(v_d))$ is the lowest. After selecting the next hop, v_c adds v_{nh} to $m.visited$ and forwards m to v_{nh}.

The following special cases occur:

- **NHimp** - next hop impossible - a message reaches a dead end, $neigh(v_c) \setminus m.visited$ is an empty set, the message is dropped
- **TTL0** - TTL zero - each message has a time-to-live counter decremented with every hop, when it reaches zero the message is dropped

2.3 Overlay Maintenance

The maintenance in our overlay is driven by routing. To ensure eventual delivery, each routing hop should advance messages closer towards to the destination in the space defined by d. What is more, this advancement should occur at a certain minimal rate of progression to provide efficient overlay message delivery, otherwise new connections have to be created to ensure that this happens. We base our overlay maintenance algorithm on this simple maintenance rule.

For a given next hop v_{nh} from the current node v_c towards the destination $m.dest$ we define the *routing convergence rate* as $cvg(v_c, v_{nh}, m.dest) = \frac{d(id(v_c), id(p.dest))}{d(id(v_{nh}), id(m.dest))}$. Let γ be the minimum required routing convergence rate. Routing convergence rate is the measure of how much the next hop shortens the distance to the destination. If a hop $(v_c \rightarrow v_{nh}, m.dest)$ does not satisfy the condition $cvg(v_c, v_{nh}, m.dest) \geq \gamma$ then that hop is *weak*, otherwise it is *strong*.

When a weak hop is encountered while routing a message m, the maintenance protocol sends a connection request $cr = (v_c, p.dest)$ with the destination set to $m.dest$ indicating the origin of the request as v_c, the current node. The connection request is routed towards the destination normally as other messages in the greedy self-avoiding way. When some node v_{resp} receives a connection request $cr = (v_o, dest)$ and if the hop $(v_o \rightarrow v_{resp}, dest)$ is not weak then v_{resp} responds to v_o with connection acknowledgement and the connection between v_o and v_{resp} is established and the routing of cr stops. Otherwise if $(v_o \rightarrow v_{resp}, dest)$ is weak, the cr continues to be routed.

When a timeout happens while sending on one of the connections then the sender closes that connection and removes the recipient from its neighbor set.

2.4 Maintenance Suppression

Every node keeps track of the connection requests that it has sent until either the corresponding connection response arrives or a timeout occurs. This request-response tracking has the following purpose. Consider the time between two events: (1) the sending of a connection request $c(v_c, dest)$ by node v_c and (2) the receipt of the corresponding connection response. Assume additionally that there are no connection requests being sent or responses arriving during that time. However, there may be many messages with the same destination $dest$

arriving at v_c. According to the proposed maintenance algorithm each of these messages takes a weak next hop and triggers a connection request, which is identical to the one already sent earlier. This would lead to the generation of many unnecessary connection requests.

To prevent this from happening, whenever some $cr = (v_c, dest)$ is about to be sent the list RL of requests currently awaiting their responses is checked. If any of the potential responders to the connection requests on RL is also a valid responder to cr, then cr is not sent. Let $cr' = (v_c, dest') \in RL$ be some connection request awaiting its response. Let v_r be the potential responder to cr', v_r must satisfy the condition (1) $cvg(v_c, v_r, dest') \geq \gamma$. If v_r is also a valid responder for cr, then it also satisfies (2) $cvg(v_c, v_r, dest) \geq \gamma$. We also know that the three identifiers of the three nodes $(v_r, dest, dest')$ must satisfy the (3) triangle inequality. Combining (1), (2) and (3) we obtain (4) $\gamma d(id(dest), id(dest')) < d(id(v_c), id(dest)) + d(id(v_c), id(dest'))$. If there is any $cr' = (v_c, dest') \in RL$ for which (4) is true then $cr = (v_c, dest)$ is not sent.

The operation of our overlay routing and maintenance algorithm has been summarized in Algorithm 1. Note that for clarity the handling of timeouts, TTL0 and NHimp events has been omitted.

3   Simulation Results

In this section we examine the behavior of our routing and overlay maintenance algorithms experimentally.

3.1   Experimental Setup

We use ProtoPeer [12], an event-driven simulator. Each node in the simulated network generates messages in a Poisson process. The generation rates are identical across all the nodes. A node v_i generates a message m with the destination $v_j \neq v_i$ selected uniformly randomly.

The average message generation rate at all nodes is set to one message every second. All messages, connection requests and connection responses are delivered with a latency uniformly distributed between 100ms and 200ms. The time-to-live for overlay messages is set to 100 hops. We do not explicitly simulate the underlying network topology. This simple network model is sufficient for verifying the correctness of our algorithm and the structural properties of the overlay, performance testing in a more realistic setting is left as future work (Section 4).

The bootstrap process is as follows. Each simulation begins with a set of 30 nodes interconnected uniformly randomly with an average degree of 5. The size of this initial network is large enough to ensure that it remains a connected graph even if some of the initial nodes depart at the beginning of the simulation. Each new joining node connects to 5 uniformly randomly selected neighbors from the network. These bootstrap connection requests and responses are delivered directly and are not routed in the overlay. In a concrete implementation the peers would be bootstrapping from a known host list (e.g. downloaded from the

---

Algorithm 1. Overlay routing and maintenance algorithm for arbitrary peer identifier spaces

---

initialize
   $RL \leftarrow \emptyset$
   $neigh \leftarrow$ initialize with peers via bootstrap mechanism

```
// application calls this function to send messages
```
function sendMessage(payload,dest)
   forwardMessage(Message(payload,self,dest,$\emptyset$))

function forwardMessage(Message(payload,src,dest,visited))
   $next\_hop = argmin_{x \in neigh \setminus visited} d(id(x), id(dest))$
   send $Message(payload, src, dest, visited \cup self)$ **to** $next\_hop$
   if $payload$ is $ConnectionRequest$ **then**
      return
   end
   ```
// maintenance is triggered by weak hops
// only for payload that is not a ConnectionRequest
```
   if $\frac{d(id(self),id(dest))}{d(id(next\_hop),id(dest))} < \gamma$ **then**
      ```
// check the maintenance suppression condition
```
      if $\neg \exists_{cr'(self,dest') \in RL} \gamma d(id(dest), id(dest')) <$
      $d(id(self), id(dest)) + d(id(self), id(dest'))$ **then**
         $RL \leftarrow RL \cup cr(self, dest)$
         ```
// route the connection request as a new message
```
         forwardMessage(Message(ConnectionRequest(self,dest),self,dest,$\emptyset$))
      end
   end

receive $Message(payload, src, dest, visited)$
   if $payload$ is $ConnectionRequest(origin, dest)$ **then**
      ```
// check if can accept the connection request
```
      if $\frac{d(id(origin),id(dest))}{d(id(self),id(dest))} \geq \gamma$ **then**
         send $ConnectionResponse(self, dest)$ **to** $origin$
         return
      end
   else if $dest{=}self$ **then**
      deliver Message to application
      return
   end
   forwardMessage(Message(payload,src,dest,visited))

receive $ConnectionResponse(responder, dest)$
   $RL \leftarrow RL \setminus cr(self, dest)$
   $neigh \leftarrow neigh \cup responder$

Web), we do not simulate this process in detail since our overlay is not sensitive to the choice of initial neighbors for the peer.

To demonstrate that the results are independent of the chosen identifier space we select five representative spaces for the experiments:

- 1D - one dimensional ring, as in Chord[1],
  $I = [0,1)$, $d(a,b) = min_{k \in \{-1,0,1\}} |a - b + k|$
- 2D - two dimensional spherical coordinates, the identifiers are placed on the sphere $I = [0, 2\pi)^2$ and the shortest distance is measured along the great circle crossing the two identifiers
- 3D - three dimensional Euclidean space with wraparound (surface of a 4D hypertorus).
- PFX - prefix routing as in Pastry[4] with the identifier space of 128bit vectors, assume we are computing $d(a,b)$, bits in a and b are compared from the highest order bit to the lowest order bit, if i is the index of the first bit which differs between a and b, then $d(a,b) = 2^i$
- XOR - XOR distance metric as in Kademlia [5], an XOR of two identifiers is computed and the result is taken as an integer distance value with 160 bits

For all the identifier spaces the nodes select their identifier uniformly randomly out of the set of all possible identifiers.

The arrivals and departures of the nodes (churn) are simulated. Arrivals are a Poisson process with a default average rate of 0.002 nodes per second. The lifetime of the nodes is power-law distributed [13] with the minimal lifetime of 10s and the exponent of -1.2.

During the simulation we track the number of TTL0 and NHimp events (Section 2.2). The churn is low in most simulations. We devote section 3.3 to the study of routing failures under high churn conditions.

3.2   Scaling

To test the fundamental scaling properties of the overlay we let it increase in size over time by setting the minimal node lifetime to a higher value of 50s. For different network sizes and identifier spaces we measure the average routing path length and the average and maximum degrees. The average path length is measured over all the messages that have reached their destinations. This excludes the connection requests and responses. The node degrees are measured on the snapshot of the overlay topology at the end of a measurement epoch.

The measurements are plotted on Figure 1. Both the average path length and the average degree scale logarithmically in terms of the number of nodes as in the state-of-the-art overlays. In addition the logarithmic scaling of the maximum degree is evidence for a balanced degree distribution, which is crucial for balancing the message forwarding load among the peers.

The case of the 3D identifier space is an outlier in our scaling experiments. In contrast to other identifier spaces, the average degree is rising much more rapidly. This is caused by the "curse of dimensionality" problem, which we discuss in Sect.3.5.

3.3 Maintenance Overhead and Failures in Extreme Churn Conditions

Apart from scaling, another important characteristic of an overlay is its maintenance overhead and resilience to node departures or failures. To measure these characteristics we switch from the power-law node lifetime model to Poisson arrivals and departures, the rate of arrivals and the rate of departures are gradually increased. While this happens the network size is kept around 1000 nodes. To stress test the overlay we set the churn rates to values that are considerably higher than those typically seen in peer-to-peer network deployments.

Figure 2summarizes the results. Our overlay maintenance algorithm keeps the routing failures under 0.2% even when 40% of the nodes are replaced with new ones every minute. For small churn rates the maintenance traffic increases faster with the increasing churn, when the churn is higher the massive parallelism in connection request sending lowers the number of connection requests a newly joined node needs to open as it is more likely to receive connection requests from other newly joined nodes. As the churn rate increases the average degree decreases, there are more missing connections in the topology caused by churn. Even though some poorly connected nodes might lay on the routing path, its average length increases only slightly in high churn conditions.

The overall robustness of our overlay is high, the overlay does not loose connectivity, high routing efficiency and low failure rates are maintained.

3.4 Varying the γ Parameter

The routing convergence parameter γ is crucial in our algorithm. We explore experimentally how the changes of this parameter influence the performance of the overlay. For *gamma* values varying from 0 to 4 and for the different identifier spaces we grow the network until it reaches 1000 nodes. For the resulting network we measure the average number of hops and the average node degree.

The results show (Figure 1) that adjusting the γ parameter allows for precise control of the path length vs. degree tradeoff. Distinct operational regimes can be defined:

- **low degree** - $\gamma \ll 1.0$ - most of the hops are strong and only a few new connections are opened, message routing relies more on the self-avoidance property of the routing algorithm, many nodes need to be visited as the routing gradually and mostly randomly converges towards the destination, messages are frequently dropped due to the TTL0 and NHimp events (Section 2.2)
- **high degree** - $\gamma \gg 1.0$ - most of the hops are weak and a large number of connections needs to be opened to form strong hops to the different areas of the identifier space, the convergence of a message is guaranteed by the high γ value, the distance to the destination exponentially decreases (at least by the γ factor in each hop), self-avoidance rarely has to be used and the average path length is small

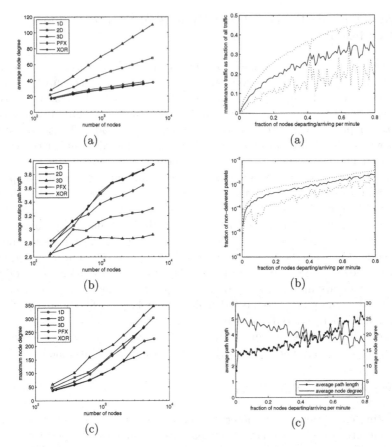

Fig. 1. The scaling of the average path length and the node degree. Each data point is an average of 20 measurements. Standard is under 10%, omitted for clarity.

Fig. 2. Maintenance cost and failure rates under extreme churn conditions. Results from 20 independent experiments. Standard deviations marked with the dotted lines.

– **balanced** - $\gamma \approx 1.0$ - in this regime the scaling of both the average degree and the average number of hops are logarithmic in terms of the number of nodes as demonstrated in section 3.2.

In the next section we discuss the high degree regime further and provide a way for selecting the γ parameter such that the overlay operates in the balanced regime.

3.5   Routability

For a given identifier space only for some γ values the overlay is in the balanced regime and we clearly need a way of determining these values. To achieve this we define the concept of routability. The *routability* $R(\gamma, I, d)$ of the identifier

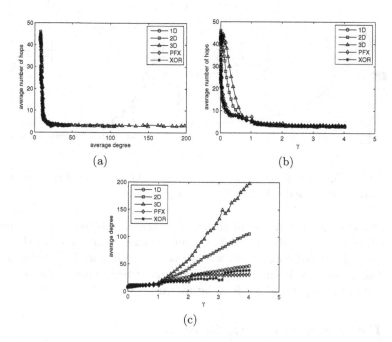

Fig. 3. The influence of varying the γ parameter on the performance of the overlay. Each point represents a measurement on a separate overlay that was independently run.

space (I, d) under the convergence rate γ is the expected probability of finding a strong next hop taken over all possible sources and destinations. Let X be the random variable describing the distance between two random identifiers in the (I, d) space, then $R(\gamma, I, d)$ is $\int_0^{x_{max}} \mathbb{P}(X = x)\mathbb{P}(0 < X < \frac{x}{\gamma})dx$. The values of R are computed numerically for the different spaces and values of γ (Figure 4). Low routability values indicate that the network will operate in the high degree regime and vice versa high routability is a good predictor of the low degree regime (compare these results to Figure 3).

To provide an extreme case we have included a highly dimensional space EUC50, the surface of 51-dimensional hypertorus with Euclidean distance metric. The value of routability for EUC50 at $\gamma = 2.0$ is very close to 0.0 and for $\gamma = 1.1$ it is 0.17 which we have verified experimentally to be enough to provide logarithmic scaling. This result also demonstrates that highly dimensional spaces are not good a good choice for routing due to their "curse of dimensionality" [14]. Only when the node degree is very high can efficient routing be achieved.

The routability concept can be conveniently used to find the ranges of γ values and identifier spaces for which the network operates in the balanced regime, i.e. with routability values in the mid-range, close to 0.5. It has to be noted that this is a necessary condition for good scaling of the network, not a sufficient one.

Routability depends on the variable X which among other factors depends on the distribution of the identifiers in the identifier space, which thus far was

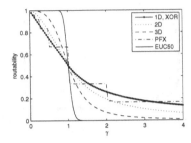

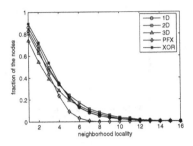

Fig. 4. Routability for the different identifier spaces and values of γ

Fig. 5. The fraction of nodes with a given locality

assumed to be uniform. If that distribution is skewed, this may greatly influence the routability value. The routability formula should also factor in the actual message traffic distribution which we assumed to be uniformly distributed over the set of all source-destination pairs. Exploring these dependencies is beyond the scope of the paper and is left as future work.

3.6   The Emergence of Local Structures

Some overlays maintain a completely deterministic set of neighbors, others create random links [8]. However, all of the current structured approaches maintain at least one deterministic connection, usually to its closest neighbor. Those connections are eagerly maintained and are crucial to reliable routing, especially at the very last hop. In our routing and maintenance algorithm, the creation of this type of connections is not explicitly a part of the algorithm. However, nodes following the simple local rule for sending the connection requests create a dense network of short-range links.

To measure this effect we define the value of *locality* for each node. With a vertex $v_i \in V$ we associate the series of vertices $l_{i1}, l_{i2}, ... l_{i(n-1)}$ where $n = |V|$, and $\forall_{i,j,k} d(id(v_i), id(l_{ij})) < d(id(v_i), id(l_{ik})) \Rightarrow j < k$. Then locality of v_i is equal to m if $\forall_{k=1..m} l_{ik} \in neigh(v_i)$. Informally, the locality of v_i is the number of nodes closest to v_i in the identifier space such that all of these closest nodes are connected to v_i.

We take the topology snapshots of the overlays constructed in different identifier spaces with 1000 nodes. We measure the fraction of nodes with the given value of locality (Figure 5). The topology snapshot is taken while the overlay is churning and thus some of the nodes may have just joined the overlay and have not opened a sufficient number of connections or some of the nodes may have lost a connection due to a neighbor's departure and have not replaced this missing connection with another one. Churn decreases the measured fractions of nodes. Despite churn, 80-90% of nodes are connected to their closest neighbor, and 70-80% of the nodes are connected to two of their closest neighbors. In a 1D ring identifier space if all of the nodes have locality 2 then there exists a global ring spanning all of the nodes, it is possible to visit all the nodes by hopping along that ring in one of the two directions. In overlays based on the 1D ring identifier space (e.g. Chord[1]) this global spanning ring is explicitly maintained. In the case of our algorithm, the ring emerges as a result of the simple connection opening rule in the maintenance algorithm. This rule is independent of the identifier space used and similar local structures are universally created in identifier spaces other than the one-dimensional ring.

4   Discussion of Results and Future Work

We have shown how to use the concept of routability to find the range of γ parameters that ensure that the network stays in the balanced operating regime for a given identifier space. We verified experimentally that this regime exhibits optimal overlay scaling properties, however one might ask a question why do the values of γ close to 1.0 still produce networks with logarithmically increasing routing path length, despite the fact that each hop is not required to shorten the distance to the destination. This is explained by the fact that each node has a non-negligible probability of opening a long range link which acts as an effective shortcut in the identifier space. Though there are few of those long range links in the case when γ is close to 1.0 they significantly lower the average path length. We plan to complement our extensive simulations with an analytical treatment of the routing and maintenance algorithms to explore this phenomenon in detail and relate it to other work in small-world networks, such as the results of Kleinberg [15].

Throughout the paper we have assumed that γ is a system-wide constant. However, it may vary in the following ways:

 - per node - each node might locally decide at what routing convergence rate it forwards the messages, indirectly this gives the node a way of controlling its degree
 - over time - γ can change to adapt to changing network conditions, e.g. churn
 - per message - some types of traffic may be prioritized, e.g. traffic with a specific destination might have a higher routing convergence rate associated with it

Fine-grained control over γ gives a considerable degree of flexibility in shaping the overlay, adjusting the length of the routing paths for different types of traffic and deciding which nodes the traffic traverses through. This may be particularly useful in cases when the source-destination distribution of the message traffic is skewed.

In our maintenance algorithm the connection requests are triggered by messages and are normally sent immediately after the message itself towards the same destination as the message. It is very likely that both the connection request and the message that triggered it follow the same routing path. This can be exploited to piggyback the connection requests on the actual application messages. This can be easily implemented since each message already contains the visited list, the connection request can then be a single bit flag attached to a node in the visited list. We plan to implement connection request piggybacking and investigate how much maintenance bandwidth can be saved in this way.

On the other hand, our protocol adds additional overhead to each of the messages by storing the visited list. The size of this list equals to the number of hops and scales logarithmically in network size. The list is used to prevent self looping while routing. However, once the overlay is stable there are no loops and the visited lists are not used. For $\gamma > 1$ Each routing loop has at least one weak hop, this suggests that it may be sufficient to store visited nodes only when the next hop is weak. Moreover, instead of explicitly storing the whole list of visited nodes, Bloom filters on node identifiers could be used to drastically decrease the space needed for storing the list. These optimizations await experimental investigation.

If our overlay is used as the routing substrate for the distributed hash tables, the emergent local structures (Section 3.6) can be used to manage the replicas of the key-value pairs. Furthermore, the node identifier space can be selected such that it better suits the needs of a particular DHT key space and the application that uses the DHT. We plan to implement a DHT on top of our overlay and investigate the advantages brought by flexible identifier spaces.

5   Related Work

Our approach is most similar to Freenet [10], which follows simple rules for opening new connections. Moreover, in Freenet, just as in our approach, routing is loop avoiding. However, in Freenet loop avoidance is implemented by keeping the routing state at the nodes, while in our algorithm we keep it in the message, which is a stronger form of loop avoidance. Another major difference is that in Freenet new connections are opened by performing a random walk when the node joins a network. This leads to a scale-free topology with a power-law node degree distribution with a small number of nodes having a very high degree. In our case the maximum degree increases only logarithmically with the overlay size.

We have shown how adjusting γ can be used to trade off between the node degree and the routing path length. In a real network implementation this corresponds to the maintenance bandwidth vs. routing latency tradeoff. This problem has been studied in the context of Accordion [16], where given a bandwidth budget the protocol adjust the number of maintained connections to the current churn levels. The obtained tradeoff curves are similar to the ones in figure 3. A simplified version of this algorithm can be implemented in our overlay by making γ a function of the current churn level and the given bandwidth budget.

A widely studied topic in the domain of complex networks are the small-world graphs. They are commonly defined [17, 18] as graphs having both a small diameter and high clustering coefficient. Our overlay satisfies both of these properties, except to quantify the clustering we employed our own measure - the node locality (Section 3.6). Our overlay can be viewed as a dynamic model for small-world growth. The first dynamic model capable of generating networks having small-world properties was proposed by Watts and Strogatz [19] in 1998. In their approach they start with a regular graph and then modify its structure through random rewiring. The probability of rewiring can be controlled. As the probability increases, the network goes through three topologically distinct stages. First, the topology is highly regular with high average path length, then it is small world and finally completely random with short path lengths but low clustering coefficient. Kleinberg [15] places the nodes on regular multi-dimensional lattices and addresses the problem of connection length distribution that would ensure efficient routing. We generalize on this work further by considering a wider family of spaces in which nodes are embedded and proposing a concrete routing and network growth algorithm that is able to achieve logarithmic routing path scaling. We performed measurements of the distance distribution between connected nodes in topologies generated by our overlay maintenance algorithm. The distributions exhibit very consistent power-law characteristics identical to the ones observed by Kleinberg. However, we do not include these results as they lay outside the scope of the paper.

6   Conclusions

In Gnutella [11], requests are flooded through the network until they reach their destination. The nodes are not embedded in any space. What gives the structured overlays their structure is the space the nodes are embedded in. Once the space is added the flow of messages is given directionality and they no longer have to move in all directions simultaneously to find their destinations but only towards directions that decrease their distance to the destinations. A form of this greedy routing is used in all state-of-the-art structured overlays. Routing decisions are made solely based on local measurements of gradients in the identifier space. The knowledge of the properties of the whole space is not necessary. We have proposed a generic algorithm that for any metric identifier space is able to maintain the overlay and route messages based only on local information and simple rules.

Each overlay runs a maintenance algorithm that opens connections to ensure that for every node every forwarded message is brought closer to the destination in terms of the space distance. Our overlay maintenance algorithm generalizes this rule. The proposed algorithm is parameterized by γ which allows for precise control of the path length vs. degree tradeoff while generating relatively small maintenance traffic even under high churn.

Normally, the overlay maintenance algorithm keeps the network prepared for efficient traffic routing from any source to any destination. In our algorithm maintenance is tied to routing and connections are created on demand and only cover the set of destinations that actually appear in the traffic without opening unnecessary connections. This may be controlled at a finer granularity by associating different γ values with different forwarding nodes and different types of routed traffic.

The main insight of our work is that the identifier space is flexible and the properties of the individual identifier spaces used in structured overlays are not significant as long as the spaces provide a consistent local gradient for the greedy routing algorithm to follow. Moreover, the topology of the overlay does not have to be an eagerly maintained pre-defined rigid structure but can emerge in a self-organized way from simple local maintenance rules that adapt the topology to the identifier space. This departure from the structural rigidity of the overlays opens new possibilities of handling skewed overlay traffic distributions, prioritizing traffic, adapting to in-homogeneous allocation of node identifiers and customizing the identifier space to the needs of the application.

References

[1] Stoica, I., Morris, R., Karger, D., Kaashoek, F., Balakrishnan, H.: Chord: A scalable Peer-To-Peer lookup service for internet applications, pp. 149–160.
[2] Aberer, K., Cudré-Mauroux, P., Datta, A., Despotovic, Z., Hauswirth, M., Punceva, M., Schmidt, R.: P-Grid: a self-organizing structured P2P system. SIGMOD Record 32(3), 29–33 (2003)
[3] Zhao, B.Y., Kubiatowicz, J.D., Joseph, A.D.: Tapestry: An infrastructure for fault-tolerant wide-area location and routing. Technical Report UCB/CSD-01-1141, UC Berkeley (April 2001)
[4] Rowstron, A., Druschel, P.: Pastry: Scalable, decentralized object location, and routing for large-scale peer-to-peer systems. In: Guerraoui, R. (ed.) Middleware 2001. LNCS, vol. 2218, pp. 329–337. Springer, Heidelberg (2001)
[5] Maymounkov, P., Mazieres, D.: Kademlia: A peer-to-peer information system based on the xor metric (2002)
[6] Bharambe, A.R., Agrawal, M., Seshan, S.: Mercury: supporting scalable multi-attribute range queries. In: SIGCOMM '04: Proceedings of the 2004 conference on Applications, technologies, architectures, and protocols for computer communications, pp. 353–366. ACM Press, New York (2004)
[7] Ratnasamy, S., Francis, P., Handley, M., Karp, R., Schenker, S.: A scalable content-addressable network. In: Proceedings of the 2001 conference on applications, technologies, architectures, and protocols for computer communications, pp. 161–172. ACM Press, New York (2001)
[8] Manku, G., Bawa, M., Raghavan, P.: Symphony: Distributed hashing in a small world (2003)
[9] Aberer, K., Alima, L.O., Ghodsi, A., Girdzijauskas, S., Hauswirth, M., Haridi, S.: The essence of P2P: A reference architecture for overlay networks. In: P2P2005, The 5th IEEE International Conference on Peer-to-Peer Computing (2005)
[10] Clarke, I., Sandberg, O., Wiley, B., Hong, T.W.: Freenet: A distributed anonymous information storage and retrieval system. In: Federrath, H. (ed.) Designing Privacy Enhancing Technologies. LNCS, vol. 2009, pp. 46–53. Springer, Heidelberg (2001)
[11] Ripeanu, M.: Peer-to-peer architecture case study: Gnutella network (2001)
[12] Protopeer: http://lsirpeople.epfl.ch/galuba/protopeer
[13] Bustamante, F., Qiao, Y.: Friendships that last: Peer lifespan and its role in (2003)
[14] http://en.wikipedia.org/wiki/curse_of_dimensionality
[15] Kleinberg, J.: The Small-World Phenomenon: An Algorithmic Perspective. In: Proceedings of the 32nd ACM Symposium on Theory of Computing, ACM Press, New York (2000)
[16] Li, J., Stribling, J., Morris, R., Kaashoek, M.F.: Bandwidth-efficient management of DHT routing tables. In: Proceedings of the 2nd USENIX Symposium on Networked Systems Design and Implementation (NSDI '05), Boston, Massachusetts (May 2005)
[17] Newman, M.: The structure and function of complex networks (2003)
[18] Albert, R., Barabási, A.: Statistical mechanics of complex networks.
[19] Watts, D.J., Strogatz, S.H.: Collective dynamics of "small-world" networks. Nature 393, 440–442 (1998)

A Common Architecture for Cross Layer and Network Context Awareness

Manolis Sifalakis[1], Michael Fry[2], and David Hutchison[1]

[1] Lancaster University, Computing Dept., Infolab21
LA1 4WA Lancaster, UK
{m.sifalakis,d.hutchison}@lancs.ac.uk
[2] The University of Sydney, School of Information Technologies
NSW 2006, Australia
Michael.Fry@usyd.edu.au

Abstract. The emerging Internet and non-Internet environments have renewed interest in flexible and adaptive communication subsystems residing in end and intermediate systems, which utilise cross layer and wider network context information. To date most cross layer solutions have been very application and/or network specific, and lack re-usability. Here we propose a common architecture to support autonomic composition of functions using generic views of information derived from lower level primitives. At its heart is a distributed Information Sensing and Sharing framework. A combination of key features of this framework are the decoupling of information collection from information use, its capability to multiplex information sources, its operational independence from any specific protocol configuration, and its use outside a node context.

1 Introduction

The current TCP/IP-based architecture of the Internet has served very well for the last twenty years, underpinning network growth and global information sharing at a scale that was unimaginable at the time it was created. However, the one-size-fits-all, end-to-end model of communication embedded in the TCP/IP architecture is now under pressure through the emergence of pervasive wireless technologies, in new networking paradigms. In the emerging Internet there is an increasing disparity between fast, wired core networks and highly heterogeneous access networks [1]. The TCP/IP stack is ill suited to this environment of highly heterogeneous networks and devices, causing poor performance and operational instability. As Internet technologies and protocols are the basis for the next generation of converged, multi-service networks, the rigidities of the wired and data-centric TCP/IP world needs to be relaxed.

A key idea gaining momentum in recent "clean slate" approaches is that the pure end-to-end architectural model of the Internet no longer works. This model has already been broken in reality by the proliferation of NATs, firewalls and proxies, so it is time to recognise this architecturally. The new Internet will consist of more loosely connected "compartments" of networks, some implementing full TCP/IP and some not, consisting of IP and non-IP devices. At the interstices of these compartments will

D. Hutchison and R.H. Katz (Eds.): IWSOS 2007, LNCS 4725, pp. 103–118, 2007.
© Springer-Verlag Berlin Heidelberg 2007

be application-aware interconnect points which will manage "network impedance" and tackle inter-communication issues between network compartments such as addressing, routing, protocol translation, security/trust, performance etc [1,2]. For example, this view underpins the work of the current EC 6th FP Autonomic Network Architecture (ANA) Project [3].

A further key idea is cross-layering, which has been shown to be more appropriate in newer, non-traditional environments, e.g. [4]. Cross layering breaks down layer boundaries, sharing information about network and application state, which allows performance to be optimised. However cross layering has to date been used in a problem and/or network specific manner. To ensure stability of operation and ongoing interoperability in the wider Internet environment, a generic, engineered approach is required.

Finally, as the Internet becomes the global network of convergence, there is a need for greater application adaptivity. Cross layering supports such adaptivity. However, there is also a need for better feedback from across the network, in terms of currently available bandwidth, delay, jitter etc, providing richer context for adaptivity decisions such as codec choice and content placement.

To solve these critical problems we propose dynamic composition, which will match application requirements to the transient state of the network, using information sensed across traditional protocol layers and from the wider network context, to determine choice of protocol functions, mechanisms and parameters for optimal performance. Our approach is to fuse ideas of cross layering and of active and passive measurement into a generic framework, to provide unified views of network context. These can be used at different levels of abstraction to choose protocol functions, and dynamically compose them into a communication system for the particular application, stream and context.

At the heart of our scheme is a generic Information Sensing and Sharing (ISS) framework which provides a network-wide knowledge plane. The framework can be 'programmed' in order to abstract and aggregate information primitives, and to provide event notification at the appropriate level. While a key use of this framework is dynamic protocol composition to ensure optimal performance, it can also underpin network-wide functions such as autonomic network management or support for network resilience metrics.

This paper is organised as follows. In section 2 we expand on the motivations for our approach. Section 3 describes a high level approach for leveraging functional adaptation in a system, so as to highlight the context of this work. As the primary focus of this work is on information sharing, in section 4 we describe the ISS framework in more detail. Section 5 then validates the ISS framework by proof-of-concept scenarios of cross layering and network context awareness. In section 6 we discuss related work before concluding in section 7 with a summary of future work.

2 Motivation

Recent years have seen a proliferation in the development of new network access technologies, especially wireless data networks in the form of Mobile Ad Hoc Networks (MANETs), Wireless Sensor Networks (WSNs), and Wireless Mesh Networks

(WMNs). Some are predicting an imminent "wireless explosion", whereby use of mobile devices, equipped with multiple, heterogeneous wireless network interfaces, will become a pervasive and predominant form of Internet access – the so-called "HetNet" [1]. However research to date into these networks has almost universally focused on the physical, MAC or routing layers. There has been very little attention paid to upper layer protocols [5,6]. These networks have been studied as self-contained networks, with little consideration of them being interconnected to the wider Internet. However with developments like field deployment of real WSNs, eg [7], and prospective roll-outs of metropolitan wide WMNs, end-to-end protocol issues are emerging as reality.

At the same time there is widespread consensus in the networking research community that the conventional TCP/IP stack is not suited to these emerging environments. For example, wireless networks display substantial variance in reliability, speed, error rates etc, and TCP/IP does not deal effectively or efficiently with such environments. The canonical example of this is the behaviour of the TCP congestion control mechanism, which equates time-out events with packet loss caused by network congestion, and then uses "multiplicative decrease" of the sending window to throttle transmission rate [8]. TCP's assumed equation of packet loss with network congestion is typically correct for the wired Internet. However on wireless sub-nets packet loss is often caused more by transmission errors due to noise, and then the TCP congestion control action is unnecessary and unhelpful. Consequently there has been significant research activity regarding TCP over wireless links, but with mixed results [9], while a solution to this problem could produce significant gains in performance [10]. This problem, and others related to the rigidities and assumptions of the TCP/IP architecture, have not been solved and require a fresh approach.

Congestion control exemplifies a more fundamental problem in the TCP/IP stack, which is the opaqueness of layering. While problems with layering have previously been recognised [11], these have become much more pressing with the widespread advent of wireless. This has renewed interest in *cross layering*, which embodies the principle that layers expose information to other layers, which is then used to influence protocol behaviour.

There has been much recent research aimed at optimising performance within some limited application domains such as video streaming, using information exposed by the physical and MAC layers of wireless networks, eg [12,13]. However a limitation of much of this work is that it addresses a particular application-level problem, using techniques specific to that problem. Furthermore, it is now being realised that one-off, uncoordinated optimisations may have competing interests, and thus have unintended side-effects that jeopardise system stability and operational correctness [14]. There is a great danger that a mish-mash of special purpose "tweaks" of protocol stacks will proliferate in response to frustrations over wireless Internet performance, causing rampant instability.

We therefore propose a framework that has a number of key goals. It must permit efficient sharing of information across all protocol layers, independent of any specific family of protocols. It should support the derivation of multiple views of information, including both simple aggregation of information sources and the capability to automatically generate higher level abstractions in the context of the different protocol layers. Different views of information should be discoverable and reusable. Such cross layer information is used to dynamically generate optimisations. However, there is also a need to maintain a global view and control of the existing optimisations so as to enable accountability, operational correctness and stability of the system.

What also propose a unified framework and mechanisms for managing cross layer and network context information sharing and optimisations. This fuses previously disparate work on cross layering and network measurement. Architecturally, we explicitly break the end-to-end model, assuming a loosely connected set of heterogeneous network compartments. Our framework may be deployed on client and server end systems, and also on intermediate proxy machines acting as gateways. We envisage a controlled deployment by network providers and/or enterprise network owners. Not all end systems will be able to support our framework, eg a WSN node. In such cases the managing proxy node for the sub-net will compose and deploy appropriate protocol systems to the nodes, in order to optimise both communication and "bits per watt" node performance. While not within the scope of our current work, which is focused on performance optimisation and resilience, this architecture could more generally be used to address wider issues in the new Internet such as routing, authentication and accountability [15], controlling "docking" or "vertical hand-offs" of roaming users and systems.

3   Enabling Autonomic Functional Composition

We envisage a clean slate approach for enabling adaptation and autonomicity in a wide range of cases and independently of any single network architecture. This is illustrated in Figure 1.

The main components of this abstract representation are the following:

- **The function composite.** This is where the complete data path functionality of a network node resides. A graph interconnecting functional blocks essentially dictates the different data paths that information may flow across, depending on the role the node in a network. Which protocol functions or layers are present (or enabled) in the composite is based on any sort of inference (however simple or complex) that resides in a component external to the composite. Composition in an autonomic network can be a highly dynamic process that can occur at bootstrap time or at runtime.

- **The logic environment.** This is decision making logic for determining what functions to include in the composite as well as how to interface them. It may be either monolithic or distributed across a number of independent modules. The decisions are based on heuristics as well as on input provided by the subsequent component. In principle, it should be of no importance whether the input originates in the external environment or the internal state of the node as long as it is semantically consistent with the expected by the modules information.

- **The information sensing and sharing framework.** This framework, which is the main focus of this paper, is the heart of information exchange and awareness in the network node. It provides the input for the logic that drives the functional composition. The essence of this component lies in an event-notification mechanism that allows interested parties to exchange information when certain events occur. The multiplexing capacity of this event mechanism enables both the combination of information from various information sources to be made available to the serviced modules as well as the dispatching/de-multiplexing of events and information to multiple interested parties. One of the main objectives in this framework (as already mentioned) is the decoupling of the information collection process from the use of this information. This permits features such as multiplexing, aggregation and way of combining the collected (shared) information before it is supplied to its user. Furthermore, it enables flexibility in the information exchange over the network, between instances of this framework.

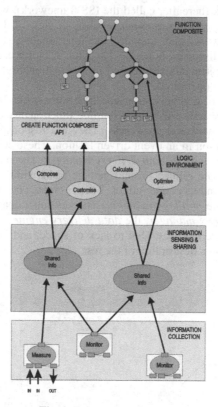

Fig. 1. Framework Overview

- **The information collection modules.** This component includes the entities responsible for the collection of the information that may be shared. These may include measurement elements distributed across the network, monitor hooks inside a host operating system, logic modules that want to share their state, or even protocol functional blocks from the function composite, ie cross-layer information.

This model provides an abstract and unifying approach for acquiring, sharing and using (state) information, which addresses many of the shortcomings of current stand alone cross-layer solutions. This approach of exchanging information using an abstracted mechanism may also leverage tasks such as network and systems monitoring and management (monitor agents and other management components would be simple sinks of the sensory input in the logic environment) in legacy systems. At the same time it can leverage the design of more dynamic future network systems. Its current deployment with the ANA network node as part of the respective EU funded project [3] is providing a challenging crash-test for its feasibility.

4 Information Sensing and Sharing (ISS) Framework

In this section we focus on the design and implementation of the information sensing and sharing framework (hereafter called the ISS framework), which is the major contribution of this paper. Figure 2 presents an abstract view of the building blocks that comprise the framework and support its functionality, together with a scenario/example that is discussed in the next section. A closer examination of the main building blocks (those semantically related to its functionality) follows, along with a description of the basic operation.

There are two sets of components in the ISS framework. The first set of components (on-line components), provide the runtime functionality of the ISS framework, namely system (node/network) awareness. They are responsible for collecting and disseminating information in an event driven fashion. These components are the *multiplexor*, the *remoting*, and the *data delivery manager*. The other set of components (off-line components) are related to management tasks within the framework or between the framework and its client entities. These components include the *authenticator*, the *registry* and the *ontology manager*. The key functionality is provided by the *multiplexor*, the *ontology manager*, the *data delivery engine*, and the *remoting* components, which we describe next. Due to lack of space, and because they are not of core relevance, we omit a description of the rest.

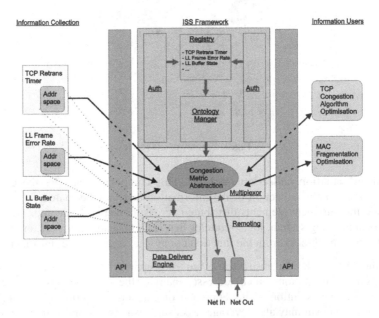

Fig. 2. ISS Framework

- **The multiplexor** is the heart of the ISS framework as it is responsible for multi-plexing events triggered by one or more event sources and demultiplexes them to one or more interested client entities. Any entity sharing information of any type acts as an event source, and any entity interested on information acts as an event sink. Any ISS client may act as either an event source (providing information), or event sink (consuming information) or both. The event multiplexing functionality is founded on Boolean algebra to describe the processing that takes place and to guarantee that at from any canonical state, the system will progress to another also canonical state (closure property) and therefore retain stability and predictability. Boolean logic operators combining event sources, are therefore both extensible and easily combined to generate multiplexing capabilities (satisfying the requirements specification of a sink). Figure 2 exemplifies three event sources or monitors on the left, operating at different protocol layers. Two clients (or event sinks) on the right side have registered higher level abstractions of interest that specify how the event sources are to be multiplexed.
- **The data delivery manager** is responsible for the sharing of the data and delivery to interested parties (event sinks) when a (combination) of event(s) is triggered. An analysis of the recent literature in cross layering [39] regarding different types of information that is typically shared across the network or across layers, leads us to classify information sources into three main categories: those that serve simple no-tifications (binary), those that provide single value information (scalar), and those that provide a larger volume of (spatially or temporally) collected data. In the first case, if an information consumer is only interested in the occurrence of an event then the multiplexor provides the complete functionality. However, in the latter two cases, where a larger volume of information is shared, the data delivery com-ponent will decide on an scheme for delivering the information as well as how the information may be arranged depending on semantic heuristics such timeliness and volume. For instance aggregation operations may be instructed and copy of the data to a location appointed by the information consumer or combine the informa-tion in dynamically generated data structures as prescribed by the information con-sumer and buffered in a queue.
- **The remoting facility** is used to extend the ISS functionality beyond the node's scope across the network, in order to enable network driven context awareness. It augments the functionality of the mutliplexor and the data delivery component across the network in a uniform way, thus promoting a universal view of the in-formation collection and dissemination process towards the clients of the ISS framework. As it simply extends the corresponding APIs, it is not bound to any specific network transport and may deploy any available transport protocol. A sig-nificant limitation of course is the existence of delays or errors during the propaga-tion of information across the network and this has to be taken into account in the construction of a multiplex. However, in cases where this is an inevitable condition the ability to localize processing and aggregate information/events before transmit-ting them over the network is actually a benefit. The main additional flexibility of-fered through the ISS framework is the ability to combine separate or different event sources across the network. These issues are still an open issue in the pro-posed design.

- **The ontology manager**. One of the main feasibility challenges in the ISS framework is the task of understanding the client/sink requirements and translating them to a multiplexing of event/information sources. The abstraction that ISS provides between information collection and information use relies on this capability. This requires some formal means for expression of requirements (from the information users), and an inference process for associating them correctly with the capabilities and services provided by the entities that generate the events and provide the information. A domain ontology backed by a knowledge base of user provided "experience" is deployed to leverage this process. The knowledge base stores information of how combinations of events associate with high level abstractions that the information consumers use to express their requirements. Some simple examples are shown below.

A summary of the operation of the ISS framework is as follows. Entities, which are able to collect and share information, register with ISS as information providers, while entities that want to acquire information register as information consumers. Information providers are essentially event sources for the ISS framework, while information consumers are event sinks. Any single client module of the ISS framework may register as an information provider, consumer or both. During the registration process the client is first authenticated with the framework and acquires an ID token (hash key) which presents to the framework thereafter in all transactions. If the client is an information consumer module the ontology manager parses its requirements specification, consults the knowledge base for translating to the appropriate

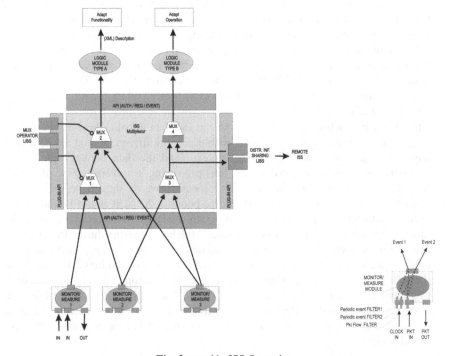

Fig. 3. a and b: ISS Operation

bindings of event sources, and generates the event multiplex description, which is passed to the multiplexor for instantiation. The multiplexor then instantiates the 'wiring' and registers the events with the event sources. Thereafter, operation begins, and whenever the appropriate combination of events occur the collected information is delivered to the event sink, in the form determined by the data delivery component. This is exemplified in Figure 3.

In Figure 3.a , *mux1* combines events from *monitor1* and *monitor2* and propagates an event to *mux2* which in turn combines with an event from *monitor3* and upon trigger fires a notification to *moduleA* (which presumably computes some logic). Similarly, *mux3* combines events from sources *monitor2* and *monitor3* and propagates an event to a *mux* at a remote ISS instance in the network as well as to *mux4*. Finally *mux4* combines the output from *mux3* and a remote event source from the network, and upon trigger sends a notification to logic *module2*. The plug-in interface on the left enables simple processing capabilities for the multiplexing elements so as to allow them to perform "aggregation" functions (used for internal processing of the combined input signals). Example of such operations might be simple data aggregation, averaging of values, min-max operations, etc.

Figure 3.b at the right corner, illustrates the event API of an information provider modules by which they are able to deploy input filter corresponding to the events to be fired. For example when filter mask 1 and 3 are matched event 2 is fired, while when filter mask 2 and 3 are matched event 1 is fired.

One concern regarding the ISS framework is obviously the performance overhead that is introduced. Although, initial evaluation results are not yet available, this concern has led to a number of implementation strategies that aim to minimize overhead. First, the service provided by the ISS framework does not rely on a running server process but rather is implemented as a set of dynamically loaded shared libraries that maintain a shared memory allocation for the common parts of all user processes. A second improvement is that once the initial event multiplex has been produced for an information consumer, a number of algebraic optimizations take place to reduce the Boolean operations required. The optimizations benefit from techniques in the literature (Mealy/Moore machines), and the expected effect is the reduction of the number of computations, the number of intermediate processing steps in the propagation of the event notifications, and the amount of memory allocations required for the data structures in the multiplex. In most test cases that we have considered, after simple optimizations the number of steps is reduced to 2, and the same stands for the number of data structures.

5 Framework Scenarios

We now provide examples of operation of the ISS framework by way of an initial validation.

5.1 Congestion Signals

This scenario exemplifies the use of higher layer abstractions provided by the framework to properly indicate instances of congestion, which solves the canonical TCP congestion problem described earlier. It uses two cross-layer optimisations derived from the literature [16,17]. Figure 4 illustrates 5 different notifications encoded under a *congestion* abstraction (ontology class), multiplexing through Boolean muxes, three raw metric event sources (TCP retransmission timer monitor, link layer error rate monitor and link layer transmission buffer state). These events can trigger different optimisations at the MAC layer or the TCP layer of a current TCP/IP stack according to Table 1. The notifications can be sent as callbacks to two (or more) client consumer (logic) modules that perform the optimizations.

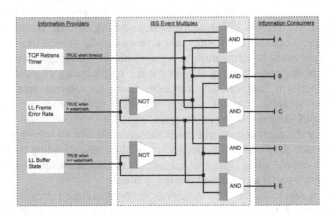

Fig. 4. Raw Metrics Multiplexing

Table 1. Notifications and optimisations

TCP Retrans Timer	LL Frame Error Rate	LL Buffer State	Signal	Interpretation – Action
TRUE	FALSE	FALSE	A	Congestion in the network (e2e path). Enable TCP congestion algorithm and possibly do resource reservation.
TRUE	FALSE	TRUE	B	Congestion at the immediate next hop. Enable congestion relaxation at the MAC layer.
TRUE	TRUE	X	C	Errors at the LL. Freeze congestion control at the TCP level. Change fragmentation scheme at LL.
X	FALSE	TRUE	D	Congestion at the LL. Enable channel reallocation.
X	TRUE	TRUE	E	Errors due to interference at the LL. Change MAC fragmentation scheme.

5.2 Dynamic MAC Error Control at an Intermediate System

We require a logic module that enables or disables error correction at an interface (link-layer) on a per transport flow basis. For instance imagine a small device that has two wireless interfaces: one WWAN (that supports GPRS) and one WLAN (that supports IEEE 802.11x). The MAC layer of the WWAN interface supports error correction while that of the WLAN does not. As the transmission speeds of the two interfaces also vary substantially, the emerging problem is that if an application that requires reliable end to end transmission (e.g. TCP) takes place of the WLAN interface it may experience considerable problems with throughput due to the lack of error control at the wireless hop. On the other hand an unreliable (e.g. UDP) time critical transmission (such as a media stream), would similarly encounter substantial performance problems over the GPRS interface where ARQ-based error control is provided by default. Therefore, being able to enable or disable MAC error control on a per-flow per-interface basis is a useful optimisation.

A logic module to carry out such and optimisation will typically use input from two event sources. One that notifies the presence of link layer errors at an interface above a watermark level, and a second that notifies the appearance of data-flows on an interface that needs reliable delivery (e.g. TCP) and which experiences increased RTTs due to frame errors. As the wireless hop may exist anywhere inside an end-to-end path, the appearance of an end of a reliable flow can be detected by means of the SYN and FIN TCP packet interception on the interface (TCP layer information). However, as both are typically memoryless events, one needs to remember the presence of a SYN to assume the existence of the TCP stream thereafter until the arrival of a FIN packet. The operation of the event multiplex in the ISS framework is illustrated in Figure 5.

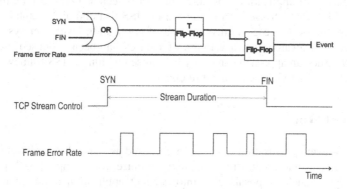

Fig. 5. Dynamic Error Control

When a SYN or FIN happens they trigger a notification that simply reverses the state of the T flip-flop. Since the SYN would always precede the FIN, the D flip-flop will be activated by the presence of the SYN (beginning of TCP stream) and deactivated by the presence of the FIN (end of the TCP stream). If during the active period of the D flip-flop an increased frame error rate is observed the D flip-flop will fire the event to enable error control. Even if the increased error rate appears before the stream, it will still be stored in the D flip-flop and trigger the event for enabling error control when a TCP stream appears. While even if the error rate is temporarily reduced when the TCP flow appears, the error correction mechanism will still be employed, since the recent presence of increased errors is remembered in the D flip-flop. Furthermore this mechanism will prevent the deployment of error control for unreliable (UDP) flows.

5.3 Network Context Awareness

Finally we present some high level description of how the ISS framework can support wider network context awareness. Adaptive multimedia applications have a fairly long history on the Internet. Early audio and video conferencing tools, such as *vat* and *vic*, pioneered the use of performance monitoring to provide feedback for adaptation. Thus, measures of throughput and error rate were used to dynamically adapt the choice of codec used, or to adapt frame rates. Skype is an example of a more modern adaptive application [18]. Our framework can assist such applications by embedding the appropriate active and passive monitors and providing a more generic event notification facility. Such facilities can then be re-used by new applications.

As a further example, the ISS framework may be used to optimise overlay routing via RONs. RONs (Resilient Overlay Networks) are overlays where each overlay member cooperates to provide a view of the resilience of the underlying routing infrastructure. These nodes operate as reference points ("lighthouses") in the routing infrastructure for finding alternative paths in the presence of routing failures or fluctuations. The ISS framework can assist in assessing the resilience of network nodes and links based on a variety of network metrics, thus allowing for more consistent and reliable selection of RON members as well as the coexistence of multiple RONs for different classes of applications. Routing protocols such as OSPF can be instantiated as logic modules which use monitoring via the ISS to build and maintain link state overlay topologies. More generally, the use of application level overlays to support network context awareness via a range of metrics has been proposed and demonstrated [19]. Such an approach can easily be realised within the ISS framework, which will provide abstract views of context to clients.

6   Related Work

Many cross-layer optimisations have been proposed as stand alone solutions targeting specific protocols and applications that they optimise, for example [20-23] to name a few. Rarely do such proposals take into account application requirements is [24]. Many of these proposed cross-layer optimisations e.g. [25,26] lack an architectural perspective, that would enable them to be seen as architectural extensions instead of architectural violations. They do not contemplate impact on other applications or conflicting interactions with other optimisations. The benefits of using a cross-layering framework such as ISS lie in providing a basis for resolving these shortcomings. It enables sharing of information used by optimisations as well as state information in a uniform way that can be used to ascertain stable operation [14].

A number of frameworks have appeared in the literature in recent years from areas such as autonomic networks, context aware computing, and the communities studying the convergence of heterogeneous networks [3,34,35,36,37,38,40]. Many of these framework proposals have been designed for specific protocol configurations and suites, e.g.[28,29] rely on the existence of ICMP packets or IPv6 extension headers, while[30] assumes the existence of specific TCP/IP packet formats for the piggyback-ing and sharing of information across the network. [31,32] propose the extension of the existing TCP/IP protocol interfaces, and [33] relies on protocol specific data struc-tures and types. Additionally, most proposed cross layer frameworks (all but [30,34] to our knowledge) are limited to node-local scope, rather than network-wide context. Finally the incentives and benefits of an event-based model have been claimed in a number of similar application domains, mainly (but not exclusively) related to sys-tems architecting using middleware [40,41,42].

Our approach is differentiated from most previous work in a number of ways. By decoupling the information collection process it enables re-use of optimisations in application domains that have similar abstract requirements (the use of a domain ontology maps abstractly expressed requirements to mechanisms in protocol setups). A further issue is replication of information collection, e.g. [22,26,27]. The ability of the ISS framework to multiplex information sources for various information users reduces the performance penalties of repeating the same measurements unnecessarily. This improvement is consolidated by the ability to perform localised aggregation or processing of the collected information before propagating an event notification. In addition it pushes cross-layering out of the node context to a network context in a uniform way. Finally its functionality is not tied to any specific protocol features.

7 Conclusion

We have presented a generic architecture for managing and making available cross layer and network context information. We have demonstrated the feasibility of the architecture via examples. The development of the ISS framework is currently work in progress. The main focus in the immediate future is on completing the functionality of the multiplexor subsystem, which is the heart of the framework. The challenge is to produce a lightweight subsystem that does not introduce significant delays during the event propagation process and maintains scalability (that is delays are not propor-tional to the size of the event multiplex), while at the same time retaining flexibility. Initial results are anticipated in the near future.

A second major area of current focus is the abstractions that the ISS framework supports between the information collection modules and the information consumers. In this respect, the challenge we face is the development of a sufficiently descriptive, yet simple, ontology for (i) expressing the requirements of the information user mod-ules in terms of general protocol semantics, (ii) describing the services provided by the information collector modules, and (iii) parsing and mapping these to autonomi-cally generate multiplexes of event sources to event sinks. Finally a third area, which will initially explore monitoring and measurement solutions from the literature, is the development of the remoting service.

Acknowledgments

This work was funded in part by the European Union Information Society Technologies Framework Programme 6 (EU IST FP6), under the auspices of the Autonomic Network Architecture project (EC-0174489), where Lancaster University steers research on resilience and cross-layering led by J. Sterbenz and D. Hutchison.

References

1. Mapp, G., Cottingham, D., Shaikh, F., Vidales, P., Patanapongpibul, L., Balioisian, J., Crowcroft, J.: An Architectural Framework for Heterogeneous Networking. In: Proceedings of International Conference on Wireless Information Networks and Systems (August 2006)
2. Clark, D., Sollins, K., Wroclawski, J., Faber, T.: Addressing Reality: An Architectural Response to Real-World Demands on the Evolving Internet. In: Proceedings of ACM SIGCOMM, pp. 247-257 (2003)
3. Autonomic Network Architecture, Situated and Autonomic Communications - EU IST FP6, ACM Computer Communications Review vol 36-2 (April 2006), http://www.ana-project.org
4. Borgia, E., Conti, M., Delmastro, F.: MobileMAN: Design, Integration and Experimentation of Cross-Layer Mobile Multihop Ad Hoc Networks. IEEE Communications 44(7) (2006)
5. Akyildiz, I.F., Su, w., Sankarasubramaniam, Y., Cayirci, E.: A Survey on Sensor Networks. IEEE Communications (August 2002)
6. Akyildiz, F., Wang, X., Wang, W.: Wireless Mesh Networks: a Survey. Computer Networks 47 (2005)
7. Hartung, C., Han, R., Seielstad, C., Holbrook, S.: FireWxNet: a Multi-Tiered Portable Wireless System for Monitoring Weather Conditions in Wildland Fire Environments. In: Proceedings of ACM MobiSys. ACM, New York (2006)
8. Jacobson, V.: Congestion Avoidance and Control. In: Proceedeings of ACM SIGCOMM. ACM Press, New York (1988)
9. Balakrishnan, H., Padmanabhan, V., Seshan, S.: A Comparison of Methods for Improving TCP Performance Over Wireless Links. IEEE/ACM Transactions on Networking 5(6) (1997)
10. Krishnan, R., Sterbenz, J., Eddy, W., Partridge, C., Allman, M.: Explicit Transport Error Notification (ETEN) for Error-Prone Wireless and Satellite Network. Computer Networks 46(3) (2004)
11. Wakeman, I., Crowcroft, J., Wang, Z., Sirovica, D.: Layering Considered Harmful. IEEE Network , 7–16 (January 1992)
12. Liu, Q., Zhou, S., Giannakis, G.: Cross Layer Scheduling with Presribed QoS Guarantees in Adaptive Wireless Networks. IEEE Journal on Selected Areas in Communication 23 (May 2005)
13. Khan, S., Peg, Y., Steinbach, E., Sgroi, M., Kellerer, W.: Application Driven Cross Layer Optimisation for Video Streaming Over Wireless Networks. IEEE Comms. 44(1) (2006)
14. Kawadia, V., Kumar, P.: A Cautionary Perspective on Cross Layer Design. IEEE Wireless Communication , 3–11 (February 2005)
15. FIND (2006), http://www.nets-find.net/

16. Ci, S., Sharif, H., Noubir, G.: Improving the Performance of a MAC Layer by Using Congestion Control/Avoidance Methods in Wireless Networks. In: Proceedings of ACM Symposium on Applied Computing, Las Vegas, ACM, New York (2001)
17. Kang, J., Nath, B.: Resource Controlled Mac Layer Congestion Control Scheme in a Cellular Packet Network. In: Proceedings of 59th IEEE Conference on Vehicular Technology, IEEE Computer Society Press, Los Alamitos (2004)
18. Baset, S., Schulzrinne, H.: An Analysis of the Skype Peer-to-Peer Internet Telephony Protocol. In: Proceedings of IEEE INFOCOM, Barcelona (April 2006)
19. Fry, M., MacLarty, G., Wakeman, I.: Using Overlays to Support Context Awareness. In: Proceedings of Third Workshop on Context Awareness for Proactive Systems, Surrey, UK (June 2007)
20. Holland, G., Vaidya, N., Bahl, P.: A Rate-Adaptive MAC Protocol for Multihop Wireless Networks. In: Proc.7th Annual Int'l. Conf. Mobile Comp. and Net, ACM Press, New York (2001)
21. Misic, J., Shafi, S., Misic, V.: Cross-Layer Activity Management in an 802.15.4 Sensor Network. IEEE Communications Magazine (January 2006)
22. Haratcherev, I., Taal, J., Langendoen, K., Lagendijk, R., Sips, H.: Optimised Video Streaming over 802.11 by Cross-layer signaling. IEEE Communications Magazine (January 2006)
23. Ksentini, A., Naimi, M.: Toward an Improvement of H.264 Video Transmission over IEEE 802.11e through a Cross-Layer Architecture, IEEE Communications Magazine (January 2006)
24. Khan, S., Peg, Y., Steinbach, E., Sgroi, M., Kellerer, W.: Application driven Cross-Layer Optimisation for Video Streaming over Wireless Networks. IEEE Comm Magazine (January 2006)
25. Kliazovich, D., Granelli, F.: A Cross-layer scheme for TCP Performance Improvement in Wireless LANs, Technical Report DIT-04-025, Informatica e Telecomunicazioni, University of Trento (2004)
26. El Batt, T., et al.: Power Management for Throughput Enhancement in Wireless Ad-hoc Networks, IEEE ICC, 2000, pp. 1506-1513
27. Liu, Q., Zhou, S., Giannakis, G.: Cross-layer scheduling with Prescribed QoS Guarantees in Adaptive Wireless Networks. IEEE JSAC 23 (May 2005)
28. Sudame, P., Badrinath, B.: On Providing Support for Protocol Adaptation in Mobile Wireless Networks. Journal of Mobile Networks and Applications 6 (2001)
29. Wijting, C., Prasad, R.: A Generic Framework for Cross-Layer optimisation in Wireless personal Area Networks. Wireless Personal Communications Journal 29 (2004)
30. Winter, R., Schiller, J., Nikaein, N., Bonnet, C.: CrossTalk: Cross-Layer Decision Support Based on Global Knowledge. IEEE Communications Magazine (January 2006)
31. Kompella, R., Greenberg, A., Rexford, J., Snoeren, A., Yates, J.: Cross-Layer Visibility as a Service. In: proceedings of Hotnets Workshop (2005)
32. Wang, Q., Abu Ragheff, M.A.: Cross-layer signaling for next-generation wireless systems. IEEE Wireless Communications and Networking Conference (WCNC) (2003)
33. Chinta, M., Helal, A., Hernandez, E.: ILC-TCP: An Interlayer Collaboration Protocol for TCP (2003)
34. Razzaque1, M., Dobson, S., Nixon, P.: A Cross-Layer Architecture For Autonomic Communications. In: proceedings of Int'l Workshop on Autonomic Communications, Paris (September 2006)

35. Hasswa, A., Nasser, N., Hassanein, H.: Tramcar: A Context-Aware Cross-Layer Architecture for Next Generation Heterogeneous Wireless Networks. In: proceedings of IEEE International Conference on Communications (ICC), Istanbul, Turkey (June 2006)
36. E2RII Project, Motorola Labs: http://e2r2.motlabs.com/
37. IST-UNITE Project: http://www.ist-unite.org/
38. Haggle Project: Situated and Autonomic Communications - an EC FET European Initiative (EU IST FP6). ACM Computer Communications Review, vol. 36-2, (April 2006), http://www.haggleproject.org/index.php/Main_Page
39. Sifalakis, M., Hutchison, D., Sterbenz, J., Zseby, T., Salamatian, K.: Functional Composition Framework, Autonomic Network Architectures, Deliverable D2.2 (February 2007), http://www.ana-project.org/images/deliverables/D.2.2.-Func-Comp.pdf
40. Paolo, C., Coulson, G., Gold, R., Lad, M., Mascolo, C., Mottola, L., Picco, G.P., Sivaharan, T., Weerasinghe, N., Zachariadis, S.: The RUNES Middleware for Networked Embedded Systems and its Application in a Disaster Management Scenario. In: Proceedings. of 5th IEEE International Conference on Pervasive Computing and Communications (Percom07), White Plains, NY, IEEE Computer Society Press, Los Alamitos (2007)
41. Blair, G., Coulson, G., Andersen, A., Blair, L., Clarke, M., Costa, F., Duran-Limon, H., Fitzpatrick, T., Johnston, L., Moreira, R., Parlavantzas, N., Saikoski, K.: The Design and Implementation of OpenORB v2. IEEE DS Online, Special Issue on Reflective Middleware 2(6) (2001)
42. Chan, A.T.S., Siu-Nam, C.: MobiPADS: a reflective middleware for context-aware mobile computing. Software Engineering, IEEE Transactions 29(12) (December 2003)

Network Topology Reconfiguration Against Targeted and Random Attack

Kosuke Sekiyama[1] and Hirohisa Araki[2]

[1] Nagoya University, Furo-cho, Chikusa-ku, Nagoya 464-8603 Japan
sekiyama@mein.nagoya-ac.jp
[2] University of Fukui, 3-9-1, Bunkyo, Fukui, 910-8507, Japan
araki@robo.mein.nagoya-ac.jp

Abstract. The issue on optimality and robustness has become a major concern in large-scale network systems. While a star-like centralized network surtcture is optimal in terms of the average path length, it is vulnerable to the breakdown arising in the central node. Scale-free network (SFN) is known to be effective topology in terms of both the average path length and robustness against random breakdown. However, if the hub nodes are intentionally attacked, SFN is found vulnerable. In this paper, we propose an evolutionary network model which reconfigures a network suructure according to the various types of breakdown or intentional attacks while maintaining the system performance. The local evaluation indice and control parameters are introduced to regulate a balance between efficiency and robustness. Simulation results suggest that the proposed approach is promising.

Keywords: Evolutionary Network, Adaptive Network Reconfiguration, Intentional Attacks.

1 Introduction

A great deal of information and traffic circulate over complex and large-scale networks. Such examples abound in an airline route or road map, and the Internet web [1]. These are important social infrastructures and desired to possess sufficient robustness against unpredictable breakdown in order to maintain expected function consistently. Including redundant nodes and links to the network is expected to make the system more robust, however it would lead the system to more expensive and less efficient in the ordinary situations. Therefore, a well balanced network topology taking into account both optimality and robustness is of great importance [2]. Recently, it has come to be known that a number of real world networks configure the scale-free network (SFN), such as seen in the airline route map, electrical power network, and web of the Internet [3,4,5]. One of the main features in SFN is that the network topology exhibits a power-law degree distribution: $P(k) \sim k^{-\gamma}$, where k is the number of link attached to a randomly chosen node in the network and γ is the scaling exponent. SFN is named for the fact that the power-law distributions do not have a median, which indicates

D. Hutchison and R.H. Katz (Eds.): IWSOS 2007, LNCS 4725, pp. 119–130, 2007.

the typical size of the system [6]. It is also known that the average path length between the nodes are surprisingly small in SFN [7], hence the efficient transport is expected over the network. However, despite the fact that SFN is robust against random breakdown or removal of nodes, it is rather fragile to intentional attacks against hubs, in which links are concentrated [8,9,10,11]. In this paper, we present an autonomous reconfiguration model of network topology to maintain the balance between efficiency and robustness under the intentional attack to nodes. In a huge network, localized attacks to the network or breakdown are more likely to happen rather than the uniform and random breakdowns. In order to cope with the problem, an evolutionary network approach is applied in an attempt to strengthen the reliability and efficiency under the intentional attack. The intentional attack means that the hubs (i.e., centralized nodes) are more likely to be targeted by attacks according to the degree. Such an attack will lead disintegration of the network, hence more flat and redundant network topology is preferred, while a centralized network structure is more effective in a steady and secure condition. In this paper, modified preferential linking is introduced such that each node reconnect to the other nodes based on the evaluation of breakdown rate. Steady condition without node breakdown will lead to more centralized and cost-efficient network topology, while SFN or random network are obtainable in the case that intentional attack is present.

2   Optimal Network Topology by Rewiring Process

2.1   Model Outline

Suppose a network composed of N nodes, which is depicted as a directed graph in fig.1(a). The solid line in the figure indicates an active connection from a sender to other nodes, while the dotted line indicates the acknowledge to the active connection. Where, a pair of the active connection and the acknowledge is regarded as *one undirected link* between the corresponding two nodes, supposing a situation in a communication network. The number of directed link that is allowed to be operated for each node is assumed to be L, then the number of corresponding acknowledge link from the linking node is also L, hence the average link number for each node over the network is $\langle k \rangle = 2L$. We exclude the single loop and double link as shown in fig.2. From what follows, the number of link for node is referred to as degree.

Figs.1(a)-(d) show an outline of network reconfiguration process, where we presume a situation of breakdown and recovery in the comunication network. From the initial condition (fig.1(a)), a random breakdown or intentional attack occurs in a node, and the links to the breakdown node are disconnected (fig.1(b)). Then, the surviving nodes will attempt to rewire the connection from the breakdown nodes to another alive node. A node to be rewired is selected probabilistically from all available nodes in the network, avoiding the single loop and the double link structure (fig.1(c)). After a certain recovery time, the breakdown nodes are supposed to recover from a failure and included to the network, hence the network size is assumed to be constant over long time interval (fig.1(d)).

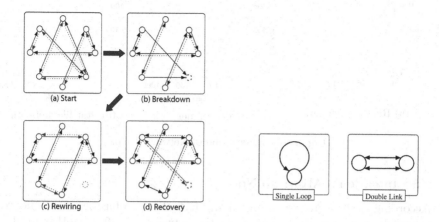

Fig. 1. Rewiring process due to node breakdown   **Fig. 2.** Single loop and double link

2.2 Rewiring Process of Network Topology

The rewiring link process is defined to reconfigure the primary network topology, which is extension from the conventional preferential linking model [12,13]. The preferential linking will select a node by the probability which is proportional to the degree of the node. This means that centralized nodes are more likely selected by the other nodes. Let k_i be the degree of node i, then modified preferential linking from node i to j is given by the following equation,

$$p_{ij} = \frac{(\bar{k}_j)^\beta}{\sum_{h \in S_i}(\bar{k}_h)^\beta}. \tag{1}$$

Where, S_i is the node set available for rewiring from node i. For simplicity, S_i is assumed to include all nodes except for the nodes which will conflict with the linking conditions of simple loop and double link in fig.2. Also, $\bar{k}_j$ denotes time average of degree for the past T time step,

$$\bar{k}_j = \frac{1}{T}\sum_{\tau=t-T}^{t} k_j(\tau). \tag{2}$$

Reconfigured network topology is statistically characterized by degree distribution of the network. Some typical degree distributions are illustrated in fig.3. Also, β in eq.(1) indicates the extent of preferential strength. In the case of $\beta = 1$, eq.(1) reduces to the conventional preferential linking, and $\beta = 0$ means random selection, and a random network is obtainable as shown in fig.3(a). If β takes larger value, the reconfigured network topology approaches to a star-like network (fig.3(c)) through a scale-free network structure (fig.3(b)). Therefore, parameter β becomes a control parameter for a basic network topology configuration.

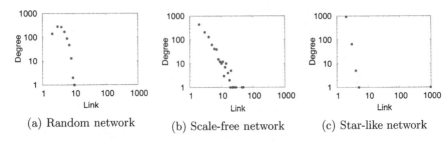

(a) Random network   (b) Scale-free network   (c) Star-like network

Fig. 3. Reconfiguration of Network Topology

2.3 Intentional Attack to Node

According to the conventional research [14,15,16,17], robustness of the network is examined by deleting a node from the network one after another until the network is disintegrated. An intriguing result is that SFN is robust in that the network can remain integrated until nearly 50 % of nodes are removed from the network. However, it is unlikely that such a large part of the network is randomly deleted in reality. Therefore, we assume that a breakdown for node i occurs independently by the following probability,

$$q_i = \min(vk_i + d, 1.0). \tag{3}$$

Where, d denotes the natural breakdown probability regardless of the degree $k \in [L, N-1]$ and v denotes the coefficient that is related to intentional attack according to the node degree k. While $v = 0$ means uniform random attack, if v takes a larger value, the node with large degree is exposed to higher risk of breakdown. Such a breakdown pattern is referred to as an intentional attack. The hub nodes are more likely to be attacked with an increase of the probability q_i. Thus, the optimal network topology has to be attained corresponding to the attack pattern.

2.4 Evaluation Index for Network Topology

To evaluate efficiency of the network topology, three evaluation indices are introduced.

Efficiency Index based on local estimation
The first index is as to transition efficiency over the network. Let N_i be the number of reachable node from node i within n-hop transitions. In particular, this paper deals with the case of n=2 (2-hop). Then, normalizing reachable node over the network, the efficiency index E is defined as the average reachable area of the network, which is expressed as

$$E = \frac{1}{N} \sum_{i=1}^{N} \frac{N_i}{N}. \tag{4}$$

Where, E has similar meaning to an average path length.

Robustness Index

The second index is related to robustness evaluation against natural or intentional breakdown over the network. Suppose that the number of reachable node from node i is N_i, then a breakdown of the peripheral node would cause disconnection of links and reduce the reachable node number from N_i to $\hat{N}_i$. An example is shown in fig.4 where the number of reachable nodes from node i within 2-hop is $N_i = 6$ and the consequence of breakdown in node j leads to disconnections of links and reachable node is reduced to $\hat{N}_i = 2$. Such an influence is normalized over the network. The normalized robustness index R is defined in a similar manner to eq.(4) as follows,

$$R = \frac{1}{N} \sum_{i=1}^{N} \frac{\hat{N}_i}{N_i}. \tag{5}$$

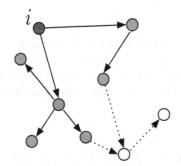
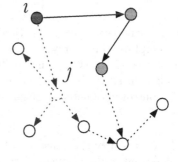

(a) Initial condition around nodes$i$ (b) After breakdown of node j

Fig. 4. Reachable node within 2-hop after node breakdown

Hybrid fitness index

The third index provides a hybrid fitness of the efficiency and the robustness. According to different breakdown patterns over the network, the aim here is to realize a well-balanced network in terms of the efficiency and the robustness defined by eqs.(4) and (5).

$$H = ER = \frac{1}{N^2} \sum_{i=1}^{N} \frac{N_i}{N} \sum_{j=1}^{N} \frac{\hat{N}_j}{N_j}. \tag{6}$$

The efficiency and the robustness indices eqs.(4) and eq.(5) are updated in a regular time step T as following,

$$\bar{R}_i(t) = \frac{1}{T} \sum_{\tau=t-T}^{t} R_i(\tau), \tag{7}$$

$$\bar{E}_i(t) = \frac{1}{T} \sum_{\tau=t-T}^{t} E_i(\tau). \tag{8}$$

And they are employed for adaptive network reconfiguration for the rest of the paper.

3   Adaptive Network Toplogy Reconfiguration

3.1   Adaptive Rewiring Process Based on Evaluation Index

Topology reconfiguration of the network is executed by rewiring links according to observation of breakdown pattern in the peripheral nodes. The modified preferential linking is employed, which is extension from the conventional preferential linking model [12,13]. Firstly, the observed condition of node j is denoted by $b_j \in \{0,1\}$ which takes $b_j = 1$ in the case that a node breakdown occurs and otherwise $b_j = 0$. The average of recent T step, $\bar{b}$ is given by

$$\bar{b}_j = \frac{1}{T} \sum_{\tau=t-T}^{t} b_j(\tau). \tag{9}$$

Eq.(9) denotes the observed time average breakdown rate for the node j and it is employed in the rewiring process. The rewiring probability is based on the form of eq.(1), however the coefficient β is extended so as to incorporate the breakdown situations. Therefore, such an extension is given by introducing a new control parameter as given by

$$\eta_{ij} = \beta \exp[-\gamma_i \bar{b}_j]. \tag{10}$$

Where, β is now the constant which is specifically predefined so that the given preferential probability can regulate various network topologies: random, scale-free, and star-like structures. γ_i is a parameter to be adjusted as rewiring process proceeds. The new preferential linking rule is now given by

$$p_{ij} = \frac{(k_j)^{\eta_{ij}}}{\sum_{h \in S_i} (k_h)^{\eta_{ih}}}. \tag{11}$$

The rewiring process, however, is driven by two different purposes. One is to repair the network topology for the node breakdown, and this process is referred to as *passive rewiring*. Another rewiring process is to optimize the network efficiency when serious breakdown is absent. Such a rewiring process is referred to as *active rewiring*.

(1) *Passive rewiring.* The passive rewiring process aims at reproducing the directed link which was lost due to a breakdown of the connected node. Then the rewiring probability should be updated so that the network will evolve to more distributed and robust structure against the intentional attack, which is realized by increasing the value of γ_i in eq.(10).

(2) *Active rewiring.* In the absence of serious breakdown, a redundant network structure should be optimized. By the active rewiring process, some links are to be rewired once per T step, and the value of γ_i in eq.(10) is reduced in an attempt to realize more efficient network structure, i.e., centralized form.

The update rule of γ_i is as follows. Let t_n denote the discrete time step which is incremented when either of the active or passive rewiring process is executed, then γ_i is updated with eqs.(7) and (8) as follows;
For passive rewiring:

$$\gamma_i(t_n) = \left(1 + \frac{e^{-\bar{R}_i(t_{n-1})}}{T}\right)\gamma_i(t_{n-1}).\tag{12}$$

For active rewiring:

$$\gamma_i(t_n) = \left(1 - e^{-\bar{E}_i(t_{n-1})}\right)\gamma_i(t_{n-1}).\tag{13}$$

The increasing rate of 12 is attenuated by $1/T$ to fit the scale of variation between increasing and decreasing rate. The optimal γ_i is acquired based on the balance of both processes.

3.2   Preliminary Simulation Results

In what follows, preliminary simulation results are shown to examine that the self-regulation of γ is able to cope with unknown intentional attack pattern. Some intentional attack patterns are imposed on the network in tab.1, where v and d are parameters in eq.(3). The simulation is run until $t = 0$ 20000, and the attack pattern is changed in every 5000 step. The network setting is also defined in tab.2. The initial network topology is set to a random network. In the

Table 1. Intentional attack pattern

	v	d	time step
Natural Breakdown	0.0	0.01	$0 \sim 5,000$
Light Attack	0.001	0.01	$5,001 \sim 10,000$
Heavy Attack	0.01	0.01	$10,001 \sim 15,000$
Light Attack	0.001	0.01	$15,001 \sim 20,000$

Table 2. Simulation setting 1

N	The number of node	1000
L	The number of directed link	2
T	Time step for active rewiring	100
β	Basic preferential linking parameter	2

case of the natural breakdown, where intentional attack proportional to the link degree is absent and moderate breakdowns occur randomly, the star-like network topology emerged as shown in the degree distribution fig. 5(a). It can be seen that only the single node has 1000 links and this is huge hub in the network. Also, as the attack pattern becomes intensive from natural breakdown to light attack, and to heavy attack, the network topology evolved to more distributed structure as shown in fig. 5(b) and (c) to enhance the robustness. On the other hand, when

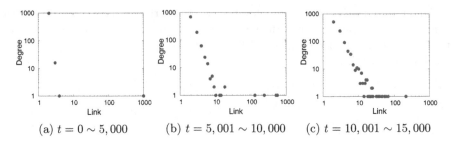

(a) $t = 0 \sim 5,000$ (b) $t = 5,001 \sim 10,000$ (c) $t = 10,001 \sim 15,000$

Fig. 5. Network topology corresponding to the attack pattern

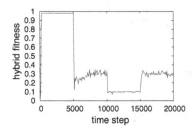

Fig. 6. Hybrid fitness corresponding to the network topology

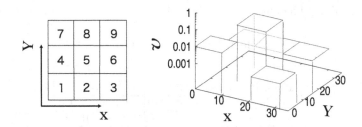

Fig. 7. Left:Area division, Right:Attack pattern on each area

the attack pattern changed from Heavy Attack to Light Attack at $t = 15,000$, the network topology was reconfigured to a more centralized form to recover the efficiency. From these results, it can be said that the proposed adaptive network reconfiguration method functions to acquire a desired network structure, which is well-balanced between efficiency and robustness corresponding to the attack patterns. Fig.6 shows time evolution of the hybrid fitness according to the network rewiring. During $t = 0$ 5000, the hybrid fitness takes maximum value because most efficient network is organized. Under the heavy attack, it declined but regained soon its performance as the network improved its topology as attack was lightened. From these results, it can be concluded that the proposed adaptive network reconfiguration method functions to acquire a desired network structure, which is well-balanced between efficiency and robustness corresponding to the attack patterns.

Table 3. Simulation parameter set 2

N	The number of node	$1089(33 \times 33)$
L	The number of directed link	2
T	Time step for active rewiring	100
O	Observation range of node	15
β	Basic preferential linking parameter	2

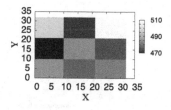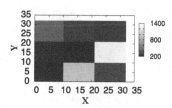

Fig. 8. Left:Link distribution of initial condition, Right:Link distribution at $t = 5000$

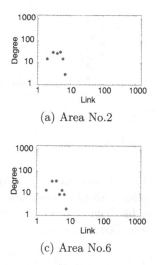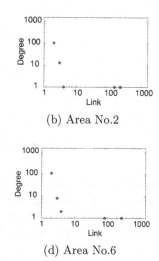

(a) Area No.2

(b) Area No.2

(c) Area No.6

(d) Area No.6

Fig. 9. Left:Degree distribution at initial state, Right:Degree distribution at $t = 5000$

4   Localized Topology Reconfiguration

4.1   Rewiring Including Distance Between Nodes

In a huge network, the attack pattern and breakdown frequency of the node will not be equivalent over the network, on the contrary, some specific part of the network may suffer from more seirious damage. Therefore, topology reconfiguration process should be localized according to the frequency of node breakdown. In the following, local adaptation method of network reconfiguration is discussed. In reality, it is not appropriate to assume that each node can recognize the

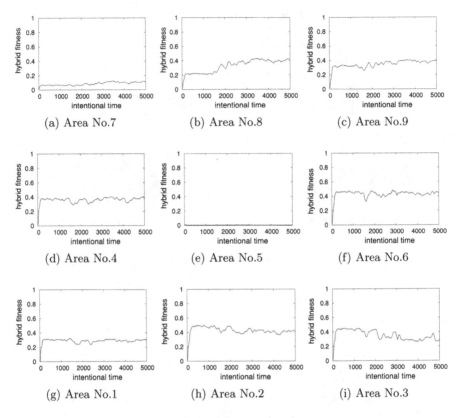

Fig. 10. Hybrid fitness of each area

entire node conditions, hence we assume that the node that can be linked is also restricted to a neighboring node set, which is placed whithin observation range of radius O from node i. Let d_{ij} be the distance between node i and j, then the preferential linking probability from node i is modified as follows;

$$p'_{ij} = \frac{(k_j)^{\eta_{ij}}/d_{ij}}{\sum_{h \in S_i}(k_h)^{\eta_{ih}}/d_{ih}}. \tag{14}$$

Where, the exponetinal η_{ij} is equivalent to eq.(10).

4.2 Simulation Results

Simulations are conducted to examine that presented topology reconfiguration process can cope with a localized intentional attack. Basic network configuration is given in tab.3. The region in the network is decomposed into 9 numbered areas as shown in fig.7 (left). Also, the strength of localized attack to each area is defined in fig.7(right) where vertical axis is the value of v used for eq.(3). In this simulation setting, area 5 receives most heavy attack, and area 1 and 3, 7, 9 receive heavy and light attack respectively. Simulation is run until $t = 5000$ step. Simulation results in fig.8 show that starting from a uniform link distribution in each area, most link is directed to the safety area such as area No.2 and No.5 escaping from No.5 being attacked. Also, the evolution of the degree distribution in area No.2 and No.6 suggest that network topology has been optimized from the random network to the star-like structure because of absence of breakdown in these areas.

On the other hand, figs.10 show time development of the hybrid fitness for respective areas. Area No.2 and No.5 exhibit best performance, however the other areas also exhibit the balanced performance except for the area No.5 which is exposed by the most heavy attack. This preliminary results show that localized adaptation of network topology is suggested, however more in-depth analysis has to be made.

5 Conclusion

In this paper, we presented a dynamical reconfiguration process of the network topology that realizes coordination between the optimality and the robustness against intentional attack to the network. Three primary indices to indicate efficiency, robustness and these hybrid evaluation criteria were discussed. With these indices, the extended preferential linking rule was applied. Simulation results suggest that the well-balance network topology can be reconfigured accoriding to various intentional attack patterns. It is also shown that if such an intentional attack pattern is localized, the localized adaptation is attainable. In the future work, we deal with the influence of cascade failure caused by the intentional attack to the network and aim to realize alleviation of caused failure.

References

1. Echenique, P.: J.g. Gardenes, Y. Moreno. Dynamics of jamming transitions in complex networks. cond-mat/0412053 (2004)
2. Valente, A.X.C.N., Sarkar, A., Howard, A.: Two-peak and three-peak optimal complex networks. Physical Review Letters 92(2), 118702 (2004)
3. Huberman, B.A., Adamic, L.A.: Growth dynamics of the world wide web. Nature 401, 130 (1999)
4. Faloutsos, C., Faloutsos, M., Faloutsos, P.: On power-law relationships of the internet topology. ACM SIGCOMM '99 29, 251–263 (1999)
5. Strogtz, S.H.: Exploring complex networks. Nature 410, 268–276 (2001)
6. Barabasi, A.L.: The New Science of Networks. Perseus Books Group (2002)
7. Cohen, R., Havlin, S.: Scale-free networks are ultrasmall. Physical Review Letters 90(5), 58701 (2003)
8. Albert, R., Jeong, H., Barabasi, A.-L.: Error and attack tolerance of complex networks. Nature 406, 378–382 (2000)
9. Cohen, R., Erez, K., ben Avraham, D., Havlin, S.: Resilience of the internet to random breakdowns. Physical Review Letters 85(21), 4626–4628 (2000)
10. Cohen, R., Erez, K., ben Avraham, D., Havlin, S.: Breakdown of the internet under intentional attack. Physical Review Letters 86(16), 3682–3685 (2001)

11. Zhao, L., Park, K., Lai, Y.-C.: Attack vulnerability of scale-free networks due to cascading breakdown. Physical Review E 70, 35101 (2004)
12. GoLmez-Gardenes, J., Moreno, Y.: Local versus global knowledge in the barabalsi-albert scale-free network model. Physical Review E 69, 37103 (2004)
13. Kalisky, T., Sreenivasan, S., Braunstein, L.A., Buldyrev, S.V., Havlin, S., Stanley, H.E.: Scale-free networks emerging from weighted random graphs. cond-mat/0503598 (2005)
14. Motter1, A.E., Lai, Y.-C.: Cascade-based attacks on complex networks. PHYSICAL REVIEW E 66, 65102 (2002)
15. Crucitti, P., Latora, V., Marchiori, M., Rapisarda, A.: Error and attack tolerance of complex networks. PHYSICA A 340, 388–394 (2004)
16. Guillaume, J.-L., Latapy, M., Magnien, C.: Comparison of failures and attacks on random and scale-free networks. In: Higashino, T. (ed.) OPODIS 2004. LNCS, vol. 3544, Springer, Heidelberg (2005)
17. Lee, E.J., Goh, K.-I., Kahng, B., Kim, D.: Robustness of the avelanche dynamics in data-paket transport on scale-free networks. Physical Review E 71, 56108 (2005)

A Self-organizing Control Plane for Failure Management in Transparent Optical Networks*

Nina Skorin-Kapov[1,2] and Nicolas Puech[1]

[1]GET / Telecom Paris - LTCI - UMR 5141 CNRS, Networks and Computer Science Department, École Nationale Supérieure des Télécommunications, Paris, France
[2]Department of Telecommunications, Faculty of Electrical Engineering and Computing, University of Zagreb, Zagreb, Croatia
nina.skorin-kapov@fer.hr, npuech@enst.fr

Abstract. Self-organizing systems are present in many areas of nature and science, and have more recently been increasingly applied to telecommunications. These systems often exhibit common structural properties, such as the small-world property, and can react to changes in their environment with no centralized control. With ever-increasing capacity requirements, Transparent Optical Networks (TONs) have been established as the enabling technology for future long-haul high-speed backbone networks. Designing fast security mechanisms is critical, particularly due to the high speeds and transparency inherent in TONs. In this paper, we propose a self-organizing small-world control plane for failure management in TONs, which can improve scalability and adapt to changes in the network.

Keywords: Self-organization, small-world phenomenon, transparent optical networks, control plane, failure management.

1 Introduction

Self-organization is a phenomenon where low-level interactions between individual entities spontaneously emerge in certain global properties. These so-called 'emergent' properties, which are spontaneously achieved through the selfish actions of individuals, have certain functionality, i.e., fulfill a purpose beneficial for the system as a whole. Common structural properties have been observed in many such systems [1]. One of the most important is the 'small-world' property [2], a term coined to describe networks which are highly clustered with short average path lengths. Self-organizing systems and concepts have been observed in many areas of life and science, from fireflies flashing in perfect synchrony to the interconnection of web pages on the World Wide Web [3]. Although self-organizing concepts have not yet been fully exploited in the design and functioning of telecommunication networks, applying these concepts to various areas

---

* This work was supported by a Postdoctoral Research Fellowship from École Nationale Supérieure des Télécommunications, Paris, France. The authors are also grateful to the French and Croatian Governments who supported their work by funding their joint COGITO project HONeDT.

D. Hutchison and R.H. Katz (Eds.): IWSOS 2007, LNCS 4725, pp. 131–145, 2007.

in communications is currently being intensively researched. Examples include applications in peer-to-peer networks [4], as well as ad hoc and cellular wireless networks [5]. However, to the best of our knowledge, these concepts have not yet been systematically applied and explored in the context of transparent optical networks.

In Transparent Optical Networks (TONs), the physical network consists of an interconnection of optical fibers employing Wavelength Division Multiplexing (WDM). WDM is a technology which can exploit the large potential bandwidth of optical fibers by dividing it among different wavelengths. TONs are dynamically reconfigurable networks where a virtual topology is created over the physical optical network by establishing all-optical connections, called *lightpaths*, between pairs of nodes. These connections can traverse multiple links in the physical topology and yet transmission via a lightpath is entirely in the optical domain making them transparent. In order to provision, establish, maintain, tear down, or reroute lightpaths due to new connection requests, changing traffic, and/or unexpected failures in the network, an optical control plane is maintained employing various routing and signalling protocols [6]. Control information is sent on a separate wavelength than data signals on each link and is electronically processed at each node.

Although the transparency of TONs offers many advantages, such as speed and insensitivity to data rate and protocol format, it makes monitoring much more difficult since it must be performed in the optical domain. Some of the optical monitoring (OPM) equipment and techniques available today include optical power monitors, optical spectrum analyzers, OTDRs (Optical Time Domain Reflectometer), eye monitors, BER (Bit-Error-Rate) estimation techniques, pilot tones, and others [7]. A survey of optical monitoring techniques can be found in [8]. Most OPM equipment generates alarms upon observing suspicious behavior. These alarms can be used to detect certain failures, but by no means all of them. Furthermore, due to the high cost of monitoring equipment, it is not realistic to assume all nodes are equipped with full monitoring capabilities. Thus, obtaining monitoring information from nodes with high monitoring capabilities efficiently is necessary to ensure reliable network operation.

A failure management system is employed by the TON to deal with various failures, including both component malfunctions and deliberate attacks. Attacks can be particularly malicious since they can propagate through the network and appear sporadically. Attacks most often include jamming and/or tapping legitimate data signals by exploiting component weaknesses, such as gain competition in amplifiers and crosstalk in switches. Various failures have been described in [9], [10], and [11]. Failure management consists of preventing, detecting, and reacting to such failures. *Prevention* mechanisms, such as strengthening and/or alarming the fiber, are measures taken to prevent failures from occurring. *Detection* mechanisms are responsible for identifying and diagnosing failures according to the alarms received from monitoring equipment (via the control plane), locating the source, and generating the appropriate notification messages to ensure successful reaction. Methods to locate and recover from various component faults are proposed in [12]. Localization algorithms to help locate the source of various attacks are given in [7], [13], and [11]. Finally, *reaction* mechanisms restore the proper functioning of the network by isolating the source of the failure, reconfiguring the connections, rerouting, and updating the security status of the network [10]. In the presence of attacks, reaction mechanisms should quickly isolate the source to preclude further attacks. Moreover, the source and destination nodes of failed lightpaths need to be notified quickly so they can launch their restoration mechanisms before triggering higher layer restoration. Additionally, efficiently restoring failed lightpaths is crucial due to the hight data rates involved which could potentially lead to huge data loss.

In this paper, we are concerned with the control mechanisms enabling efficient detection and fast restoration in the presence of failures. Namely, when a failure occurs, optical monitoring equipment sends alarms via the control plane to be analyzed by the failure management system. Lightpaths affected by the failure are then restored as quickly as possible, while failure management works on locating, isolating, and repairing the failure. Here, we do not discuss the specific routing or signalling protocols involved, but present a general model for the optical control plane. Namely, we propose a self-organizing scheme to maintain an optical control plane whose structure implicitly enables fast monitoring information exchange for both detection and restoration purposes. The algorithm self-organizes the control plane into a 'small world'. The motivation for this is to reduce the average path length of the control plane to speed up the flow of control information, while maintaining high clustering to improve resiliency to false alarms and the resolution power of true alarms. Simulations show that the proposed scheme significantly reduces the average path length while maintaining fairly high clustering, and can adapt to changes in the network in a self-organized manner.

The rest of this paper is organized as follows. First, Sec. 2 gives an overview of the 'small-world' concept. Then in Sec. 3, we propose a self-organizing control plane which is supported by the simulation results presented in Sec. 4. Finally, Sec. 5 concludes the paper.

2   Small Worlds

Up until the 1990's, complex systems were generally modeled using regular and random graphs. However, many real-world self-organizing networks, from the collaboration of film actors to biological ecosystems, lie in between these extremes of order and randomness. Such complex networks have been successfully described using the small world [2] model. The term *small world* is used to describe networks that are highly clustered with short average path lengths. The average path length, L, is a global property describing the typical separation between any two nodes in the network. It is defined as the average hop distance between all pairs of nodes. The clustering coefficient, C, is a local property describing the typical cliquishness of a local neighborhood. For each node, we find the ratio of edges in its immediate one-hop neighborhood (including itself) to the total possible number of edges in this neighborhood[1]. These values, averaged over all the nodes in the network, define the clustering coefficient, C.

While regular lattices are highly clustered with long average path lengths, and random graphs exhibit low clustering with short average path lengths, small world structures are somewhere in between. Watts and Strogatz [2] proposed a 'rewiring' method, referred to as the WS algorithm, to generate such structures. The procedure initially starts with a ring lattice and then randomly replaces, or *rewires*, existing links with random ones with probability p. It has been shown that even for very small p, i.e., a tiny bit of rewiring, a small world is born. An example of a small world network generated in this manner is shown in Fig. 1.

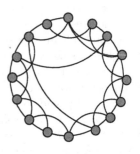

Fig. 1. A small world network generated by the WS procedure where $0 < p << 1$

Applying the small-world concept to communication networks could prove beneficial [14], helping to improve information flow and propagation speed in the Internet, ad hoc networks, and possibly transparent optical networks. Intuitively, high-speed shortcuts between distant parts of a network could enable faster system-wide communication, thus aiding dynamic processes such as synchronization, control, and management.

3   The Proposed Self-organizing Control Plane

The physical topology of the transparent optical network is far from being a random graph since geographic location and installation cost considerations play a major role. The physical topology of the mesh core network is usually more clustered and lattice-like. As already mentioned, a control plane is maintained in the network on a separate supervisory channel on each link. Thus, the control plane topology is equivalent to the physical topology, with point-to-point control lightpaths in each direction between every two physically neighboring nodes. Such a topology can have a fairly high average path length between distant parts of the network, making control information exchange relatively slow. Adding some 'shortcuts' to create a small world can help reduce the average path length.

It is not realistic to add physical long-range links between distant nodes due to the cost of installing fiber and the inherent need for optical regenerators. However, establishing some long-range control *lightpaths* between distant nodes over the existing physical topology is feasible. Basically, the control plane would be a hybrid control plane composed of point-to-point control lightpaths on each physical link and a set of directed long-range control lightpaths. An example of such a hybrid control plane for a reference European core topology from [15] is

---

[1] It is assumed that there can be at most a single edge between a pair of nodes.

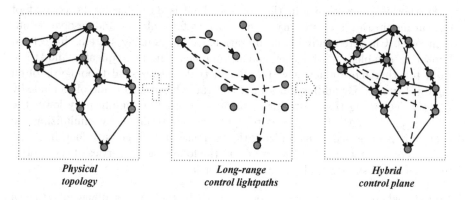

| Physical
topology | Long-range
control lightpaths | Hybrid
control plane |

Fig. 2. An example of a hybrid control plane on a reference European core topology from [15]

shown in Fig. 2. This idea was first introduced in [9] and further developed in [16]. In this paper, we propose a self-organizing scheme to create such a hybrid structure and maintain it in the presence of changes.

To create a small-world control plane topology in a self-organized manner, each node must choose to which distant nodes it wants to be connected to via lightpaths in such a way that their selfish behavior emerges in the desired global structure. Although the physical neighbors are fixed, each node is free to choose its distant neighbors, called 'informants', from which it obtains additional information about other parts of the network. These extra lightpaths are directed, originating at the informant and terminating at the node which chose it.

3.1   The Desired Global Structure

The motivation for creating a small world control plane is to be able to exchange monitoring and control information quickly, particularly in the context of failure management. It is desirable that the management system receive alarms generated from monitoring equipment (via the control plane) as quickly as possible to ensure fast failure detection and localization. In the meantime, it is of utmost importance that lightpaths affected by the failure be restored quickly due to the very high data rates inherent in TONs which can potentially lead to critical data loss causing severe service disruption. Additionally, fast restoration is necessary to ensure that lightpaths are restored before higher layers trigger their own restoration procedures creating a race condition. Failed lightpaths can be restored by utilizing preplanned back-up paths or reactive rerouting strategies. In both cases, the end nodes of the failed lightpath must efficiently be signalled to handle the failure [17]. Since it is not realistic to assume that extensive optical monitoring is available at each node, failures along a particular lightpath trigger alarms only at a subset of optical monitoring nodes which the lightpath traverses. Thus, it is desirable that the source and destination nodes of lightpaths be well connected to the monitoring nodes they traverse.

Furthermore, clustering in the control plane is desirable in the context of optical monitoring and security to help detect false alarms and resolve redundant ones. Clustered individuals in various self-organizing systems have been known to establish trust easier and communicate more frequently and, thus, work together more efficiently [18]. Recall that the physical topology is often highly clustered. By adding long-range control lightpaths, a trade-off is made by slightly decreasing the clustering coefficient in order to significantly lower the average path length. Our goal is to optimize this trade-off by minimizing the drop in clustering while maximizing the decrease in average path length.

In accordance with all of this, we deem the following properties of the control plane as the desired global structural properties.

Low L, where L is the average path length in the control plane in terms of hops. (A hop is considered to be a control lightpath.)

High C, where C is the clustering coefficient as described in Sec. 2. (Since the clustering coefficient is defined for an undirected graph, the directed long-range control lightpaths are considered undirected in the calculation of C.)

Low $L_{mon\_to\_s}$ and $L_{mon\_to\_d}$, where $L_{mon\_to\_s}$ and $L_{mon\_to\_d}$ are the average path lengths in hops from each monitoring node to the source and destination nodes, respectively, of all data lightpaths passing through it, averaged over all the monitoring nodes in the network.

3.2   Local Behavior Rules

Our goal is to create and maintain a control plane topology in a self-organized manner where the selfish behavior of individual nodes emerges in the desired global properties. In addition to its fixed physical neighbors, each node can choose distant 'informants' from which it obtains additional information about other parts of the network. Not all nodes are equally attractive to use as informants. Naturally, each node prefers to connect to nodes with access to more information relevant to it. For example, suppose node j has certain monitoring equipment available to monitor lightpaths passing through it. Furthermore, suppose node i happens to be the source node of a lightpath routed via node j. Node i would benefit from having j as an informant because if the monitoring equipment at node j detects a failure, node i could be informed very quickly (in a single hop) and could, thus, launch its restoration mechanism faster.

It is also important that the control plane self-maintains and self-organizes to adapt to changes in its environment. Namely, nodes can change over time causing a shift in the attractiveness of informants. In the presence of traffic changes and/or failures, several data lightpaths could be reconfigured. New monitoring equipment could also be acquired or existing equipment could fail. Furthermore, informants could acquire a bad reputation after sending false information. Nodes in our control plane can choose new informants, in light of these changes, subject to certain constraints.

All nodes in the network have certain local information available. Each node is aware of all the lightpaths originating at it, terminating at it, and passing through it (called transient lightpaths). A node maintains the following information regarding each lightpath: its source node, its destination node, the wavelength it utilizes, the input port on which it arrives (unless it originates at the node), and the output port on which it is transmitted (unless it terminates at the node). Each node is also aware of the monitoring information available to it. It can have various optical monitoring equipment to monitor passing lightpaths, such as spectrum analyzers or power monitors.

To create and maintain our desired small world control plane, we propose the following self-organizing scheme. Initially, each node chooses one random informant and establishes a corresponding control lightpath. Periodically, each node i sends a $rating\_request$ message to a random node j in the network demanding its rating. The rating of node j, when requested by node i, represents its attractiveness as a potential informant to node i. We denote this as $Rating(j, i)$ and it depends on both i and j.

```
Node (i) Behavior Protocol
Initialization:
currentInformant := NULL ;
Rating(NULL, i) := 0;
Begin:
Periodically, send rating_request to a random node j;
if received rating_request from a node k then
    Send rating_reply to k;
end if
if received rating_reply from node j then
    Compute Rating(j, i);
    if Rating(j, i) > Rating(currentInformant, i) then
        Tear down control lightpath (currentInformant, i);
        Establish new control lightpath (j, i);
        currentInformant := j;
    end if
end if
End
```

Fig. 3. The pseudocode of the node behavior protocol

Upon receiving a rating request, node j returns a $rating\_reply$ message, whose contents will be described later on. From the information provided in the $rating\_reply$ message, node i can calculate $Rating(j, i)$. It then compares j's rating to the rating of its current informant. If j's rating is better, it tears down the lightpath connecting it to its current informant and establishes a new lightpath from node j using the signaling protocol employed by the control plane. We set a limit on the maximum number of nodes for which a node can be an informant (i.e., each node has a maximum control plane out-degree) due to the limited resources available at each node. The pseudocode of the local node behavior protocol is shown in Fig. 3.

To help describe function $Rating(j, i)$, we define the following parameters.

$Phy_{j,i}$ is a binary parameter indicating if nodes j and i are physical neighbors and, thus, already connected via one-hop lightpaths along the physical link connecting them.

$Free\_Port_j$ is a binary parameter indicating whether there are free resources at node j to handle becoming an informant for a new node.

Mon_j is an integer representing the level of optical monitoring equipment and
techniques used at node j. If there is no optical monitoring available, $Mon_j = 0$.
With increased equipment and better techniques, the level increases.

$TLPs_j^i$ is an integer which represents the number of transient data lightpaths
passing through node j whose source node is node i.

$TLPd_j^i$ is an integer which represents the number of transient data lightpaths
passing through node j whose destination node is node i.

$Hops_{j,i}$ represents the length of the shortest path in hops in the physical topology from node j to node i.

CP_j^{in} is an integer representing the in-degree of node j in the control plane
topology.

The rating function is then defined as

$$Rating(j,i) = (1 - Phy_{j,i}) \cdot Free\_Port_j \cdot [Hops_{j,i} \cdot Mon_j \cdot (TLPs_j^i + TLPd_j^i) + CP_j^{in}]. \tag{1}$$

If nodes j and i are already physical neighbors, i.e., $Phy_{j,i} = 1$, then there is
no need for a new control lightpath between them since they are already one-hop
away. Thus, rating $Rating(j,i) = 0$. The same is true if node j does not have
any free resources (i.e., a free output port) to establish a new control lightpath
originating at it. Otherwise, the rating depends on the information that can be
obtained from node j which is relevant to node i.

Node j monitors all its transient lightpaths in accordance with the level of
optical monitoring capabilities available to it, i.e., Mon_j. If node j detects a
failure, it sends an alarm to failure management and the source and destination
nodes of the corresponding lightpaths via the control plane. The more lightpaths
that pass through node j that happen to have their source or destination at node
i, and the better the optical monitoring performed at node j, the more attractive
j is as an informant to i.

Furthermore, node i will receive alarm(s) from j in the presence of failure
(provided j's monitoring equipment detects it) via the shortest path in the current control plane topology. Thus, the longer this path, the more desirable it is
for node i to employ node j as an informant in order to reduce this path. In the
$Rating(j,i)$ function, however, the parameter $Hops_{j,i}$ represents the shortest
path in the physical topology and not the control plane. The motivation for this
is as follows. As the control plane changes over time, the shortest paths between
nodes in the control plane also change. Thus, if the shortest path between j and
i in the control plane were included in function $Rating(j,i)$, the rating could
change due to a shift in the control topology even if there are no significant
changes in the network with respect to traffic flows, data lightpaths, monitoring
equipment, etc. Since each change in the control plane requires certain signalling
overhead to tear down and establish a new informant, it is not desirable to have
frequent modifications. We aim to optimize the trade-off between the stability
of the control plane and its ability to adapt to changes in the network. By considering the shortest physical path between nodes j and i in the rating function,
the protocol initiates fewer changes and yet often gives a good indication of the

distance between the nodes. Essentially, it is a tradeoff between updated information and control overhead. Since the shortest path between two nodes in the physical topology is the longest possible shortest path in the control topology (i.e., adding informants can only lower this path), $Hops_{j,i}$ considers the worst case scenario for the node. Furthermore, preliminary testing indicated that considering the physical shortest path in the rating function, instead of the shortest path in the control plane, lowered L for most cases while performing the same with respect the remaining criteria.

Since nodes in the network maintain only local connectivity information, they do not have knowledge of the shortest paths to all other nodes in either the physical or the control plane topology. Thus, a counter is included in the *rating_reply* message which counts the number of hops for the message to get from node j to node i. In this message, node j provides all the elements required to calculate $Rating(i, j)$, except for $Hops_{j,i}$. Once the message arrives at node i, the final $Rating(i, j)$ is calculated by node i using the information held in the counter and the *rating_reply* message. Since it is not crucial that the periodic updates performed at each node be extremely fast, we send *rating_request* and *rating_reply* messages using only the point-to-point lightpaths in the control plane (and not via informants). The 'shortcuts' in the control plane are reserved only for crucial monitoring information when a failure occurs and are not used up by other less-important signalling and control overhead. This way the counter would calculate the shortest path in the physical topology $Hops_{j,i}$. If we were to define $Hops_{j,i}$ as the shortest path in the current *control plane* topology, then the *rating_request* and *rating_reply* messages could be sent over any link in the control plane.

The last element in the rating function is simply the control plane in-degree of node j. For the case when j has a high monitoring level and many transient lightpaths relevant to node i, this parameter will not significantly affect the rating. However, if two nodes have similar ratings with respect to monitoring transient lightpaths, the node with a higher control in-degree is considered more attractive since it has access to more one-hop control information.

In the approach, we suppose that every node has exactly one 'informant'. This assumption is made for simplicity but need not be so for the general case. Furthermore, we assume nodes have global knowledge of the existence of all other nodes in order to send random *rating_request* messages. Since the physical topology is for the most part fixed[2], this is feasible but limits scalability. We are currently investigating various modifications of the model to deal with these issues.

4   Numerical Results

In order to evaluate the proposed self-organizing scheme, we developed an event-driven simulator in C++. For simplicity, we assumed that the periodic updates of nodes are performed synchronously. We tested the algorithm on a reference

---

[2] Changes in the physical topology do not occur very frequently due to the difficulties involved in laying down fiber.

topology of a pan-European basic network from the COST Action 266 project [15] with 30 nodes and 48 bidirectional edges, shown in Fig. 4. We assumed two levels of monitoring, differentiating between non-monitoring nodes ($Mon_j = 0$) and nodes which are equipped with at least some optical monitoring equipment ($Mon_j = 1$). To decide which nodes have optical monitoring equipment, we used the monitoring placement policy described in [13]. According to this policy, if a node is non-monitoring, all its neighbors must be monitoring nodes. Furthermore, if a node is of degree one, its neighboring node must be a monitoring node.

Before running the simulation, an initial virtual topology was created for the data plane as follows. First, a traffic matrix was generated using the method suggested in [19] where a fraction F of the traffic is uniformly distributed over $[0, C/a]$ while the remaining traffic is uniformly distributed over $[0, C * \Upsilon/a]$. The values were set to $C = 1250$, $a = 20$, $\Upsilon = 10$, and $F = 0.7$ as in [19]. To establish the initial virtual topology, we set up lightpaths between pairs of nodes in decreasing order of their corresponding traffic, with at most 5 lightpaths originating and 5 light-

Fig. 4. The European basic network topology, from [15]

paths terminating at each node, i.e., we assumed 5 transmitters and receivers were utilized per node[3]. Lightpaths were routed on their shortest physical paths, in terms of hops, and we assumed that there were enough available wavelengths on all links.

In the first simulation scenario, referred to as Scenario 1, requests to tear down the lightpaths comprising the initial virtual topology described above, arrived according to a Poisson process with rate $\lambda = 5$. New lightpath requests also arrived according to a Poisson process with rate $\lambda = 5$, with exponentially distributed holding times with mean $b = 10$. We assumed that the monitoring equipment at nodes was fixed. In this scenario, the values of $TLPs_j^i$ and $TLPd_j^i$ in the informant *Rating* function can change over time while the remaining parameters remain constant.

Simulations were run for 3 cases. In the first case, the control plane topology was kept equivalent to the physical topology with no long-range shortcuts. This is denoted as Phy_CP. In the second case, a hybrid control plane was created at simulation start time by choosing a random informant for each node in the network, establishing the corresponding directed lightpath, and super-positioning it onto the physical topology. This control plane, denoted as $Random\_CP$, was then kept constant throughout the simulation. The third case ran the

---

[3] At most one lightpath was established between the same pair of nodes.

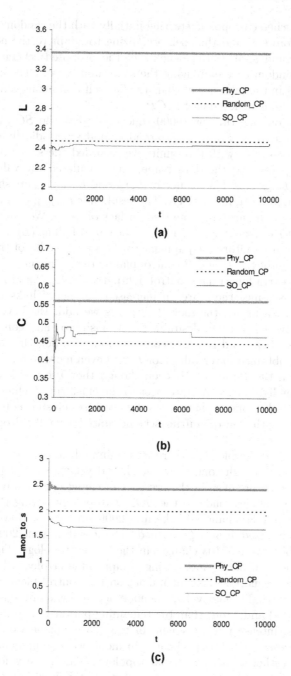

Fig. 5. The average path length in the control topology (a), the clustering coefficient (b), and the average path lengths from monitoring nodes to the source nodes of their transient lightpaths (c) for Scenario 1

self-organizing scheme proposed, starting initially with the random control plane topology $Random\_CP$ and then self-organizing to adapt to the network state. This way we could analyze the benefits of the proposed scheme in comparison with the random case employing the same number of 'shortcuts' but self-organizing itself in the presence of changes. The self-organizing control plane for the third test case is denoted as SO_CP.

Each simulation was run for 10000 time units. For the SO_CP algorithm, nodes sent $rating\_request$ messages to random nodes periodically every 10 time units. Furthermore, every 10 time units we recorded the structural properties of the control plane and the data plane, and calculated the values for L, C, $L_{mon\_to\_s}$, and $L_{mon\_to\_d}$. The results for L, C, and $L_{mon\_to\_s}$ are shown in Fig. 5 in plots (a), (b), and (c), respectively. The results for $L_{mon\_to\_d}$ are analogous to those of $L_{mon\_to\_s}$ and are, thus, omitted for lack of space. We can see from plots (a) and (c) that the average path length of the control plane (L) and the average path lengths from monitoring equipment to the source nodes of transient lightpaths ($L_{mon\_to\_s}$) of the Phy_CP control plane are significantly decreased with the addition of extra long-range control lightpaths (SO_CP and $Random\_CP$). This makes sense since there are an increased number if links in the control plane topology. Naturally, the more lightpaths we add, the lower the average path length. However, it is not desirable to establish too many control lightpaths due to extra overhead and resource consumption. The Self-Organizing Control Plane SO_CP, obtained lower values for L, and even more so for $L_{mon\_to\_s}$ (and $L_{mon\_to\_d}$), than the $Random\_CP$ even though they use the same number of extra long-range lightpaths. With respect to the clustering coefficient C, adding random edges to the control plane naturally decreases clustering to some extent. However, applying the self-organizing scheme caused a smaller drop in clustering than the random case.

Note that it is desirable that there be a minimal number of changes in the control plane due to high control overhead, and yet we want it to achieve the desired global structure even in the presence of changes. To analyze our model, we recorded all changes made to the SO_CP topology during the simulation. Initially, there were 80 changes in the first 2000 time units. However, once the control plane stabilized, it only performed 2 changes from time 2000 until 10000, even though there were 55103 changes in the virtual topology. This shows that learning the location of the monitoring equipment and physical distances between nodes has a more significant impact on the control plane topology than changes in the virtual topology, i.e. the node protocol is more sensitive to variations in $Hops(j, i)$ and Mon_j than the remaining parameters in the $Rating(j, i)$ function. Thus, intense rearrangement of the control plane would more likely occur in the presence of drastic changes in monitoring equipment or the physical topology, rather than the virtual topology. This is very fortunate since monitoring equipment at nodes and the physical topology are generally fairly constant and change slowly over time. Thus, the control plane would be quite stable.

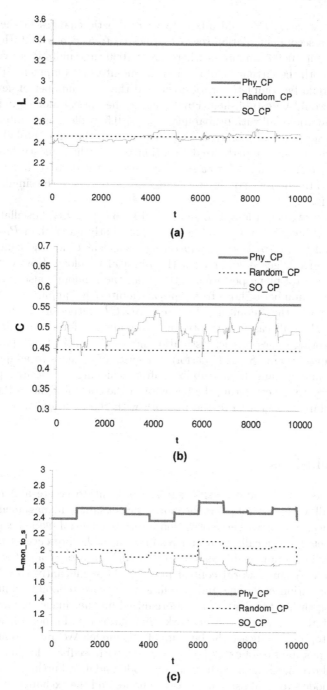

Fig. 6. The average path length in the control topology (a), the clustering coefficient (b), and the average path lengths from monitoring nodes to the source nodes of their transient lightpaths (c) for Scenario 2

To see how SO_CP would adapt to more drastic changes in the network, we created a second simulation scenario, referred to as Scenario 2. Here we ran simulations for 10000 time units where every 1000 time units there were major changes in both the virtual topology and the monitoring equipment. The virtual topology would be completely torn down and the same number of new random lightpaths would be established. Furthermore, the optical monitoring available at each node would fail with probability $P_{mon} = 0.5$, while non-monitoring nodes would gain new optical monitoring equipment with the same probability. This is, of course, a much hyperbolized situation but can help us see how SO_CP can adapt and self-organize into a stable state with the desired global structural properties in the presence of drastic changes. The results of the simulations[4] are shown in Fig. 6.

For the average path length L shown in Fig. 6.(a), SO_CP oscillates around $Random\_CP$ but both remain close and significantly lower than Phy_CP. We can see from Fig. 6.(b) that the clustering coefficient C for the control planes with long-range edges is lower than the physical topology. However, the self-organizing control plane performs better than the random constant one. With respect to the number of hops from optical monitoring equipment to the source of the lightpaths they monitor (Fig. 6.(c)), SO_CP outperformed $Random\_CP$ and Phy_CP in all cases. The situation is analogous for the number of hops from monitoring nodes to the destination nodes of transient lightpaths. When drastic changes occur, SO_CP performs a series of changes to adapt in a self-organizing manner and then stabilizes after achieving the desired properties. We are currently investigating the behavior of the control plane in the presence of node failure and growth of the network with the addition of new nodes or links.

5   Conclusions

In this paper, we propose a self-organizing scheme to create and maintain a hybrid small world control plane for more efficient failure management in transparent optical networks. The motivation for such a control plane lies in the fact that fast detection, localization and restoration in the presence of failures are particularly important in TONs due to very high data rates and their inherent transparency. A small world control plane could significantly speed-up monitoring information exchange and potentially improve reliability. Furthermore, maintaining such a topology in a self-organized manner makes it more scalable and robust to changes in the network. Simulations performed on a reference European topology indicate the benefits of this model. We are currently investigating the possibilities of extending this model with feedback loops to minimize the control overhead incurred by periodic node updates. Furthermore, developing trust models to establish trust between nodes and the exchange of reputation information could prove beneficial.

---

[4] The results for $L_{mon\_to\_d}$ are again omitted since they are analogous to those of $L_{mon\_to\_s}$.

References

1. Strogatz, S.H.: Exploring Complex Networks. Nature 410, 268–276 (2001)
2. Watts, D.J., Strogatz, S.H.: Collective Dynamics of 'Small-World' Networks. Nature 393, 440–442 (1998)
3. Flake, G.W., Pennock, D.M., Fain, D.C.: The Self-Organized Web: The Yin to the Semantic Web's Yang. IEEE Intelligent Systems 18(4), 75–77 (2003)
4. Hales, D., Arteconi, S.: SLACER: A Self-Organizing Protocol for Coordination in Peer-to-Peer Networks. IEE Intelligent Systems 21(2), 29–35 (2006)
5. Dixit, S., Yanmaz, E., Tonguz, O.K.: On the Design of Self-Organized Cellular Wireless Networks. IEEE Communications Magazine 43(7), 86–93 (2005)
6. Li, G., Yates, J., Kalmanek, C.R., Wang, D.: Control Plane Design for Reliable Optical Networks. IEEE Communications Magazine 40(2), 90–96 (2002)
7. Mas, C., Tomkos, I., Tonguz, O.: Failure Location Algorithm for Transparent Optical Networks. IEEE Journal on Selected Areas in Communications 23(8), 1508–1511 (2005)
8. Kilper, D.C., et al.: Optical Perfprmance Monitoring. Journal of Lightwave Technology 22(1), 294–304 (2004)
9. Skorin-Kapov, N., Tonguz, O., Puech, N.: Self-Organization in Transparent Optical Networks: A New Approach to Security. In: The 9th International Conference on Telecommunications (Contel 2007), Zagreb, Croatia, pp. 7–14 (invited paper) (2007)
10. Médard, M., Marquis, D., Barry, R., Finn, S.: Security Issues in All-Optical Networks. IEEE Network 11(3), 42–48 (1997)
11. Bergman, R., Médard, M., Chan, S.: Distributed Algorithms for Attack Localization in All-Optical Networks. In: Network and Distributed System Security Symposium (NDSS'98) (session 3, paper 2), San Diego, Cal., USA (1998)
12. Li, C.-S., Ramaswami, R.: Automatic Fault detection, isolation, and Recovery in Transparent All-Optical Networks. Journal of Lightwave Technology 15(10), 1784–1793 (1997)
13. Wu, T., Somani, A.: Cross-talk Attack Monitoring and Localization in All- Optical Networks. IEEE/ACM Transactions on Networking 13(6), 1390–1401 (2005)
14. Collins, J.J., Chow, C.C.: It's a Small World. Nature 393, 409–410 (1998)
15. Inkret, R., Kuchar, A., Mikac, B.: Advanced Infrastructure for Photonic Networks. In: Extended Final Report of COST Action 266, Faculty of Electrical Engineering and Computing, pp. 19–21. University of Zagreb, Zagreb (2003)
16. Skorin-Kapov, N., Tonguz, O., Puech, N.: A 'Small World' Hybrid Control Plane for Reliable Transparent Optical Networks. IEEE Journal of Selected Areas in Communications (submitted)
17. Sivakumar, M., Shenai, R.K., Sivalingam, K.M.: A Survey of Survivabilty Techniques for Optical WDM Networks. In: Sivalingam, A.M., Subramaniam, S. (eds.) Emerging Optical Network Technologies: Architectures, Protocols and Performance. Springer Science+Media, Inc., ch. 3, pp. 297–332 (2005)
18. Buchanan, M. (ed.): Nexus: Small Worlds and the Groundbreaking Theory of Networks, pp. 199–204. W. W. Norton & Company, Inc, New York (2002)
19. Banerjee, D., Mukherjee, B.: Wavelength-Routed Optical Networks: Linear Formulation, Resource Budgeting Tradeoffs, and a Reconfiguration Study. IEEE/ACM Transactions on Networking 8(5), 598–607 (2000)

A Self-organizing Approach to Tuple Distribution in Large-Scale Tuple-Space Systems

Matteo Casadei[1], Ronaldo Menezes[2], Mirko Viroli[1], and Robert Tolksdorf[3]

[1] Università di Bologna, DEIS
Cesena (FC), Italy
{m.casadei,mirko.viroli}@unibo.it
[2] Florida Tech, Computer Sciences
Melbourne, Florida, USA
rmenezes@cs.it.edu
[3] Freie Universität Berlin, Institut für Informatik
Berlin, Germany
tolk@inf.fu-berlin.de

Abstract. A system is said to be self-organizing if its execution yields temporal global structures out of simple and local interactions amongst its constituents (e.g agents, processes). In nature, one can find many natural systems that achieve organization at the global level without a reference to the status of the global organization; real examples include ants, bees, and bacteria. The future of tuple-space systems such as LINDA lies on *(i)* their ability to handle non-trivial coordination constructs common in complex applications, and *(ii)* their scalability to environments where hundreds and maybe thousands of nodes exist. The Achilles heel of scalability in current tuple-space systems is tuple organization. Legacy solutions based on antiquated approaches such as hashing are (unfortunately) commonplace. This paper gets inspiration from self-organization to improve the status quo of tuple organization in tuple-space systems. We present a solution that organizes tuples in large networks while requiring virtually no global knowledge about the system.

1 Introduction

Self-organization is certainly a buzzword in many science fields today, which culminates in a misusage of the term. Sure self-organization has interesting characteristics but it may also be a hindrance since it makes it harder to explain/understand its consequences, causal properties, etc. If used correctly, one can exploit its characteristics for the benefit of better solving complex problems in Computer Science. So, why only recently has self-organization caught the attention of computer scientists? It is quite simple: *(i)* computer scientists are used to having the total control over their systems' workings, and *(ii)* the scale of the problems they face does not require "unconventional" solutions. The consequence is that an antiquated way of thinking floods today's applications leading to solutions that are complex, unreliable, unscalable, and hard to maintain.

Although far from the level of complexity of the real world, computer applications exhibit complexity that is hard to deal with even by today's most powerful computers.

D. Hutchison and R.H. Katz (Eds.): IWSOS 2007, LNCS 4725, pp. 146–160, 2007.

Data organization is one of the problems that appears again-and-again in many contexts; Cardelli [1] argues that data organization via mobility is the only way we can overcome physical limitations in networks such as bandwidth limitations and latency. Hence, if data are kept close to where they are needed, one can optimize access time. A further hypothesis we introduce based on Cardelli's statement is that rarely a single piece of data of a particular kind is required alone during the execution of a system. That is, processes deal with collections of data which tend to be similar in format. Therefore, data should be not only kept close (as per Cardelli's arguments) but also clustered by their kind.

The achievement of solutions that could satisfy both the locality principle and the clustering principle is not trivial, particularly if we assume dynamic environments. Self-organized phenomena in nature, such as brood sorting, appear to solve the afore-mentioned problems in a very elegant way, since no global decisions are required. Rather, the solution is an emergent property of the ants' local interactions. Our purpose in this paper is to borrow from these good ideas and employ them to provide a solution to the data organization problem in distributed tuple-space coordination systems.

Coordination systems are constantly being pointed as a good mechanism to deal with some of the complex issues in large-scale systems. The so-called separation between computation and coordination [2] enables a better understanding of the complexity in distributed applications, but the success of these systems depends on *(i)* their ability to handle non-trivial coordination constructs common in complex applications, and *(ii)* their scalability to environments where hundreds and maybe thousands of nodes exist. While *(i)* has been successfully solved in tuple-space systems (with the introduction of more expressive models), tuple-space systems still have difficulties overcoming the hurdles imposed by *(ii)*.

One approach becoming popular for dealing with *(ii)* is the use of emergent coordination mechanisms [3]. Examples of this approach include mechanisms proposed in models such as SwarmLinda [4] and TOTA [5]. This paper explores one mechanism proposed in SwarmLinda referring to the organization of data (tuples) in distributed environments, using solutions borrowed from natural forming multi-agent swarms, more specifically based on ant's brood sorting behavior [6].

In tuple-space coordination systems, the coordination itself takes place via generative communication using tuples. Tuples are stored and retrieved from distributed tuples spaces. As expected, the location of these tuples is very important for performance— if tuples are maintained near the processes requiring them, the coordination can take place more efficiently. In this paper we devised a mechanism to implement the approach originally suggested in SwarmLinda to solve the tuple distribution problem. Then, we evaluated the performance of our approach using a metric based on the spatial entropy of the system. We showed that the organization pattern of tuples in nodes emerges from the configuration of nodes, tuple templates, and connectivity of nodes. Then, using definitions taken from Camazine *et al.* [7], we demonstrated that our approach is truly self-organizing.

2   Characteristics of Self-organization

The use of self-organization has increased in the last decade. A quick search using Google Scholar [8] shows that about 92,000 articles since 2000 include either the term "self-organization" or "self-organizing" (accounting also for British spelling). Amongst these, nearly 20,000 are in the fields of Computer Science, Mathematics and Engineering. How can this be explained? We believe there are three reasons for this increase:

1. There is clearly a hype around this term. It is fashionable and subject of many scientific and non-scientific books, conferences, and workshops.
2. It is interesting. One cannot deny that self-organization can offer new insights on how to deal with complex problems.
3. The complexity of today's problems forces us to look elsewhere for solutions. Nature, and in particular insect colonies (where self-organization is ubiquitous), deals well with complexity.

On one hand, we have Item 1 leading to the unfortunate situation in which authors use the term without a clear justification that their approaches are indeed self-organizing. On the other hand, we have Items 2 and 3 that are acceptable incentives for the use of self-organization in tuple-space organization.

Camazine *et al.* [7] argue that even though self-organization is not a simple term to define, it can be viewed as a process referring to a plethora of mechanisms that, working together, cause the emergence of patterns, structures and order in systems. Moreover, these mechanisms are limited to local interactions amongst the individuals that compose the system. Bonebeau *et al.* [9] prefer to provide the ingredients to self-organization rather than define the term. They refer to mechanisms such as positive and negative feedback, amplification of fluctuations (eg. random walks), and the existence of multiple interactions.

But what could be an acceptable definition of self-organization? In the context of this paper, self-organization can be loosely defined as *a process where the entropy of a system decreases without the system being guided or managed by external forces*. It is a phenomenon quite ubiquitous in nature, particularly in natural forming swarms. These systems exhibit a behavior that seems to surpass the one resulting from the sum of all the individuals' abilities. By self-organizing their activities, they achieve goals beyond their individual capabilities [9,10].

3   Linda and Self-organization

The LINDA coordination model is based on the associative-memory communication paradigm. LINDA provides processes with primitives enabling them to store and retrieve tuples from tuple spaces. Processes use the primitive *out* to store tuples. They retrieve tuples using the primitives *in* and *rd*; these primitives take a template (a definition of a tuple) and use associative matching to retrieve the desired tuple—while *in* removes a matching tuple, *rd* takes a copy of the tuple. Both *in* and *rd* are blocking primitives, that is, if a matching tuple is not found in the tuple space, the process executing the primitive blocks until a matching tuple can be retrieved.

LINDA is arguably the most successful coordination model as it has been the basis for many research projects [11,12] and commercial products [13,14,15]. Yet, all the implementations of this model suffer from limitations when they need to face large-scale environments (based on the number of tuple spaces, number of nodes in the network, etc.).

One of the major scalability problems in LINDA is the idea of tuple organization— related to the data-organization issues described in Section 1. The LINDA community has tried many approaches to tuple distribution before turning to self-organization. LINDA is a model that has been proposed 20 years ago and since then has undergone many extensions, implementations, and studies. However, only recently with the proposal of SwarmLinda [4], self-organization has started to be a choice for implementing scalable LINDA systems.

In this paper we aim at optimizing distribution and retrieval of tuples by dynamically determining storage locations on the basis of the template of a particular tuple. It should be noted that we do *not* want to program the clustering, but rather make it emerge from algorithms implemented by exploiting self-organizing mechanisms.

To achieve self-organization we see a network of nodes as the terrain where *out*-ants roam. These ants have to decide at each hop on the network, if the storage of the carried tuple should take place or not. The decision is made stochastically but biased by the amount of similar tuples around the ant's current location. There should also be a guarantee that the tuple will eventually be stored. This is achieved by having an aging mechanism associated with the *out*-ant.

The similarity function is another important mechanism. Note that it may be too restrictive to have a monotonic scheme for the similarity of two tuples. Ideally we would like to have a function that says *how similar the two tuples are* and not only if they are exactly of the same template.

In this paper, our approach is based on the brood sorting algorithm proposed by Deneubourg *et al.* [6]. We demonstrate that it is possible to achieve a good level of entropy for the system (level of tuple organization) without resorting to static solutions or centralized control.

4   A Solution for Tuples Distribution

Several approaches for tuple distribution in LINDA systems can be found in the literature [16,17,13], but none of them have proven to be usable in the implementation of scalable systems. In order to find a solution to scalability, SwarmLinda went to find inspiration in self-organization and adopted the concept of *ant's brood sorting* as proposed by Deneubourg *et al.* [6].

We consider a network of distributed nodes each containing exactly one tuple space. New tuples can be inserted using the LINDA primitive *out*; tuples are of the form $N(X_1, X_2, ..., X_n)$, where N represents the tuple name and $X_1, X_2, ..., X_n$ represent the tuple arguments. As discussed in Section 1, a good solution to the distribution problem must guarantee the formation of clusters of tuples in the network. More precisely, tuples belonging to the same template should be stored in the same tuple space or in neighboring tuple spaces. Note that this should be achieved in a dynamic and decentralized fashion, since deterministic solutions based on central decisions are unfit to be used in large-scale scenarios.

In order to decide how similar tuples are, we need a *similarity function*. The similarity function can be defined as $\delta(tu, t)$, where tu and t are input arguments: tu and t represent tuples. It is generally expected that a similarity function would return values in a range between 0 and δ_{max}, where δ_{max} is the value of the maximum similarity between two tuples, usually $\delta_{max} \leq 1$. However, δ_{max} is a control parameter of the proposed approach and can be used to regulate the attractiveness level of similar tuples.

Upon the execution of an *out(tu)* primitive, the network, starting from the tuple space in which the operation is executed, is visited in order to decide where to store the tuple tu. According to *brood sorting*, the *out* primitive can be viewed as an ant, wandering in the network searching for a good tuple space to drop tuple tu, that represents the carried food.

The proposed solution to the distribution problem is composed of the following phases: *(i)* a *Decision Phase* to decide whether to store tu in the current tuple space or not; *(ii)* a *Movement Phase* to choose the next tuple space if the decision taken in *(i)* is not to store tu in the current tuple space. The process then starts from *(i)*.

4.1 Decision Phase

During the decision phase, the *out*-ant primitive has to decide whether to store the carried tuple tu-food in the current tuple space. This phase involves the following steps:

1. Calculate the concentration F of tuples having a template similar to tu by using a *similarity function* defined for the system.
2. Calculate the probability P_D to drop tu in the current tuple space.

The concentration F is calculated by considering the similarity of tu with all the other tuples t in tuple space TS, given by:

$$F = \sum_{\forall t \in TS} \delta(tu, t) \tag{1}$$

According to the original brood sorting algorithm used in SwarmLinda, the probability P_D to drop tu in the current tuple space is given by:

$$P_D = \left(\frac{F}{F + K} \right)^2 \tag{2}$$

with $0 \leq P_D \leq 1$. Differing from the original brood sorting algorithm, the value of K is not a constant and represents the number of tuple spaces that an *out*-ant can visit. When an *out* operation is executed, the value of K is set to the maximum number of tuple spaces each *out*-ant can visit (*Step*).

Each time a new tuple space is visited by an *out*-ant without storing the carried tuple tu, K is decreased by 1. When K reaches 0, P_D becomes 1 and tu is automatically stored in the current tuple space, independently of the value of F. K is adopted to implement the *aging mechanism*, since we want to avoid to have an *out*-ant wandering forever without being able to store the carried tuple tu. If $K > 0$ and the tuple tu is not stored in the current tuple space, a new tuple space is chosen in the movement phase from the neighbors of the current one.

4.2 Movement Phase

The *movement* phase occurs when the tuple carried by an *out*-ant is not stored in the current tuple space. This phase has the goal of choosing, from the neighborhood of the current tuple space, a good neighbor for the next hop of the *out*-ant. This neighbor is a tuple space with a high concentration F of tuples equal or similar to the carried tuple tu.

If we denote by n the total number of neighbors in the *neighborhood* of the current tuple space, and F_j the concentration of tuples similar to tu in neighbor j (obtained by Equation 1), we can then say that the probability P_j of having an *out*-ant move to neighbor j, is calculated by the proportional-selection equation below:

$$P_j = \frac{F_j}{\sum_{i=1}^{n} F_i} \qquad (3)$$

Adopting this equation for each neighbor, we obtain $\sum_{i=1}^{n} P_i = 1$. Moreover, the higher the value of P_j, the higher the probability of choosing neighbor j as the next hop of the *out*-ant. After a new tuple space is chosen, the whole process is repeated starting from the decision phase (as described in Section 4.1).

5   Applying SwarmLinda to Scale-Free Networks

In order to verify the applicability of the SwarmLinda distribution mechanism to large networks, we chose to perform simulations on scale-free topologies [18]. This choice was mainly driven by the consideration that almost every real network of computers features a scale-free topology—e.g. the WWW [18,19]. Next section reports a brief description of the scale-free networks used for our experiments, and the algorithm used to generate these networks.

5.1   Sample Scale-Free Networks

The networks used in our experiments were generated using the original *B-A Scale-Free Model Algorithm* presented by Barabási and Albert in [18]. This algorithm is briefly recalled in the following description. Given an initial small number m_0 of tuple spaces:

- at each step of the algorithm, a new tuple space is added and connected to $m < m_0$ already existing tuple spaces.
- The higher the degree k_i of an already-existing tuple space i, the higher the probability of connecting the newly-introduced tuple space to i.

The probability P_i to have the added tuple space connected to i is:

$$P_i = \frac{k_i}{\sum_j k_j}$$

so that already existing tuple spaces with a large number of connections have a high probability to get new connections. This phenomenon is also called *rich get richer*.

In order to generate our sample scale-free networks we chose $m_0 = m = 2$, and started with an initial network of two tuple spaces linked to one another. Then, each new tuple space was connected to two already-existing ones according to the *B-A Scale-Free Model Algorithm*. We generated two scale-free networks composed of 30 and 100 tuple spaces.

5.2 Methodology

We want our distribution mechanism to achieve a reasonable organization of tuples. Tuples having the same template should be clustered together in a group of tuple spaces near to each other. The concept of *spatial entropy* is an appropriate metric to describe the degree of order in a network. Denoting by q_{ij} the amount of tuples matching template i within tuple space j, n_j the total number of tuples within tuple space j, and s the number of templates, the entropy associated with tuple template i within tuple space j is

$$H_{ij} = \frac{q_{ij}}{n_j} \log_2 \frac{n_j}{q_{ij}} \qquad (4)$$

and it is easy to notice that $0 \le H_{ij} \le \frac{1}{s} \log_2 s$. We want to express now the entropy associated with a single tuple space

$$H_j = \frac{\sum_{i=1}^{s} H_{ij}}{\log_2 s} \qquad (5)$$

where the division by $\log_2 s$ is introduced in order to obtain $0 \le H_j \le 1$. If we have r tuple spaces, then the *spatial entropy* of a network is

$$H = \frac{1}{r} \sum_{j=1}^{r} H_j \qquad (6)$$

where the division by r is used to normalize H, so that $0 \le H \le 1$. The lower the value of H, the higher the degree of order in the considered network. Moreover, a value of H_j equals to 1 corresponds to a situation of complete chaos in tuple space j, since we have $q_{ij} = \frac{n_j}{s}$ for $1 \le i \le s$. Oppositely, a value of H_j equal to 0 corresponds to a situation of complete order, since all the n_j tuples in tuple space j are of the same template. For each simulated network, we performed a series of 20 simulations, using each time different values for the $Step$ parameter that represents the maximum number of steps a tuple can take. One run of the simulator consists of the insertion of tuples in the network—via *out* primitive—until there are no pending *out*s to be executed in the entire network.

After the execution of a series of 20 simulations for a given network, the value of the *spatial entropy* H is calculated as the average of the single values of H resulting from each simulation; we call this value *average spatial entropy* (H_{avg}). For each network topology presented in the next section, we considered tuples of four different templates: $a(X)$, $b(X)$, $c(X)$, and $d(X)$.

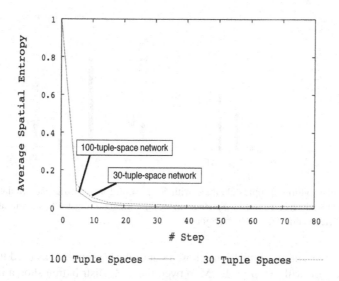

Fig. 1. Trend of *average spatial entropy* H_{avg} resulting from the simulation of the two scale-free networks used in our experiments

The following section presents the results obtained by simulating two scale-free instances. First, we performed simulations on a 30-tuple-space scale-free network in order to preliminary evaluate the performance of the distribution mechanism. Then after having proven that our approach is able to achieve a good tuple distribution, we executed simulations on the 100-tuple-space scale-free network, in order to further evaluate the performance of the distribution mechanism on larger networks.

5.3   Simulation Results

Both network instances were simulated for values of *Step* in the range from 0 to 80 steps, considering the occurrence of 60 *out* operations per tuple space—15 per tuple template. In particular, *Step* = 0 corresponds to a simulation performed without applying the distribution mechanism: indeed, in this situation, every tuple is directly stored in the tuple space in which the corresponding *out* operation occurs.

The simulation results for the 30-tuple-space scale-free network are reported in Figure 1. Observing the results, we can clearly see that if we use a value of *Step* large enough to let an *out*-ant explore the network, the value of H_{avg} becomes small, meaning that the network features a high degree of clustering. For values of *Step* greater that 20, the distribution mechanism shows a high degree of insensitivity to different values of *Step*: this is due to the fact that, even though we choose *Step* > 20, the capability of exploring the entire network of our distribution mechanism remains the same. For this reason, the trend of H_{avg} tends to an horizontal asymptote with a value approximately equals to 0.014.

Figure 1 shows only the trend of H_{avg} for different values of *Step*, but it does not provide any information about the tuple distribution in the network at the end of a simulation. Figure 2 reports a qualitative representation of the tuple distribution in the

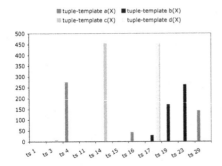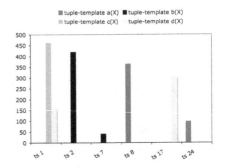

Fig. 2. Final distribution of tuples obtained with $Step = 40$ for two different simulations on the 30-tuple-space scale-free network. *Spatial entropy* corresponding to these simulation: $H = 8.51 \times 10^{-3}$ (left) and 1.42×10^{-2} (right).

network for two sample simulations chosen out of the 20 simulations used to calculate H_{avg} in the case with $Step = 40$. Moreover, the tuple distribution shown in Figure 2 (left) corresponds to a *spatial entropy* value $H = 8.51 \times 10^{-3}$, while Figure 2 (right) reports a tuple distribution featuring $H = 1.42 \times 10^{-2}$. Although these final tuple distributions are different, the corresponding values of H demonstrate that we can achieve a quasi-perfect clustering. Indeed, Figure 2 makes it clear the tendency of the distribution mechanism to organize tuples *per tuple template* amongst the tuple spaces composing the network.

Looking at the two tuple distributions reported in Figure 2 also makes it clear that even though the SwarmLinda distribution mechanism's evolution does not allow to know in advance where tuples of a certain template will aggregate, a good level of information clustering (in terms of *spatial entropy*) can be achieved in every situation. Nonetheless, as depicted in Figure 4, the final tuple distribution's pattern is sensitive not only to the values assumed by $Step$, but also to the initial conditions of the network. In fact, though the previous simulations were performed on a initially-empty network, executing simulations on a network featuring an initial configuration with one or more clusters already formed leads to a final tuple distribution in which the inserted tuples tend to aggregate around the already-formed clusters. The outcome of a set of experiments—executed considering the presence of clusters in the network—is described later.

The observed sensitivity of the distribution mechanism to the system's initial conditions does not however mean that our approach is not self-organizing. The capability to cluster information is independent from what Camazine *et al.* [7] call *external cue acting as a template for the aggregation of organisms.*

As Camazine *et al.* point out, the emergence of patterns in organism clustering is sometimes thought of as the result of a self-organizing process even though it is not. As a consequence, we need to provide indications of the true self-organizing nature of our tuple-distribution mechanism.

One first possible indication is reported in Figure 3, showing the trend of concentration F for different simulations executed using $Step = 5, 10$ and 40.

More precisely, F refers to the concentration of tuples similar to the current one in the tuple space where the tuple is stored.

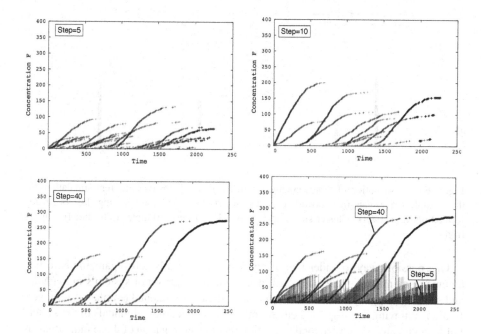

Fig. 3. Evolution of concentration F for different simulations run on the 30-tuple-space scale-free network using $Step = 5$, 10 and 40. The bottom-right graph just highlights the differences between $Step = 5$ and $Step = 40$.

Even though Figure 3 allows to know neither the tuple space where a tuple is stored nor the tuple template of that tuple, we can easily recognize the formation of several clusters. This suggests that our SwarmLinda distribution mechanism exhibits a self-organizing behavior, making clusters of similar tuples emerge from an initial state characterized by an empty network. In particular, the emergence of such clusters is only driven by the local interactions occurring between a tuple space and its neighbors. Furthermore, looking at the different charts reported in Figure 3 makes it clearly visible an increasing order arising when higher values of $Step$ are used: note in particular the comparison reported in Figure 3 between the results for $Step = 5$ and the ones for $Step = 40$.

In spite of these results, we need a stronger argument to show that the pattern resulting from our distribution mechanism arises as a result of a self-organizing process. In many cases of collective behavior, organism clustering arises as a response of the individuals to an *external cue acting as a template*, and not as a natural outcome of a self-organized pattern formation [7]. In particular, such a behavior has been observed in the stable fly of human and wood lices. In all of these cases, the aggregation process is the result of an external stimulus, an *environmental template* representing a fixed feature of the environment. Oppositely, in a self-organized aggregation the individuals respond to signals that are dynamic and affected over time from the behavior of the individuals themselves.

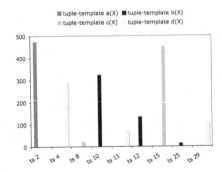 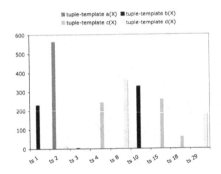

Fig. 4. Final distribution of tuples obtained with $Step = 40$ for two different simulations on the 30-tuple-space scale-free network, considering a starting condition featuring clusters already formed on the network. Cluster size used in these simulations: 10 tuples (left) and 100 tuples (right).

One possible way to understand if one's system is really self-organizing is to find an environmental template suspected to lead the aggregation of system's individuals, and see if that aggregation occurs even though the environmental template is removed. If, after removing the suspected environmental template, the individuals in the system fail to aggregate, we have a good indication that the aggregation of those individuals is based on an environmental cue and not on a self-organizing process. Furthermore, systems achieving aggregation only by environmental templates are insensitive to different initial conditions, that is to say, they come to the same final state independently of any variation in the system's initial conditions. Nonetheless, some other systems feature behavior driven by both a self-organizing process and an external cue. Here, it is desirable to find the relative contributions of these two factors to the aggregation process.

If we go back to our distribution mechanism, the role of *individual* is played by the *out*-ants wandering on a network in the attempt to store the carried tuple, while the role of possible *external cue* can be played by an initial condition featuring already-formed clusters in the network at the beginning of a simulation. The results shown previously are a first indication of the self-organizing nature of our distribution mechanism. Even though we considered a network initially empty—with no clusters already-formed in the network—the results demonstrate that our distribution mechanisms can however achieve a strong level of information clustering.

In the attempt to provide a stronger argument of the self-organizing nature of our distribution mechanism, we decided to perform further simulations considering the presence of clusters. More precisely, we ran a first set of simulations on the 30-tuple-space scale-free network considering four already-formed clusters containing 10 tuples each: *(i)* 10 $a(X)$-template tuples in tuple space 2, *(ii)* 10 $b(X)$-template tuples in tuple space 10, *(iii)* 10 $c(X)$-template tuples in tuple space 15 and *(iv)* 10 $d(X)$-template tuples in tuple space 29. As in the previous experiments, we simulated the insertion of 60 tuples per tuple space—15 per tuple template. Figure 4 (left) shows the results of one of these simulations. It is easy to see that though the size of the clusters is small—compared to the number of tuples to be inserted in the network—they act as attractors for the tuples to be stored. Moreover, we are still able to achieve information clustering, but now

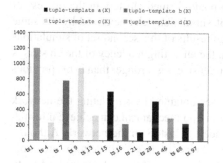
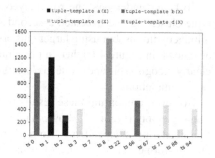

Fig. 5. Final distribution of tuples obtained with $Step = 40$ for two different simulations on the 100-tuple-space scale-free network. *Spatial entropy* corresponding to these simulation: $H = 5.43 \times 10^{-3}$ (left) and 0 (right).

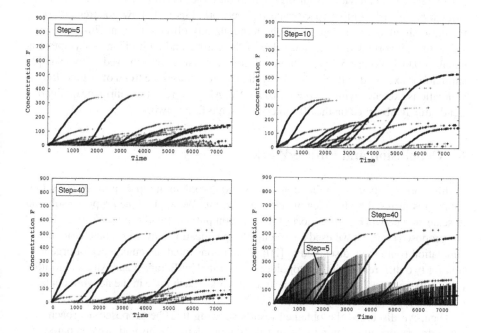

Fig. 6. Evolution of concentration F for different simulations run on the 100-tuple-space scale-free network using $Step = 5$, 10 and 40. The bottom-right graph just highlights the differences between $Step = 5$ and $Step = 40$.

the evolution of our distribution mechanism is driven not only by the time-evolving interaction between *out*-ants, but also by the presence of clusters. In fact, the distribution reported in Figure 4 (left) is characterized by the formation of clusters in the tuple spaces featuring the presence of clusters.

To better understand the role played by already-formed clusters in the process of tuple clustering, we executed a second set of simulations characterized by the same initial conditions, but using larger clusters. The results of this second set of simulations are shown in Figure 4 (right). In this situation the attracting tendency of the clusters is clearly recognizable and—due to clusters of larger size—is stronger than in the previous set of simulations.

However, comparing these results to the ones obtained by simulating the network initially empty (Figure 2) shows that the distribution mechanism can achieve information clustering independently from the presence of clusters in the network. Although already-formed clusters tend to attract the evolution of larger clusters toward tuple spaces featuring the presence of initial clusters, they are not the only cause leading to information clustering. Indeed, the dynamic and evolving interactions between *out*-ants plays also an important role in achieving a good level of clustering.

After having performed simulations on the 30-tuple-space scale-free network, in order to verify the behavior of our mechanism on larger networks, we ran a new set of simulations on a 100-tuple-space scale-free network. Figure 1 shows the trend of H_{avg} for these simulations, and confirms the trend already observed on the 30-tuple-space scale-free network. Moreover, the value of the corresponding horizontal asymptote is equal to 0.007. Figure 5 reports the final tuple concentration achieved in two sample simulations executed with $Step = 40$, while Figure 6 shows the trend of F for different simulations executed with $Step = 5, 10$ and 40. All these results confirm the qualitative trend already observed on the 30-tuple-space scale-free network.

6   Conclusion and Future Work

In this work we focused on the tuple-distribution problem in tuple-space systems. Our solution is inspired by the idea of *self-organization*. We applied the proposed solution to scale-free networks, since most of the real computer networks are scale-free. To test our proposed strategy, we used an executable specification, developed on the stochastic simulation framework discussed in [20]. We demonstrated the emergence of organization for two different scale-free topologies generated by Barabási and Albert algorithm [18]: a 30-node and a 100-node scale-free networks. Each node in the network contains only one tuple space.

Then, we discussed the obtained results showing that the proposed approach is a true self-organizing solution to the distribution problem. To this end, we compared our strategy to Camazine *et al.* argumentation about how a system has to behave to be called self-organizing. The results described in Section 5.3 clearly show that the evolution of our tuple-distribution strategy leads to the emergence of patterns in tuple organization.

In addition to this self-organizing feature, our distribution mechanism also achieves low values of *spatial entropy*—though this does not correspond to a situation of perfect clustering. However, this can be considered a good result, because the proposed mechanism works dynamically. That is, for every *out(tu)* operation to be executed, it tries to store the carried tuple tu in a tuple space with a high concentration of similar tuples. We pointed out that complete clustering may not be desirable in dynamic networks given that processes become too dependent on certain tuple spaces being available all the time. As a consequence, our approach may also be said to provide a higher degree of fault-tolerance. Note that if the tuple space containing the majority of tuples fails, the few tuples in other nodes act as "seeds" for the formation of clusters in those locations.

There are still many open issues to be treated: *(i)* we want to perform further simulations on scale-free networks with an even higher number of tuple spaces; *(ii)* to see how tuples aggregate around clusters, we are going to run simulations where the templates are not totally different from each other so they feature a certain degree of *similarity*; *(iii)* finally, *over-clustering* is a very important issue, since we may want to avoid tuple spaces containing too large clusters. Therefore, we need to devise a *dynamic similarity function* that takes into account the current concentration of similar tuples in a tuple space.

In particular, *(iii)* is very interesting in order to apply our approach in the domain of *open* networks. We believe self-organization will also play a role in the control of over-clustering. We are starting to experiment with solutions inspired by bacteria molding as they appear to have the characteristics we desire. We aim at achieving a balance in the system where clustering occurs but it is not excessive.

References

1. Cardelli, L.: Lecture notes in computer science. In: Wiedermann, J., van Emde Boas, P., Nielsen, M. (eds.) ICALP 1999. LNCS, vol. 1644, pp. 10–24. Springer, Heidelberg (1999)
2. Gelernter, D., Carriero, N.: Coordination languages and their significance. Communications of the ACM 35(2), 96–107 (1992)
3. Ossowski, S., Menezes, R.: On coordination and its significance to distributed and multi-agent systems. Concurrency and Computation: Practice and Experience 18(4), 359–370 (2006)
4. Menezes, R., Tolksdorf, R.: A new approach to scalable linda-systems based on swarms. In: Proceedings of the ACM Symposium on Applied Computing, Melbourne, FL, USA, ACM, New York (2003)
5. Mamei, M., Zambonelli, F., Leonardi, L.: Tuples on the air: A middleware for context-aware computing in dynamic networks. In: Proceedings of the 23rd International Conference on Distributed Computing Systems, vol. 342, IEEE Computer Society, Los Alamitos (2003)
6. Deneubourg, J.L., Goss, S., Franks, N., Sendova-Franks, A., Detrain, C., Chretien, L.: The dynamic of collective sorting robot-like ants and ant-like robots. In: Proceedings of the First International Conference on Simulation of Adaptive Behavior: From Animals to Animats 3, Cambridge, MA, pp. 356–365. MIT Press, Cambridge (1991)
7. Camazine, S., Deneubourg, J.L., Franks, N., Sneyd, J., Theraula, G., Bonabeau, E. (eds.): Self-Organization in Biological Systems. Princeton Univ. Press (2003)
8. Google Inc.: Google scholar. http://scholar.google.com
9. Bonabeau, E., Dorigo, M., Theraulaz, G.: Swarm Intelligence: From Natural to Artificial Systems. In: Santa Fe Institute Studies in the Sciences of Complexity, Oxford University Press, Inc., New York (1999)
10. Parunak, H.: Go to the ant: Engineering principles from natural multi-agent systems. Annals of Operations Research 75, 69–101 (1997)
11. Picco, G.P., Murphy, A.L., Roman, G.C.: Lime: Linda meets mobility. In: Garlan, D. (ed.) Proceedings of the 21st International Conference on Software Engineering (ICSE'99, Los Angeles, CA, USA, pp. 368–377. ACM Press, New York (1999)

12. Snyder, J., Menezes, R.: Using Logical Operators as an Extended Coordination Mechanism in Linda. In: Arbab, F., Talcott, C.L. (eds.) COORDINATION 2002. LNCS, vol. 2315, pp. 317–331. Springer, Heidelberg (2002)
13. Wyckoff, P., McLaughry, S.W., Lehman, T.J., Ford, D.A.: T Spaces. IBM Systems Journal Special Issue on Java Technology 37(3) (1998)
14. Freeman, E., Hupfer, S., Arnold, K.: JavaSpaces Principles, Patterns and Practice. The Jini Technology Series. Addison-Wesley, Reading (1999)
15. Ltd., G.T.: Gigaspaces platform. White Paper (2002)
16. Tolksdorf, R.: Laura — A service-based coordination language. Science of Computer Programming 31(2–3), 359–381 (1998)
17. Corradi, A., Leonardi, L., Zambonelli, F.: Strategies and protocols for highly parallel Linda servers. Software Practice and Experience 28(14), 1493–1517 (1998)
18. Barabási, A.L., Albert, R.: Emergence of scaling in random networks. Science 286, 509–512 (1999)
19. Strogatz, S.H.: Exploring complex networks. Nature 410(6825), 268–276 (2001)
20. Casadei, M., Gardelli, L., Viroli, M.: Simulating emergent properties of coordination in Maude: the collective sorting case. In: 5th International Workshop on Foundations of Coordination Languages and Software Architectures (FOCLASA'06), CONCUR 2006,, Bonn, Germany, University of Málaga, Spain, pp. 57–75 (2006)

Autonomous Optimization of Next Generation Networks

Uwe Walter

Institute of Telematics – Universität Karlsruhe (TH)
Zirkel 2, 76128 Karlsruhe, Germany
walter@tm.uka.de

Abstract. For an efficient usage of the transmission capacity of a QoS-supporting Next Generation Network, it is beneficial to influence the routing of traffic flows by the optimization of link metrics. Deploying a Network Admission Control at the network border helps to comply with assured service guarantees as it can effectively protect against overload situations, especially in times of varying traffic matrices or failures.

Since the manual adaptation of link metrics and NAC budgets is neither quick nor efficient, it makes sense to integrate these optimization algorithms into a self-configuration tool, which is able to autonomously keep the network in the best possible operational condition. This paper presents a management system that re-optimizes link metrics and NAC budgets when necessary. Different scenarios show the benefits of this approach for an increased network resilience and efficient operation.

1 Introduction

In times of increasing demands and pressure of competition, network operators have a tremendous interest in the best possible use of their transmission resources. The available capacity shall be used in the most efficient manner without negatively affecting the offered service quality.

In this context, the optimization of the intra-domain routing, e.g. when using OSPF for example, by adapting the link metrics (also known as link costs, interface costs or link weights) to the current network situation can have significant advantages [1,2,3,4,5,6]. Usually, this approach can help to reduce the maximum utilization of individual transmission links considerably. Although there are alternatives for this sort of traffic engineering (like, e.g. MPLS paths) this paper focuses on the described adaptation of IP link metrics.

It is even possible to anticipate possible link failures in the routing optimization [7,8,9,10]. In this case, link metrics are generated that help to prepare the network routing for a good performance, even under the given failure scenarios.

To reduce the negative impact of link failures and peak demands, it is common practise to keep an ample amount of transmission capacity unused in reserve (so-called overprovisioning). However, it is far more efficient to deploy Network Admission Control (NAC) mechanisms at the network borders to protect against

D. Hutchison and R.H. Katz (Eds.): IWSOS 2007, LNCS 4725, pp. 161–175, 2007.
© Springer-Verlag Berlin Heidelberg 2007

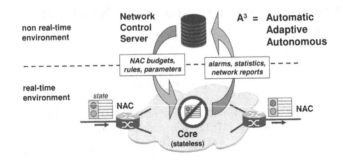

Fig. 1. Deploying a Network Control Server in a Next Generation Network

overload situations. In this case, traffic flows requesting a high-priority have to ask a NAC instance for permission before they are allowed to enter the network.

This paper is based on a NAC approach, where virtual tunnels between all ingress and egress nodes of the network are assigned with capacities called NAC budgets. These budgets are assigned to the NAC instances at the network border, where they are administered locally. Thereby, each network border node can accept or deny QoS requests autonomously without having to ask a central bandwidth broker for permission.

1.1   The Network Control Server

Both, the optimized link metrics and the NAC budgets for high-priority traffic flows should periodically or continuously get adapted to the current network and traffic situation for the most efficient operation. For this task a Network Control Server (NCS) has been developed during the KING research project [11,12].

The basic design principle of this NCS is shown in figure 1. The NCS continuously monitors the network's operating conditions and adapts its parameters if necessary, e.g. after changing traffic patterns or topology changes like link failures or re-connections.

To influence the network's behavior, the NCS is equipped with the link optimization and NAC budget adaptation algorithms, whose output can control the intra-domain routing and configure the NAC function.

The former is done by adjusting the link metrics and configuring the routers accordingly. This optimization is performed with a genetic optimization algorithm, using the link metrics vector as chromosomes and the objective function to minimize the utilization of the most heavily loaded link [9]. Since these optimizations can take quite some runtime, it is possible to include possible link failures into their calculation, as described in section 3. In case of an actual failure, this allows for some calculation time, since the network is already configured to adhere to QoS guarantees even in case of, e.g. a link failure.

The calculation of NAC budgets is described in [13] and aims to minimize the probability of high-priority flows getting blocked by the NAC instances at the network border, based on the current capacity situation observed by the

Network Control Server. The NAC budgets are upper bounds on the maximum amount of admissible traffic at each NAC instance (and per service class), where they are administered locally after being configured by the NCS.

This architecture allows to combine the advantages of a central management node and distributed functions. While the NCS is able to autonomously optimize network parameters to the current situation more efficiently than each network component on its own, the network's availability must not depend on it.

During the design and implementation of the NCS, special care has been taken that all real-time tasks, i.e., packet forwarding, QoS signalling and failure reaction, are handled autonomously by distributed network components (routers, NAC boxes). Thus, a temporary failure of the Network Control Server has no impact on the basic network operation.

In addition to keeping the network in a well-balanced operating condition, the NCS can relieve operators of routine maintenance tasks and aid in the more complex tasks, e.g. in traffic engineering for changing traffic matrices or in evaluating network upgrade options.

The benefits of the Network Control Server and its autonomous optimizations will be demonstrated in different scenarios in sections 2 to 4 of this paper. The most important results will be summarized in section 5.

1.2 Reference Networks

Shown in figure 2 are reference networks that have been used to evaluate the influence of the NCS. The COST-100G reference network consists of 11 nodes and 26 bidirectional links. The Labnet03 network has 20 nodes and 53 bidirectional links. The example network called Worldnet is constructed from 26 nodes and 54 bidirectional links. Based on the US backbone of the Tier-1 provider Sprint from 2001, a topology with 18 nodes and 40 bidirectional links has been built and is depicted in figure 3. The largest of the used reference networks is based on the US/Canada backbone of UUNet from 1997. As shown, it consists of 50 nodes and 80 bidirectional links.

2   Reaction Towards a Changing Traffic Matrix

Due to dynamic changes of the traffic that is offered to a transmission network, its data rates, sources and destinations (the so-called traffic matrix), the utilization of a network can continuously fluctuate. It is one of the biggest advantages of the NCS that it can automatically react to these variations by re-optimizing budgets and link metrics, if this is desired by the operator.

Among the first signs, which signal the necessity for a re-optimization, is an increasing rate of high-priority traffic request getting denied by the Network Admission Control instances. In other words, the ratio of blocked to admitted traffic increases because the currently active traffic budgets are no longer adapted to the new traffic situation in the best possible way.

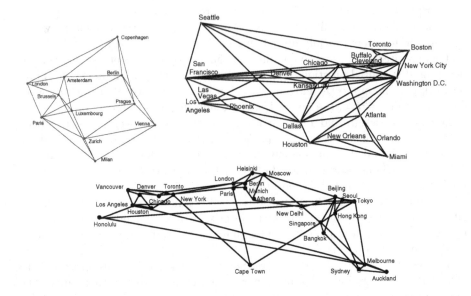

Fig. 2. Reference Network Topologies. Top left: COST-100G, top right: Labnet03, Bottom: Worldnet.

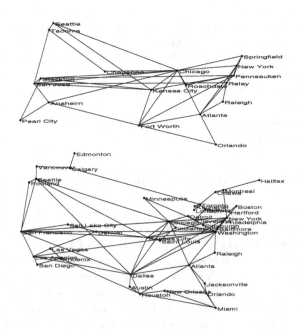

Fig. 3. Reference Network Topologies. Top: Sprint, Bottom: UUNet.

The network operator can define an acceptable threshold of the blocked traffic ratio to configure when the NCS should start re-optimizing the network parameters. There are different possibilities to define blocking rates, e.g. the maximum probability of a QoS request getting denied or the sum of the data rates of blocked high-priority request. To be able to compare blocking values among scenarios with different network utilizations, the ratio of the sum of blocked traffic rates to admitted traffic rates will be used in the following.

Instead of determing the blocking rate from NAC data about denied QoS requests, the NCS calculates blocking probabilities based on the offered traffic matrix and the currently configured NAC budgets. This is done using the offered traffic rates, the NAC budgets, a given traffic mix to compute the estimated ratio of blocked traffic b_T using the Kaufman-Roberts formula [14]:

$$b_T = \frac{\sum\limits_{k=0,1,\ldots,k} {}^b BBB_k \cdot r_m \cdot A_k}{\sum\limits_{k=0,1,\ldots,k} r_m \cdot A_k} = \frac{\sum\limits_{k=0,1,\ldots,k} {}^b BBB_k \cdot A_k}{\sum\limits_{k=0,1,\ldots,k} A_k} \tag{1}$$

In the equation b_{BBB_k} denotes the blocking probability for a (virtual tunnel) border to border budget BBB_k as a weighted average for all traffic types given in the traffic mix (consisting purely of small-bandwidth VoIP calls in the following). There are K different BBBs and A_k denotes the offered traffic (in Erlang) for budget BBB_k, corresponding to a mean offered traffic (in terms of bandwidth) of $r_m \cdot A_k$ where r_m is the mean offered bit rate of the traffic mix.

This approach offers the benefit of a quick reaction even in the range of very small blocking probabilities, since the offered traffic can be observed more steadily than very rare rejected QoS requests. It would take very long to measure rare events with sufficient precision whereas very low blocking probabilities can be easily computed analytically using the multi-rate Erlang formula [15,16].

When the Network Control Server detects an excess of rejected QoS requests an the network border, it can react in several ways. The default strategy, which is shown in figure 4, is divided into two reaction steps. The first step consists of a re-adaptation of all NAC budgets to the current traffic matrix. In many cases, these new budgets alone suffice to reduce the blocking rate to acceptable values.

If the re-adaptation of NAC budgets does not achieve a sufficient improvement of the blocking situation, the NCS escalates its reaction to the second step of the default strategy. In this case, the NCS re-optimizes all link metrics with the objective to adapt them to the current traffic matrix in the most optimal way. This is followed by the calculation of new NAC budgets, since the older ones are not valid for the new link metrics any more.

The reason for this two-step-strategy is the fact that changing link metrics (and the following convergence process of the intra-domain routing protocol) can result in temporarily degraded service quality due to packet losses or short micro-routing loops. To avoid these issues, the link metrics do only get re-optimized, if it is really necessary or –at least– leads to a significant improvement.

If blocking still remains too high after all re-optimizations, there is likely too much offered traffic. In this case, the NCS can temporarily suspend its operation and instead make suggestions how to upgrade the network [17].

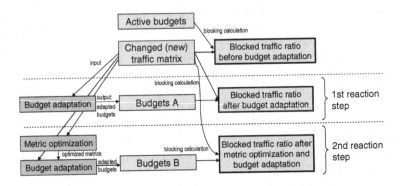

Fig. 4. Default two-step-reaction to changes of the offered traffic situation

Random traffic matrices have been generated as input parameters for the following evaluation scenarios. By nature, network optimizations show more benefits, if the network is utilized to a certain degree. Because of this reason, traffic matrices with different network utilizations were created, which result in the most heavily loaded link in the network being utilized between 30 and 80 %. This generation was done by adding or scaling random CBR flows until the maximum link utilization with optimized link metrics reached to target value. The following simulations have been done with hundreds of such traffic matrices with results being averaged, if nothing else is mentioned. As simplification, only high-priority traffic will be used. Spare bandwidth could always be consumed by lower priority traffic classes, e.g. best effort. An appropriate scheduling configuration will have to ensure the preferential treatment of high-priority traffic during forwarding by the routers.

In the evaluations of the reaction to changing traffic matrices, one traffic matrix has been selected as initial traffic matrix, based on which budgets and link metrics were optimized. Then, the traffic matrix was swapped against a randomly selected new one (with the same overall network utilization), which translates to a change of the offered traffic. Following this, the increase of the ratio of blocked traffic in this un-optimized situation was evaluated and how far the reaction strategy of the NCS was successful in reducing the blocking rate.

Exemplarily, the optimization results in the Labnet03 reference network shall be presented in the following. In the left part of figure 5 the average blocked traffic ratio of the different strategy steps is depicted (derived from 500 evaluation results). The diagram on the right side shows the average reduction of the blocked traffic ratio from the situation before the NCS reacted to the situation after a budget adaptation in the left bar. The middle bar depicts the further reduction in the second reaction step, from the situation after only the budget adaptation to the results achievable by additionally optimizing the link metrics with the traffic budgets. The bar on the right shows the overall reduction from the initial situation after a traffic matrix change to the situation after the combination of optimization steps.

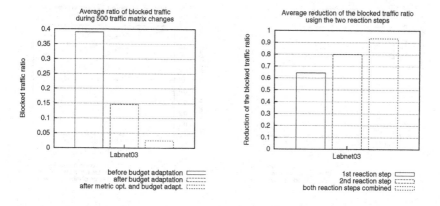

Fig. 5. Average blocked traffic ratios (left) and their reductions (right) by the NCS' optimizations in 500 random traffic matrices with a network utilization of the Labnet03 reference topology of 60 %

In these 500 simulations, the blocked traffic ratio after a changing traffic matrix lies between 14 % and 67 % at an average of 39 %. After the budget adaptation, the blocked traffic ratio drops to values between 1 % and 50 %, the average being 15 %, as shown in the middle bar of the left diagram. After the second strategy step of the NCS, the blocked traffic ratio stays at almost 20 % in the worst case of the 500 scenarios, while most results being far better. The best value is at only 0,03 %, resulting in an average of only 2,47 %.

If the optimization results are compared to the average blocked traffic ratio directly after a change of the traffic matrix amounting to 39 %, this example shows a reduction of more then 93 % as depicted by the right bar in the right diagram. The first reaction step of the NCS alone, consisting of only the budget adaptation, achieves a reduction of 64 %. The link metric optimization adds to another reduction of the blocked traffic ratio of more than 80 %.

As mentioned above, the simulations have been done not only with a network utilization of 60 % but also with traffic matrices utilizing the networks between 30 % and 80 %. Figure 6 shows the resulting blocked traffic ratio (on the left side) and the reductions the NCS was able to achieve (on the right side) in the Labnet03 reference topology.

Considering the blocked traffic ratios after metric optimization and budget adaptation in the Labnet03 with a network utilization of 50 % and 60 %, there are first occurances of blocking rates that cannot be removed by the optimizations of the NCS. On average, the network reaches its maximum capacity at these utilization values. However, it is important to note, that this threshold is valid in case of a single link failure. The budget adaptation, from which the blocking values are derived from, was performed with the objective function to include all possible single link failure into its calculations. This results in NAC budgets that remain valid in case of a single link failure, enabling the network to preserve the guaranteed service quality even during these failures.

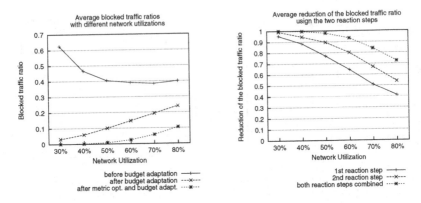

Fig. 6. Average blocked traffic ratios (left) and their reductions (right) by optimizations in 500 random traffic matrices with different network utilizations of Labnet03

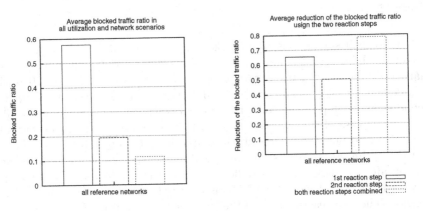

Fig. 7. Average blocked traffic ratios (left) and their reductions (right) in all reference networks and utilization scenarios

Regarding this strategy decision, the reference network Labnet03 reaches a point at utilizations of 50-60 % where individual traffic flows must be limited (or even denied) by the Network Admission Control to prevent overloads in case of link failures. This correlates nicely with the rule of thumb many providers use, who start upgrading their network when utilization reaches the order of 50 %.

Figure 7 depicts the average of the blocked traffic ratio in all reference networks with different network utilizations. The average ratio of blocked traffic after a change of the traffic matrix is 57,4 % and is decreased by 65 % after the budget adaptation to a ratio of 19,3 %. The second strategy step of the NCS further reduces the blocked traffic ratio by 50 % on average to a value of 11,4 %. Averaging all reference networks and utilization scenarios, the NCS is able to achieve a reduction of the ratio of blocked traffic amounting to more than 78 %.

3 Single Link Failures

Besides the reaction to a changing traffic situation, especially the anticipation of failure cases is one of the main strengths of the Network Control Server's optimization algorithms. This section will present some results of the evaluation of single link failures. A single link failure is defined as the bidirectional failure of a direct connection between two nodes.

During the budget adaptation, failure scenarios can be included in the calculations to influence which sort of network failures should be anticipated in advance. If the failure scenarios contain all possible single link failures, the budgets will get calculated in such a way that none of these failures will result in an overload situation inside the network. The following evaluation shall demonstrate this functionality and give an estimate about the expected consequences if no NAC budgets would be deployed and all offered traffic would be admitted.

Each simulation is carried out with one of the reference networks and randomly generated traffic matrices. For each traffic matrix optimized link metrics and NAC budgets get calculated. Both optimizations are configured to anticipate each possible single link failure. After this step, the reference network topology is used to derive all possible subset networks, where exactly one link failed. For each of these failure scenario networks, all link loads based on the original traffic matrix are determined and compared to the link loads of the fully functional reference network. These link loads represent the network situation during a link failure, if there would be no traffic limit, as if there would be no NAC.

If a link shows a load above the value 1 it is assumed to be overloaded. While such a value can result from the simulations, this would of course translate to a heavy packet loss in reality, disrupting the guaranteed service quality. The number of overloaded links (in theory) is stored as the first result for the given traffic matrix and failure scenario. For each overloaded link, the (theoretically) missing capacity is determined. The sum of these missing capacities is used as the second result.

Finally, a traffic matrix is constructed based on the originally calculated NAC budgets and used to calculate the link loads. This step gives an indication of the network situation if all NAC budgets would be completely utilized (admitting the maximum amount of high-priority traffic into the network), with no traffic being rejected by the NAC. The resulting number of overloaded links and missing capacities for this setup are used as third and fourth result. Thus, four values are calculated as results for each traffic matrix and single link failure:

- Number of overloaded links without consideration of NAC budgets
- Sum of missing capacities of the overloaded links without consideration of NAC budgets
- Number of overloaded links with traffic being limited by NAC budgets
- Sum of missing capacities of the overloaded links with traffic being limited by NAC budgets

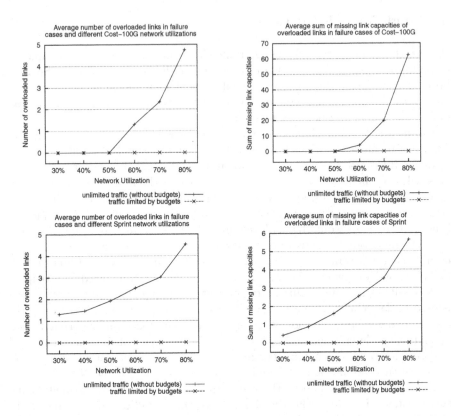

Fig. 8. Number of overloaded links and sum of missing capacities during single link failures

In the simulations performed to gather these results, 250 traffic matrices have been used for each reference network and each network utilization class along with each possible single link failures.

An exemplarily extract from the simulation results is shown in figure 8. The figure shows the average of the four described results for the reference networks Cost-100G and Sprint. In the diagrams on the left side, the mean number of overloaded links in a failure case is depicted, differing between the unlimited traffic given by the traffic matrix (no NAC) and with NAC budgets limiting the admitted traffic. On the right side, the sum of missing capacities (in generic bandwidth units) is shown.

During lower utilizations, there is no visible overload for the Cost-100G reference network in case of a single link failure. Starting with an utilization of 50 % there are overloaded links, whose number increase with the rising utilization if there are no budgets limiting the traffic. In the highest simulated utilization this leads to an average of five links being overloaded, if a single link fails.

If the incoming traffic is limited by the budgets of a Network Admission Control, there is no overload of any link in case of a single link failure.

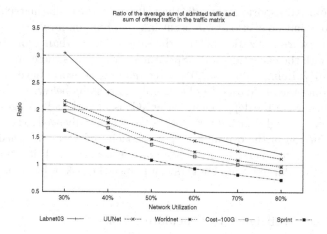

Fig. 9. Ratio of overall transport capacity in traffic matrix and allowed by NAC budgets with and without anticipation of all single link failure

The lower part of figure 8 shows the according values for the Sprint reference topology. In this example network, there are traffic matrices that lead to a link overload even in case of a low overall network utilization during link failures, if there is no limit on incoming traffic.

3.1   Impact of the Failure Anticipation

As demonstrated, the deployment of a Network Admission Control can protect against overload situations during single link failures. However, while in the best cases a NAC admits most request, it must deny and reject requests for high-priority traffic flows in certain situations. As a result, the gained benefit of being able to guarantee the service quality even during failure cases, comes at the price of a possible reduction in the overall transport capacity of a network. This can be seen as the cost for the improved resilience.

To get an insight into this reduction of transport capacity, each reference network and random traffic matrices have been used to evaluate the overall sum of transmitted traffic by adding all data rates in the traffic matrix. As a comparison, the overall sum of all NAC budgets after a metric optimization and budget adaptation has been calculated, giving an estimate of the overall transport capacity in case the QoS guarantees are protected by a Network Admission Control. The ratio of the average of both transport capacity sums is shown in figure 9 above the network utilization.

If the value of the ratio is higher than 1, this means, that the sum of the admissible traffic flows in the NAC budgets is higher than the sum of traffic flows in the traffic matrix. However, this does not mean, that there are no rejections of individual traffic flows possible, since the summarization ignores the composition and source and destination of individual flows. Simplified, in these scenarios, a network could transport more traffic than offered by the traffic matrix, even with the guarantee that no overload will occur in case of link failures.

Below a network utilization of 50 % the ratio is higher than 1 for all reference networks. With the exception of the Sprint network it stays there until a network utilization 70 %. Being utilized to 80 % two remaining networks have a ratio above 1, the Worldnet reference topology being only marginally below.

A consequence for the deployment of the NCS is, that the advantage of a higher resilience against single link failures, does not automatically need to be joined with a severe reduction of the overall transport capacity of a well-designed network. Although precise numbers heavily depend on the scenario and situation, making best use of well-adapted NAC budgets can reduce the risk of blocking too many QoS requests.

3.2   Increased Efficiency by Calculated Risks

Network operators who object even a minor reduction of the overall transport capacity for increased resilience against failures, can vote to take a calculated risk to further reduce this impact.

In this case, taking a risk means that not all possible single link failures must be taken into account when calculating NAC budgets. The failure scenarios that have the biggest hit on the overall transport capacity can be excluded. However, if such a link would really fail it would not be possible to guarantee that this would not result in the overload of a network link. In the absence of failures, this would allow a further increase of the available overall transport capacity, despite the gained protection against most possible failures.

Exemplarily such an approach was evaluated for the Sprint reference network topology. Its results were compared against the former findings (marked "Sprint" in the following). At first, taking a calculated risk, the protection against the one

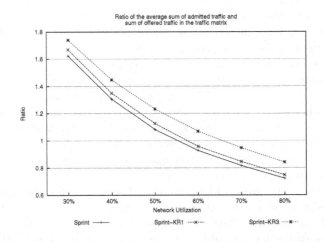

Fig. 10. Comparison of the overall sum of transport capacity

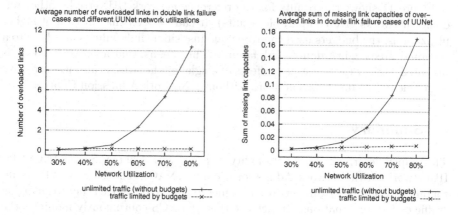

Fig. 11. Number of overloaded links and sum of missing capacities during double link failures in the UUNet reference topology

single link failure, having the biggest negative impact on the admissible traffic, was dropped, i.e. the single link failure for whose protection the NAC budgets drop the most (marked "Sprint-KR1" in the following). Then, this was extended to a scenario, were protection against the three most expensive single link failures was dropped (marked "Sprint-KR3").

Figure 10 shows the ratio of the average overall sums of admissible traffic to traffic offered in the traffic matrices for the mentioned risk cases in comparison. Dropping protection against the worst three single link failures in "Sprint-KR3" keeps the ratio above 1 even with a network utilization of 60 %. At 70 % it is only marginally below 1. As expected, "Sprint-KR3" allows a clear increase of admissible traffic compared to the "Sprint"-scenario, which protects against each single link failure.

If desired, a network operator could further mitigate the reduction of the overall network transport capacity, following the deployment of a Network Admission Control, if he accepts the risk to exclude a few failure scenarios from the resilience protection.

4   Double Link Failures

In section 3 the effects of the NCS' optimizations against single link failures have been described. It has been shown that no overload situation occurs, if the NAC budgets are deployed correctly. This section shall evaluate, how far the protection against single link failures can also lessen the negative impact of double link failures (two bidirectional links failing at the same time).

Except for the inclusion of double link failures[1], the simulations are performed exactly as the ones in section 3. Each double link failure scenario has been evaluated 250 times per utilization class.

---

[1] Failure scenarios that would separate the network are excluded.

Figure 11 shows the results for the reference network UUNet (the other reference networks give comparable results). It can be seen that in case of higher utilizations, the budgets used to protect against single link failures also help to a certain degree against double link failures. This means, that a network, which is prepared against the worst-case effects of single link failures, is also less sensitive to double link failures, as a network without a Network Admission Control.

5 Summary

This paper described the functionality of a Network Control Server (NCS) for IP networks with Network Admission Control (NAC). The NCS is able to autonomously adapt certain network parameters and re-optimize them to changing traffic or failure situations. To achieve this, the NCS continuously monitors the network and reacts when necessary.

After a change of the offered traffic matrix, an estimation is made if the new situation will lead to an increasing blocking probability of high-priority QoS requests. If necessary, new NAC budgets are calculated, shifting available NAC capacity to the best possible ingress and egress nodes. If the ratio of blocked traffic still remains undesirably high, new IP link metrics are calculated in addition to the budget adaptation.

Extensive simulations have been performed to evaluate the performance of this reaction strategy using five realistic reference network topologies and hundreds of traffic matrices. It has been verified that the deployment of NAC mechanisms protects from network overloads. NAC budgets, which are calculated taking failures into consideration, can protect from these failure scenarios.

It could be shown that over all scenarios, the re-optimization of network parameters – the adaptation of NAC budgets and link metrics – after a change of the offered traffic matrix could on average effectively cut the blocking ratio in half. Even if the NAC budgets were only calculated to protect against single link failures, they also clearly lessen the impact of double link failures compared to a NAC without failure protection.

The gained resilience and service guarantees definitely recommend the deployment of network optimization mechanisms. For increased efficiency, it seems advisable to add integrate such functionality into autonomous tools, as it has been demonstrated in the Network Control Server presented in this paper.

Acknowledgments. The simulations described in this paper used tools to optimize link metrics [9] and to adapt NAC budgets to traffic matrices [13] from the KING research project (Key components for the Internet of the Next Generation) that has been partly funded by the German Ministry of Education and Research (BMBF). Special thanks go to Uwe Riehm for his simulations and Dr. Joachim Charzinski for his contributions to this work. Thanks are also given to the reviewers for their very valuable comments.

References

1. Fortz, B., Thorup, M.: Internet Traffic Engineering by Optimizing OSPF weights. In: Proceedings of IEEE Infocom, IEEE Computer Society Press, Los Alamitos (2000)
2. Riedl, A.: A Genetic Algorithm for Routing Optimization in IP Networks Utilizing Bandwidth and Delay Metrics. In: Proceedings of IEEE Workshop on IP Operations and Management IPOM, Dallas, IEEE Computer Society Press, Los Alamitos (2002)
3. Ericsson, M., Resende, M., Pardalos, P.: A genetic algorithm for the weight setting problem in OSPF routing. Journal of Combinatorial Optimization 6(3) (September 2002)
4. Fortz, B., Thorup, M.: Optimizing OSPF/IS-IS Weights in a Changing World. IEEE Journal on Selected Areas in Communications 20(4) (May 2002)
5. Fortz, B., Rexford, J., Thorup, M.: Traffic Engineering With Traditional IP Routing Protocols. IEEE Com. Mag. 40(10) (October 2002)
6. Rexford, J.: Handbook of Optimization in Telecommunications. Kluwer Academic Publishers, Dordrecht (2005)
7. Nucci, A., Schroeder, B., Bhattacharyya, S., Taft, N., Diot, C.: IGP Link Weight Assignment for Transient Link Failures. In: Proceedings of the 18th International Teletraffic Congress (ITC-18), Berlin, Germany (August 2003)
8. Fortz, B., Thorup, M.: Robust optimization of OSPF/IS-IS weights. In: Proceedings of International Network Optimization Conference (INOC), Evry/Paris, France (October 2003)
9. Reichert, C., Magedanz, T.: A Fast Heuristic for Genetic Algorithms in Link Weight Optimization. In: 5th International Workshop on Quality of Future Internet Services (QoFIS), Barcelona, Spain (September 2004)
10. Hasslinger, G., Schnitter, S., Franzke, M.: The Efficiency of Traffic Engineering with Regard to Link Failure Resilience. Telecommunication Systems Journal 29(2) (June 2005)
11. Hoogendoorn, C., Charzinski, J., Schrodi, K., Heldt, N., Huber, M., Winkler, C., Riedl, J.: Towards the Next Generation Network. In: 12th IEEE International Conference on Network Protocols (ICNP 2004), Berlin, Germany, IEEE Computer Society Press, Los Alamitos (2004)
12. Schrodi, K.: High Speed Networks for Carriers. In: Carle, G., Zitterbart, M. (eds.) PfHSN 2002. LNCS, vol. 2334, Springer, Heidelberg (2002)
13. Menth, M.: Efficient Admission Control and Routing in Resilient Communication Networks. PhD thesis, University of Würzburg, Faculty of Computer Science, Am Hubland (July 2004)
14. Mocci, U., Virtamo, J., Roberts, J.: Broadband Network Traffic: Performance Evaluation and Design of Broadband Multiservice Networks: Final Report of Action Cost 242. Springer, New York (1999)
15. Labourdette, J.F., Hart, G.: Blocking Probabilities in Multitraffic Loss Systems: Insensitivity, Asymptotic Behaviour and Approximations. IEEE Transactions on Communications 40(8) (August 1992)
16. Nilson, A.A., Perry, M., Gersht, A., Iversen, V.: On multi-rate Erlang-B Computations. In: Proceedings of 16th International Teletraffic Congress (ITC 16), Edinburgh, Scotland (June 1999)
17. Charzinski, J., Walter, U.: Optimized Incremental Network Planning. In: Proceedings of the 13th GI/ITG Conference on Measuring, Modelling and Evaluation of Computer and Communication Systems, 27.-29.3, Nürnberg, Germany, GI ITG, VDE Verlag (March 2006) 349–362 (2006)

Bandwidth-Satisfied Multicast Services in Large-Scale MANETs

Chia-Cheng Hu

Department of Information Management, Naval Academy,
No.669, Junxiao Rd., Zuoying District, Kaohsiung City 813, Taiwan (R.O.C.)
cchu@cna.edu.tw

Abstract. Recent routing/multicast protocols in large-scale mobile ad-hoc networks (MANETs) adopt two-tier infrastructures by selecting backbone hosts (BHs) in order to avoid the inefficiency of the flooding. Further, previous MANET quality-of-service (QoS) routing/multicasting protocols determined bandwidth-satisfied routes for QoS applications. However, they suffer from two bandwidth-violation problems. In this paper, a novel algorithm that can avoid the two problems is proposed and integrated with the two-tier infrastructures to construct bandwidth-satisfied multicast trees for QoS applications in large-scale MANETs.

Keywords: *Ad-hoc network, integer linear programming, multicast protocol, bandwidth violation, quality-of-service.*

1 Introduction

A multicast group contains a special host (server) which is responsible for transmitting data packets to the other hosts (clients) in the same group. Motivated by increasing importance of real-time and multimedia applications with different quality-of-service (QoS) requirements, e.g., VoIP and video-conference, several QoS-constrained multicast algorithms for multimedia communication in wired networks have been proposed in the literature [1]-[3]. These algorithms aim to construct least-cost multicast trees with the constraints of end-to-end delay and/or bandwidth requirement.

A mobile ad-hoc network (MANET) is formed by a group of mobile hosts that can communicate with one another without the aid of any centralized point or fixed infrastructure. Because of recent provision of high-speed wireless Internet services, QoS-guaranteed applications are now crucial to new-generation wireless multimedia communication systems. To meet the QoS requirements of the applications, multicast protocols are required to construct multicast trees with QoS guaranteed. Continuing advances in semiconductor and computer technologies will soon allow large-scale MANETs to become viable and valuable in a wide variety of novel applications. Designing a robust and efficient infrastructure for distributing a large number of multicasting applications with QoS requirements in large-scale MANETs will become one of the key research challenges.

In MANETs, the network topologies may dynamically change in an unpredictable manner because hosts are free to move. Hence the admitted QoS applications may

D. Hutchison and R.H. Katz (Eds.): IWSOS 2007, LNCS 4725, pp. 176–192, 2007.
© Springer-Verlag Berlin Heidelberg 2007

suffer due to frequent route breaks, thereby requiring such applications to be re-determined over new routes. Besides, each transmission in MANETs is a local broadcast and each host shares the common radio channel with all its neighbors. An inadequate bandwidth reservation may decline network performance, which is serious in MANETS because of shared channel and limited bandwidth. Further, the above difficulties will be enlarged in a large-scale MANET. Recently, several MANET multicast protocols have been proposed in the literature [4]-[10]. They can be classified into two categories: tree-based protocols and mesh-based protocols. For both, adding a new member into an existing multicast group will cause the flooding of a join request message over the entire network. The flooding process is time-consuming and bandwidth-consuming, especially, for a large-scale MANET.

To avoid the inefficiency of flooding, two-tier infrastructures [11]-[15], [21], [24] were adopted for routing/multicasting in large-scale MANETs. Some of hosts, named backbone hosts (BHs), were selected and responsible for managing the flooding, maintaining the infrastructures and determining the routes. On the other hand, several routing protocols [12], [18], [19] and multicast protocols [15], [20] have been proposed for MANET QoS services. However, they may lead to bandwidth violation, as described below. When a new flow with bandwidth requirement is permitted, a control packet from the source is flooded in order to determine a bandwidth-satisfied route. Each host in the neighborhood of some ongoing flows may be determined as a forwarder for the new flow if the bandwidth increment does not induce bandwidth violation of it and its neighbors. However, even so, bandwidth violation may happen to its neighbors because it fails to take into account the bandwidth consumption of those hosts that are two hops distant from it. This induces a new bandwidth-violation problem in MANETs. Similarly, another bandwidth-violation problem is induced if multiple flows are permitted concurrently. The two problems are named the hidden route problem (HRP) and the hidden multicast route problem (HMRP), respectively. They will be elaborated in the next section.

The purpose of this paper is to propose a novel algorithm that can avoid HRP and HMRP. The proposed algorithm is integrated with two-tier infrastructures for QoS multicast applications in large-scale MANETs. In [21], the authors proposed a multicast protocol, named OGHAM, for large-scale MANETs. OGHAM selected BHs on-demand with the objective of minimizing the total number of hops to the other hosts, so as to shorten multicast routes. In [24], we enhance OGHAM with mobility awareness by taking host mobility into consideration. The enhanced OGHAM, denoted by M-OGHAM, is enhanced with link prediction, i.e., to estimate the amount of remaining connection time for two neighboring hosts. Link prediction is very helpful to mobility awareness, and by its aid, stable BHs and stable multicast routes can be determined.

Since bandwidth and power are limited in MANETs, one way to reduce bandwidth and power consumption is to decrease the number of hosts (i.e., forwarders) participating in packet forwarding. The problem of finding a multicast tree with minimum number of forwarders is known to be NP-hard (see [22]). If the bandwidth consumption is considered additionally, the resulting problem is also NP-hard. In this paper, we study the problem of determining a bandwidth-satisfied multicast tree with minimum number of forwarders. The problem is referred to as BSMTP in the rest of this paper, and a heuristic algorithm is proposed to provide a feasible solution to it.

2 Related Works and Problems

To meet the bandwidth requirements of bandwidth-constrained applications, a proper admission control mechanism is needed to judge if a host is allowed to forward the packets for a requested flow. In wired networks, there is a dedicated point-to-point link, denoted by $l_{i,j}$, between two adjacent nodes v_i and v_j. If v_i (v_j) transmits packets to v_j (v_i), the bandwidth consumption is bounded by the maximal available bandwidth of $l_{i,j}$. When a neighbor of v_i (v_j) transmits packets to v_i (v_j), it does not consume the bandwidth of $l_{i,j}$. So, v_i and v_j are aware of the remaining available bandwidth of $l_{i,j}$. Suppose that a new flow from the source v_s to the destination v_d is initiated and its bandwidth requirement is b_req. Let $b\_r_{i,j}$ be the current remaining available bandwidth of $l_{i,j}$. If $b\_r_{i,j} \geq b\_req$, then v_i (v_j) is allowed to forward the flow to v_j (v_i). Finally, if $v_s \rightarrow v_{f_1} \rightarrow v_{f_2} \rightarrow v_{f_3} \rightarrow \ldots \rightarrow v_{f_m} \rightarrow v_d$ is determined as the bandwidth- satisfied route for the flow, then it should have $\min\{ b\_r_{s,f_1}, b\_r_{f_1,f_2}, b\_r_{f_2,f_3}, \ldots, b\_r_{f_3,d} \} \geq b\_req$.

The MANET QoS routing/multicasting protocols, in CEDAR [12], [18], MCEDAR [15] and [20], borrowed the concept of point-to-point links from the wired networks to construct bandwidth-satisfied routes. However, a host in MANETs shares the radio channel with its neighbors so that the bandwidth is consumed not only by it, but also by its neighbors. For example, suppose that $v_s \rightarrow v_{f_1} \rightarrow v_{f_2} \rightarrow v_{f_3} \rightarrow \ldots \rightarrow v_{f_m} \rightarrow v_d$ is a bandwidth-satisfied route in MANETs. When v_{f_1} transmits packets to v_{f_2}, the bandwidths of l_{s,f_1}, l_{f_1,f_2} and l_{f_2,f_3} are consumed. Therefore, it is more difficult to determine a bandwidth-satisfied route in MANETs, because the computation of $l_{i,j}$ also relies on the neighbors of v_i and v_j.

On the other hand, AQOR [19] determined a bandwidth-satisfied route differently. The bandwidth computation is from the viewpoint of a host, instead of a link. Let b_r_i be the current remaining available bandwidth of v_i. With the same example of $v_s \rightarrow v_{f_1} \rightarrow v_{f_2} \rightarrow v_{f_3} \rightarrow \ldots \rightarrow v_{f_m} \rightarrow v_d$ above, the bandwidths of v_s and v_{f_1} are consumed when v_s transmits packets to v_{f_1}, the bandwidths of v_s, v_{f_1} and v_{f_2} are consumed when v_{f_1} transmits packets to v_{f_2}, and so on. Consequently, the total bandwidth consumptions of v_s, v_{f_1}, v_{f_2}, v_{f_3}, $\ldots$, v_{f_m} and v_d, caused by the flow, are $2 \times b\_req$, $3 \times b\_req$, $3 \times b\_req$, $3 \times b\_req$, $\ldots$, $3 \times b\_req$ and b_req, respectively.

All QoS routing/multicasting protocols described above may suffer from two bandwidth- violation problems, i.e., HRP and HMRP, because the hosts in the neighborhood of ongoing flows fail to compute the bandwidth consumptions of those hosts that are two hops distant from them.

Refer to Fig 1, where an illustrative example is shown. There are two ongoing flows from e to f and from g to h, respectively. A new flow from a to d is permitted, and the route determination for the new flow proceeds to c. Suppose that each host has the same available bandwidth, say 11 units, and the bandwidth requirements for the three flows are 2 (from e to f), 7 (from g to h) and 3 (from a to d) units, respectively. If c serves as a forwarder, then total 9-unit bandwidth of c will be consumed

(c, the predecessor of c and the successor of c each require 3-unit bandwidth). Since the ongoing flow from e to f is in the radio coverage of c, it consumes 2-unit bandwidth of c. Consequently, the remaining available bandwidth of c is $11 - 2 = 9$ units, and so c is allowed to be a forwarder.

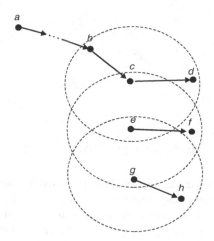

Fig. 1. An example of HRP

Now we turn our attention to the bandwidth consumption of e. Since both c and g are in the radio coverage of e, the bandwidth requirement of e is $2 + 3 + 7 = 12$ units, which exceeds its available bandwidth (11 units). The reason for the bandwidth violation is that c was not aware of the ongoing flow from g to h when it was determined to be a forwarder. In short, the bandwidth violation happened because the ongoing flow from g to h was hidden from the new flow from a to d. The problem is henceforth referred to as the hidden route problem (HRP). In contrast to the hidden terminal problem [23], which arises in the MAC layer, the hidden route problem arises in the network layer.

An illustrative example for HMRP is shown in Fig 2, where there is a multicast group and a new flow from a (server) to e and h (clients) is permitted. Suppose that each host has the same available bandwidth, say 11 units, and the bandwidth requirement of the flow is 3 units. Also note that bandwidth reservation will be made for the flow when data flow through the routes. Both c and g can be forwarders for the flow from a to e and from a to h, respectively, because their bandwidth requirement (9 units) is smaller than their available bandwidth (11 units). However, since they are in the radio coverage of each other, 3-unit bandwidth is required additionally when data flow from a to e and h. This increases their bandwidth requirement to 12 units, which causes a bandwidth violation. The bandwidth violation happens because the two multicast routes from a to e and from a to h are mutually hidden from each other. It is henceforth referred to as the hidden multicast route problem (HMRP).

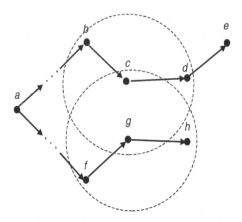

Fig. 2. An example of HMRP

Currently, two QoS routing protocols have been proposed by Yang et al. [16] and. Chen et al. [17]. They considered the bandwidth consumption of the one-hop/two-hop hosts when a new flow with bandwidth requirement was permitted. So, HRP can be avoided. However, a QoS multicast protocol should determine multiple bandwidth-satisfied routes from a server to all clients concurrently. Another bandwidth-violation problem, i.e., HMRP, is induced if multiple flows are permitted concurrently.

3   Review of OGHAM and M-OGHAM

OGHAM and M-OGHAM construct a two-tier infrastructure by selecting BHs on-demand for multicast applications. A multicast group contains a server which is responsible for transmitting data packets to the clients in the same group. To construct a multicast group, either the clients passively join the group when the server broadcasts a query packet, or they actively sends a request to the server. In this setting, a host (server or client) first broadcasts a message over a limited-range region (multicast region) to collect necessary neighboring information. Then BHs are selected in the region. And they are responsible for determining multicast routes, forwarding data packets, handling dynamic group membership, and updating multicast routes due to host movement.

When a server (client) v_i attempts to create (join) a multicast group, v_i first tries to find a BH within a region with a radius of $2r$ hops centered at v_i, where $r \geq 1$ is a predefined integer. If such a BH is found, then v_i is attached to it. Otherwise, v_i broadcasts a message over a larger region, called *multicast region*, with a radius of γ hops centered at v_i for collecting neighboring information, where $\gamma \geq 2r$ is a predefined integer. Then, v_i selects BHs for the multicast region and determines the attachment from NBHs (the hosts are not BHs) to BHs by solving a formulated 0/1 ILP. Also, v_i sends the list of all BHs and the neighboring information to each BH.

For example, suppose that a server s intends to create a multicast group and it fails to find a BH within 2 hops ($r=1$ for this example). So, s reacts by triggering the selection of BHs in a multicast region with a radius of $\gamma=4$ hops centered at s. Refer to

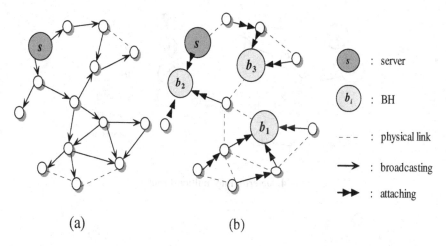

(a)   (b)

Fig. 3. BH selection in a multicast region. (a) Broadcasting. (b) Selecting BHs and attaching NBHs to BHs.

Fig 3. First, s broadcasts a message in the multicast region (refer to Fig 3(a)). Upon receiving the message, hosts in the multicast region reply their neighboring information to s. With the neighboring information, s then selects BHs and attaches NBHs to BHs (refer to Fig 3(b)).

After BHs in the multicast region are selected, clients can join the multicast group by asking the attached BHs to query the location of the server. The BH attached by the server then replies to the queries. Through the round-trip communication (querying and replying), the BH attached by the server can determine the multicast routes from the server to the clients. With the same example of Fig 3, assume that c_1 and c_2 are two clients. They join the multicast group by attaching themselves to b_1 and b_3, respectively. The server s is attached to b_2. Both b_1 and b_3 can locate s by querying b_2. At the same time, the multicast routes, i.e., $s-b_2-f_1-b_1-f_2-c_1$ and $s-b_2-f_1-b_3-c_2$, from s to c_1 and c_2, respectively, can be determined (refer to Fig 4).

As described above, the BHs attached by clients are responsible for querying the location of the server. For a client in a different multicast region from the server, the attached BH fails to locate the server, and so it floods a query message over the entire network, in order to locate the server. Through the flooding, the two multicast regions where the client and the server are positioned can be merged into a larger one. After the merging, a multicast route from the server to the client can be determined.

Following the example of Fig 4, we assume that there is a client c_3, which is outside the multicast region of Fig 4, attempting to join the multicast group created by s. The gray portion of Fig 5 shows the multicast region created by c_3. There are two BHs, i.e., b_1' and b_2', selected in the new multicast region and c_3 is attached to b_1'. In order to locate s, b_1' floods a message. Upon receiving the message, b_2' replies to b_1'. Through the message exchange, a multicast route, i.e., $s-b_2-f_1-b_1-f_2-f_4-b_1'-c_3$, from s to c_3 is then determined.

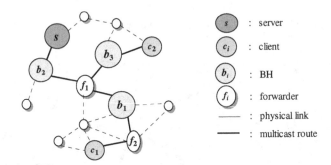

Fig. 4. Determining multicast routes

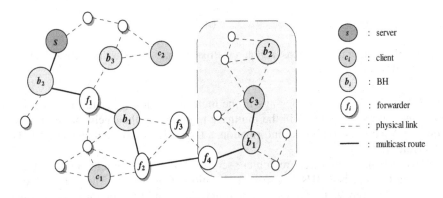

Fig. 5. Determining multicast routes across two multicast regions

Stable BHs and stable multicast routes are preferred in M-OGHAM. OGHAM and M-OGHAM differ in the computation of remaining connection time between two neighboring hosts. In M-OGHAM, the remaining connection time (dented as $t_{i,j}$) is first estimated for every pair of neighboring hosts v_i and v_j. The server (or a client) broadcasts a message over the multicast region for collecting neighboring information and the remaining connection time, in order to create (or join) a multicast group. Then, those hosts with fewer hops and longer remaining connection time to the other hosts are selected as BHs. And, the selected BHs are stable.

Recall that the BH attached by the server is responsible to determine the multicast routes from the server to the clients. In order to obtain stable multicast routes, the BHs, selected in M-OGHAM, construct an $n \times n$ matrix, denoted by N. Let $t'_{i,j}$ be the up-to-date remaining connection time between v_i and v_j, which is equal to $t_{i,j}$ minus the amount of time elapsed since $t_{i,j}$ is computed. The BH sets $N(i,j)=1$ if v_i and v_j are neighboring and $t'_{i,j} \geq \rho$, and $N(i,j)=\infty$ otherwise, where ρ is a predefined constant. Intuitively, $N(i,j)=1$ means that the link between v_i and v_j is stable. Then, stable multicast routes from the server to clients can be obtained by applying the Dijkstra's shortest-path algorithm [29] to the matrix N.

4 A Heuristic Algorithm

In this section, a heuristic algorithm that can provide a feasible solution to BSMTP is proposed. The algorithm can admit a new flow with bandwidth requirement for some multicast service by constructing a bandwidth-satisfied multicast tree. Besides, the algorithm attempts to minimize the number of forwarders for the new flow. The execution of the algorithm involves a basic procedure, named *Shortest_Routes*, which can establish shortest (minimum number of hops) routes to a particular destination. Finally, the algorithm is integrated into M-OGHAM to support bandwidth-constrained multicast services in large-scale MANETs.

Shortest_Routes has four input parameters: v_d, b_req, Δ and Φ, where v_d is a destination of the new flow, b_req is the bandwidth requirement for the new flow, and Δ, Φ are two sets of hosts. *Shortest_Routes* also has two output parameters: H and P (explained later). The execution of *Shortest_Routes* invokes another procedure, named *B_Violation*, which returns *true* if bandwidth violation happens and *false* else. *B_Violation* uses Δ and Φ for checking bandwidth violation, which will become clear later. *Shortest_Routes* is a modification of the well-known Dijkstra's shortest-path algorithm, which can compute the shortest paths from a node to the other nodes. Differently, *Shortest_Routes* establishes shortest routes in a reverse direction from all hosts to v_d. For each host v_i in the shortest routes, let h_i denote the minimum number of hops from v_i to v_d, and P_i denote the set of forwarders along the shortest v_i-v_d route. Initially, set $h_i = 1$ and $P_i = \{v_i\}$ if v_i is neighboring to v_d and the inclusion of v_i in the shortest routes will not cause bandwidth violation. Otherwise, set $h_i = \infty$ and $P_i = \{\}$ (the empty set). Host v_i will (will not) cause bandwidth violation if $B\_Violation(b\_req, \Delta, \{v_i\} \cup \Phi)$ returns *true (false)*.

The shortest routes are established iteratively by a repeat-until loop. In each iteration, a host v_x that is closest to v_d (i.e., h_x is minimum) and was not selected before is determined. Then, for each neighbor v_j of v_x, the v_j-v_d route (i.e., P_j) is replaced with the v_x-v_d route augmented with the v_j-v_x hop (i.e., $P_x \cup \{v_j\}$), if the new v_j-v_d route is shorter and does not cause bandwidth violation. The latter can be checked by invoking $B\_Violation(b\_req, \Delta, \{v_j\} \cup P_x \cup \Phi)$. Finally, the shortest routes are represented by means of H and P, which are two sets containing all h_is and P_is, respectively. *Shortest_Routes* is detailed in the following.

```
Procedure  Shortest_Routes(v_d, b_req, Δ, Φ, H, P);
    /* V is the set of all hosts. */
    X←V-{v_d};
    for  each v_i ∈ X  do
        if  v_i and v_d are neighboring and
             B_Violation(b_req, Δ, {v_i} ∪ Φ) = false
            then  h_i ← 1 and P_i ← {v_i}
            else  h_i ← ∞ and P_i ← {};
    repeat
            determine v_x∈X so that h_x=min{h_i|v_i∈X} and de-
            lete v_x from X;
```

```
if  hₓ ≠ ∞
    then  for  each neighbor vⱼ of vₓ  do
          if  hⱼ > hₓ + 1 and
              B_Violation(b_req,Δ,{vⱼ}∪Pₓ∪Φ) = false
              then  hⱼ ← hₓ + 1 and Pⱼ ← Pₓ ∪ {vⱼ}
until  hₓ = ∞ or X = {};
H ← {} and P ← {};
for  each vᵢ ∈ V - {vₔ}  do
     H ← H ∪ {hᵢ} and P ← P ∪ Pᵢ.
```

Next, we explain the execution of *B_Violation*. *B_Violation* has three input parameters: b_req, Δ and Φ', where b_req and Δ are inherited from *Shortest_Routes* and Φ' is a superset of Φ. The hosts in Δ (Φ') are destinations (forwarders) of the new flow. *B_Violation* first checks if the forwarders in Φ' cause bandwidth violation. If bandwidth violation happens, then *B_Violation* returns *true*. Let N_i be the set of hosts that are neighboring to v_i and b_max_i be the maximal available bandwidth of v_i. Also, let $b\_ongoing_i$ be the total bandwidth requirement of the ongoing flows that are forwarded by v_i and M be the set of destinations and forwarders of all ongoing flows. A forwarder v_i in Φ' causes bandwidth violation if its bandwidth consumption exceeds its maximal available bandwidth, i.e., $|\{v_i\}\cup(N_i\cap\Phi')|\times b\_req + \sum_{v_k\in\{v_i\}\cup N_i} b\_ongoing_k > b\_max_i$.

Let $v_j \in M$ be a neighbor of v_i. Similarly, v_j causes bandwidth violation if $|N_j \cap \Phi'| \times b\_req + \sum_{v_k\in\{v_j\}\cup N_j} b\_ongoing_k > b\_max_j$. By means of checking v_j, HRP can be avoided to the ongoing flows that pass through the neighborhood of the forwarders in Φ'. A destination $v_l \in \Delta$ causes bandwidth violation if $|N_l \cap \Phi'| \times b\_req + \sum_{v_k\in\{v_l\}\cup N_l} b\_ongoing_k > b\_max_l$. Finally, *B_Violation* returns *false*, if no bandwidth violation happens. *B_Violation* is detailed in the following.

```
Procedure  B_Violation(b_req, Δ, Φ');
    for  each vᵢ ∈ Φ'  do
        begin
            if  |{vᵢ} ∪ (Nᵢ ∩ Φ')| × b_req
                + ∑_{vₖ∈{vᵢ}∪Nᵢ} b_ongoingₖ > b_maxᵢ
            then  return  true;
            for  each neighbor vⱼ ∈ M of vᵢ  do
                if  |Nⱼ∩Φ'|×b_req + ∑_{vₖ∈{vⱼ}∪Nⱼ} b_ongoingₖ > b_maxⱼ
                then  return  true
        end;
```

```
for   each v₁ ∈ Δ  do
  if   |N₁∩Φ'|×b_req+  ∑      b_ongoingₖ >b_max₁
                      vₖ∈{v₁}∪N₁
    then   return   true;
return   false.
```

There are multiple shortest routes to v_d established by *Shortest_ Routes*. If the new flow is carried with a single route, then the bandwidth requirement can be satisfied. However, bandwidth violation may happen if the new flow is carried with two or more routes simultaneously. In the next section, by the aid of *Shortest_Routes*, an algorithm is proposed to construct a bandwidth-satisfied multicast tree.

The proposed heuristic algorithm, named *B_Satisfied_Multicast_Tree*, has three input parameters: v_s, D and b_req, where v_s is the source (i.e., the server) of the new flow, D is the set of the destinations (i.e., the clients) of the new flow, and b_req is the bandwidth requirement for the new flow. *B_Satisfied_Multicast_Tree* intends to establish a bandwidth-satisfied multicast tree for the new flow. Let $\Delta \subseteq D$ be a subset of the destinations and F be the set of forwarders in the multicast tree. Initially, set $\Delta = \{\}$ and $F = \{\}$.

The bandwidth-satisfied multicast tree is established iteratively by a for-loop. Without loss of generality, assume $D = \{ v_{d_1}, v_{d_2}, ..., v_{d_c} \}$, where $c = |D|$. During the ith iteration, a shortest route to v_{d_i} is selected from P (computed by *Shortest_Routes*) so that the selected shortest routes to $v_{d_1}, v_{d_2}, ..., v_{d_i}$ are bandwidth-satisfied. Moreover, F and Δ are updated. At first, when $i = 1$ (the first iteration), P_s is selected, which is a shortest route from v_s to v_{d_1}, and F (Δ) is replaced with $F \cup P_s$ ($\Delta \cup \{ v_{d_1} \}$). It should be noted that P_s is bandwidth- satisfied.

When $i = 2$, a forwarder $v_x \in F$ that is closest to v_{d_2} is first determined. Then, $P_x \in P$ is selected as the shortest route to v_{d_2}. Also F (Δ) is replaced with $F \cup P_x$ ($\Delta \cup \{ v_{d_2} \}$). Now F represents a tree that connects v_s with v_{d_1} and v_{d_2}. The tree is bandwidth-satisfied (hence can avoid HMRP), as a consequence of executing *Shortest_Routes*(v_{d_2}, b_req, Δ, F, H, P), where $\Delta = \{ v_{d_1} \}$ and $F = P_s$ (P_s was obtained when i=1). The execution for the other iterations (i.e., when $3 \leq i \leq |D|$) is very similar. Finally, when the execution of *B_Satisfied_Multicast_Tree* terminates, F represents a bandwidth-satisfied multicast tree that connects v_s with $v_{d_1}, v_{d_2}, ..., v_{d_c}$. *B_Satisfied_Multicast_Tree* is detailed in the following.

```
Procedure   B_Satisfied_Multicast_Tree(vₛ,D,b_req);
   Δ ← {} and F ← {};
   for  i ← 1 to |D|  do
     begin
         Shortest_Routes(v_{dᵢ}, b_req, Δ, F, H, P);
```

```
if   i=1
    then   if  h_s ≠ ∞
           then   F ← F ∪ P_s  and  Δ ← Δ ∪ { v_{d_i} }
           else   exit
    else   begin
           determine v_x ∈ F so that h_x=min{h_i | v_i∈ F};
           if  h_x ≠ ∞
               then   F ← F ∪ P_x  and  Δ ← Δ ∪ { v_{d_i} }
               else   exit
       end
end.
```

For example, refer to Fig 6, where $B\_Satisfied\_Multicast\_Tree(s, \{c_1, c_2, c_3\}, b\_req)$ is executed to establish a bandwidth-satisfied multicast tree that connects the server s with three clients c_1, c_2 and c_3. During the first iteration, $Shortest\_Routes(c_1, b\_req, \Delta, F, H, P)$ is invoked to establish a shortest route, i.e., $s-a-b-e-c_1$, to c_1, where $\Delta = \{\}$ and $F = \{\}$ (refer to Fig 6(a)). The set F (Δ) is updated to $\{\} \cup \{s, a, b, e\} = \{s, a, b, e\}$ ($\{\} \cup \{c_1\} = \{c_1\}$).

During the second iteration, $b-f-c_2$ is established as the shortest route to c_2 (i.e., $v_x = b$ and $P_x = \{b, f\}$) after invoking $Shortest\_Routes(c_2, b\_req, \Delta, F, H, P)$, because b is the closest forwarder in F to c_2, where $\Delta = \{c_1\}$ and $F = \{s, a, b, e\}$. The set F is updated to $\{s, a, b, e\} \cup \{b, f\} = \{s, a, b, e, f\}$, which represents a tree that connects s with c_1 and c_2 (refer to Fig 6(b)). The set Δ is updated to $\{c_1\} \cup \{c_2\} = \{c_1, c_2\}$.

Similarly, during the third iteration, $e-g-c_3$ is established as the shortest route to c_3 after invoking $Shortest\_Routes(c_3, b\_req, \Delta, F, H, P)$, because e is the closest forwarder in F to c_3, where $\Delta = \{c_1, c_2\}$ and $F = \{s, a, b, e, f\}$. A bandwidth-satisfied multicast tree that connects s with c_1, c_2 and c_3 is represented by $F = \{s, a, b, e, f\} \cup \{g\} = \{s, a, b, e, f, g\}$ (refer to Fig 6(c)).

For most of multicast services, the number (i.e., $|D|$) of clients is considered a constant, in contrast to the number of hosts in the MANET. Since different permutations of v_{d_1}, v_{d_2}, ..., v_{d_c} will result in different multicast trees established, we execute $B\_Satisfied\_Multicast\_Tree(v_s, D, b\_req)$ $|D|!$ times, each for a different ordering of clients. Then the multicast tree with minimal number of forwarders is selected.

$B\_Satisfied\_Multicast\_Tree$ can be easily adapted to M-OGHAM when the BHs determine bandwidth-satisfied multicast trees. Recall that the adjacency of hosts is necessary to $B\_Violation$ (a basic procedure of $B\_Satisfied\_Multicast\_Tree$). Also, all $b\_ongoing_j$s (the total bandwidth requirement of ongoing flows that are forwarded by v_i) and the set M (the set of destinations and forwarders of all ongoing flows) are necessary to $Shortest\_Routes$ and $B\_Violation$ (the other two basic procedures of $B\_Satisfied\_Multicast\_Tree$). Hence, they should be made available before $B\_Satisfied\_Multicast\_Tree$ is invoked by the BHs.

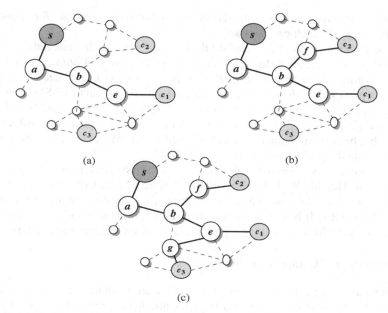

Fig. 6. A bandwidth-satisfied multicast tree. (a) A route from s to c_1. (b) A tree connecting s with c_1 and c_2. (c) A tree connecting s with c_1, c_2 and c_3.

Recall that a sever (or a client) v_i broadcasts a message over a multicast region for selecting BHs. The hosts v_js within the multicast region reply their neighboring information and remaining connection time to v_i. In this way, the BHs selected by v_i can compute the adjacency of hosts. Further, v_j also reply $b\_ongoing_j$ to v_i if v_j is a destination or forwarder of some ongoing flow. So, $b\_ongoing_j$s and M can be derived by the BHs.

5 Simulation

Simulation is implemented using the Network Simulator 2 package (ns-2) [28]. IEEE 802.11 is used as the MAC layer protocol. Data/control packets are sent using the un-slotted Carrier Sense Multiple Access with Collision Avoidance (CSMA/CA). The simulation environment models a large-scale MANET of 200 hosts which are randomly spread in a 2000m×2000m area. Each host is equipped with a radio transceiver whose transmission range is up to 250 meters over a wireless channel. The data transmission capability of each host is assumed 800 Kbps. Forty runs with different seed numbers are conducted for each scenario and collected data for these runs are averaged.

Three performance measures: receiving rate, admission rate and number of control packets are adopted. The receiving rate is the ratio of the number of data packets received by clients to the number of data packets delivered from servers. If bandwidth violation happens, it will drop drastically. The admission rate is the ratio of the number of multicast groups admitted to the number of multicast groups requested. When the admission rate goes up, the network performance increases. On the other hand, the number of control packets can reflect the overheads incurred for constructing/maintaining the multicast routes.

For convenience, we use Heu-M-OGHAM to denote the integration of M-OGHAM with *B_ Satisfied_Multicast_Tree*. Since *B_Satisfied_Multicast_Tree* is heuristic, an optimal algorithm for BSMTP is necessary in order to evaluate the performance of *B_Satisfied_Multicast_Tree*. In [27], the authors formulated BSMTP as a 0/1 integer linear programming, which can be well solved by a branch-and-bound algorithm (see [26]). We use Opt-M-OGHAM to denote the integration of M-OGHAM with such a branch-and-bound algorithm. Opt-M-OGHAM can serve as a benchmark for evaluating the performance of Heu-M-OGHAM.

The simulation is performed with two aspects. First, performance comparison is made among Heu-M-OGHAM, Opt-M-OGHAM and MCEDAR under the assumption of static hosts. Second, the same performance comparison is made for mobile hosts. MCEDAR [15] is a representative two-tier multicast protocol and an extension to CEDAR [12] which is a routing algorithm for QoS applications in MANETs.

5.1 Performance Comparison: Static Hosts

Ten multicast groups, denoted by G_1, G_2, ..., G_{10}, are randomly created. Each multicast group consists of one server and three clients. A flow requiring 50 Kbps is sent from the server to the clients. The simulation proceeds for 1000 seconds; G_1 starts first, G_2 starts after 100 seconds elapsed, G_3 starts after 200 seconds elapsed, and so on. Fig 7 and Fig 8 compare the admission rates and receiving rates of Heu-M-OGHAM, Opt-M-OGHAM and MCEDAR under the assumption of static hosts.

Refer to Fig 7. When the number of multicast groups is smaller than 5, Heu-M-OGHAM has higher admission rates than MCEDAR, as a consequence that the multicast trees constructed by it have fewer forwarders. On the other hand, when the number of multicast groups exceeds 5, Heu-M-OGHAM has lower admission rates

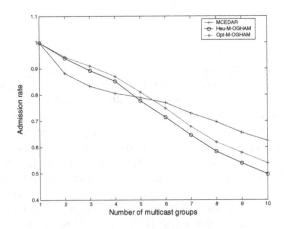

Fig. 7. Admission rate for static hosts

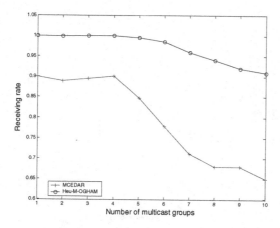

Fig. 8. Receiving rate for static hosts

than MCEDAR. Recall that the probability that MCEDAR induces HMRP jumps up when the number of multicast groups exceeds 5. Also, Fig 8 demonstrates that the receiving rate of MCEDAR drops drastically when the number of multicast groups exceeds 5. When the number of multicast groups increases to 10, the receiving rate of Heu-M-OGHAM is still high, but the receiving rate of MCEDAR drops seriously.

When there are feasible solutions to BSMTP, Opt-M-OGHAM can always determine the optimal one. However, it is likely that Heu-M-OGHAM fails to find a feasible one. Fig 7 also reveals the success rate of Heu-M-OGHAM in finding a feasible solution to BSMTP. The success rate can be estimated by means of the ratio of the admission rate of Heu-M-OGHAM to the admission rate of Opt-M-OGHAM.

5.2 Performance Comparison: Mobile Hosts

Host mobility is based on the random waypoint model [25], in which a host's movement consists of a sequence of random length intervals, called *mobility epochs*.

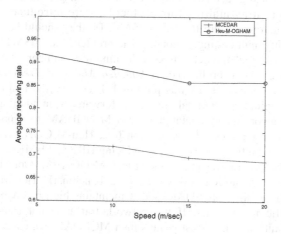

Fig. 9. Receiving rate for mobile hosts

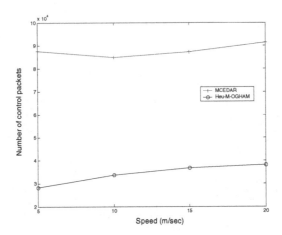

Fig. 10. Number of control packets for mobile hosts

During each epoch, a host moves in a constant direction and at a constant speed. The simulation proceeds for 1000 seconds and the speed varies from 5 to 20 meters per second (or 18 to 72 kilometers per hour). The first seven multicast groups, i.e., G_1, G_2, ..., G_7, that were created in Section 5.1 are used in the simulation.

Figure 9 compares the receiving rates of Heu_M_OGHAM and MCEDAR under the assumption of mobile hosts. Like the previous situation of static hosts, Heu_ODMRP has higher receiving rates than MCEDAR. Figure 10 shows that Heu_ M_OGHAM generates fewer control packets than MCEDAR. Control packets are generated for constructing a new bandwidth-satisfied multicast tree if the current multicast tree cannot satisfy the bandwidth requirement.

6 Conclusions

HRP and HMRP, which are two bandwidth-violation problems that may occur to previous QoS routing/multicasting protocols, were introduced in this paper. Since bandwidth and power are limited in MANETs, they should be taken into consideration in routing/multicasting protocols. The problem (i.e., BSMTP) of determining a bandwidth-satisfied multicast tree with minimum number of forwarders was studied in this paper. An algorithm (i.e., *B_Satisfied_Multicast_Tree*) that can generate bandwidth-satisfied multicast trees was proposed. It constructed multicast trees with the objective of minimizing the total number of forwarders, in addition to bandwidth satisfaction. The algorithm was integrated into M-OGHAM to support bandwidth-constrained multicast services. The integration (i.e., Heu-M-OGHAM) can construct bandwidth-satisfied multicast trees without inducing HRP and HMRP.

Performance comparison was made among Heu-M-OGHAM, Opt-M-OGHAM and MCEDAR, where Opt-M-OGHAM served as a benchmark. Heu-M-OGHAM has higher receiving rates than MCEDAR, and it maintains high receiving rates even if the network traffic is saturated. If the network traffic is not saturated, Heu-M-OGHAM can admit more multicast groups than MCEDAR, which is a consequence that the multicast trees constructed by it have fewer forwarders. Moreover, the admission rate of Heu-M-OGHAM is close to the benchmark.

References

[1] Low, C.P., Song, X.: On finding feasible solutions for the delay constrained group multicast routing problem. IEEE Transactions on Computers 51, 581–588 (2002)

[2] Kompella, V.P., Pasquale, J.C., Polyzos, G.C.: Multicast routing for multimedia communicationl. IEEE Transactions on Computers 51, 581–588 (2002)

[3] Sun, Q., Langendoerfer, H.: Multicast routing for multimedia communication. In: Proceedings of the Second Workshop on Protocols for Multimedia Systems, pp. 452–458 (1995)

[4] Corson, M.S., Batsell, S.G.: A reservation-based multicast (RBM) routing protocol for mobile networks_ initial route construction phase, ACM/Baltzer Wireless Networks 1(4), 427–450 (1995)

[5] Belding-Royer, E.M., Perkins, C.E.: Transmission range effects on AODV multicast communication. ACM/Kluwer Mobile Networks and Applications 7, 455–470 (2002)

[6] Xie, J., Talpade, R.R., Mcauley, A., Liu, M.: AMRoute: adhoc multicast routing protocol. ACM/Kluwer Mobile Networks and Applications 7, 429–439 (2002)

[7] Gupta, S.K.S., Srimani, P.K.: Cored-based tree with forwarding regions (CBT-FR), a protocol for reliable multicasting in mobile ad hoc networks. Journal of Parallel and Distributed Computing 61(9), 1249–1277 (2001)

[8] Chan, K., Nahrstedt, K.: Effect location-guided tree construction algorithms for small group multicast in MANET. In: Proceedings of the 21st International Annual Joint Conference of the IEEE Computer and Communications Societies vol. 3, pp. 1180–1189 (2002)

[9] Lee, S.J., Gerla, M.: On-demand multicast routing protocol in multihop wireless mobile networks, ACM/Kluwer Mobile Networks and Applications 7, 441–453 (2002)

[10] Garcia-Luna-Aceves, J.J., Madruga, E.L.: The core-assisted mesh protocol. IEEE Journal on Selected Areas in Communications 17, 1380–1394 (1999)

[11] Kozat, U.C., Kondylis, G., Ryu, B., Marina, M.K.: Virtual dynamic backbone for mobile ad-hoc networks. Proceedings of the IEEE International Conference on Communications 1, 250–255 (2001)

[12] Sinha, P., Sivakumar, R., Bhanghavan, V.: CEDAR: a core-extraction distributed ad-hoc routing algorithm. IEEE Journal on Selected Areas in Communications 17, 1454–1465 (1999)

[13] Sivakumar, R., Das, B., Bharghavan, V.: Spine routing in ad-hoc networks, Cluster Computing, a special issue on mobile computing 1(2), 237–248 (1998)

[14] Jaikaeo, C., Shen, C.C.: Adaptive backbone-based multicast for ad hoc networksx. Proceedings of the IEEE International Conference on Communications 5, 3149–3155 (2002)

[15] Sinha, P., Sivakumar, R., Bhanghavan, V.: MCEDAR: multicast core-extraction distributed ad-hoc routing. In: Proceedings of the IEEE Wireless Communications and Networking Conference, pp. 1313–1317. IEEE, Los Alamitos (1999)

[16] Yang, Y., Kravets, R.: Content-aware admission control for ad hoc networks. IEEE Transactions on Mobile Computing 4(4), 363–377 (2005)

[17] Chen, L., Heinzelman, W.: Qos-aware routing based on bandwidth estimation for mobile ad hoc networks. IEEE Journal on Selected Areas in Communications 23(3), 561–572 (2005)

[18] Chen, S., Nahrstedt, K.: Distributed quality-of-service routing in ad hoc networks. IEEE Journal on Selected Areas in Communications 41, 120–124 (1999)

[19] Xue, Q., Ganz, A.: Ad hoc QoS on-demand routing (AQOR) in mobile ad hoc networks. Journal of Parallel and Distributed Computing 41, 120–124 (2003)

[20] Pagani, E., Rossi, G.P.: A framework for the admission control of QoS multicast traffic in mobile ad hoc networks. In: Proceedings of the ACM International Workshop on Wireless Mobile Multimedia, pp. 3–12 (2001)

[21] Hu, C.C., Wu, E.H.K., Chen, G.H.: OGHAM: On-Demand Global Hosts for Mobile Ad-Hoc Multicast Services, accepted for Ad Hoc Networks

[22] Lim, H., Kim, C.: Multicast tree construction and flooding in wireless ad hoc networks. In: Proceedings of the ACM International Workshop on Modeling, Analysis and Simulation of Wireless and Mobile Systems, pp. 61-68 (2000)

[23] Bharghavan, V., Demers, A., Shenker, S., Zhang, L.: MACAW: a media access protocol for wireless LAN's. In: Proceedings of ACM SIGCOMM, pp. 212–225 (1994)

[24] Hu, C.-C., Wu, E.H.-K., Chen, G.-H.: Mobility-aware on-demand global hosts for ad-hoc multicast. In: Lu, X., Zhao, W. (eds.) ICCNMC 2005. LNCS, vol. 3619, pp. 375–384. Springer, Heidelberg (2005)

[25] Bettstetter, C., Resta, G., Santi, P.: The node distribution of the random waypoint mobility for wireless ad hoc. IEEE Transactions on Mobile Computing 2(3), 257–269 (2003)

[26] Geoffrion, A., Marsten, R.: Integer programming algorithms: a framework and state- of-the-art survey. Management Science 18, 465–491 (1972)

[27] Hu, C.-C., Wu, E.H.-K., Chen, G.-H.: Bandwidth-satisfied multicast trees in MANETs. IEEE International Conference on Wireless and Mobile Computing, Networking and Communications (WiMob) 3, 323–328 (2005)

[28] Network Simulator (Version 2): http://www-mash.cs.berkeley.edu/ns/

[29] Dijkstra, E.W.: A note on two problems in connection with graphs. Numerische Mathematik 1, 269–271 (1959)

Localising Multicast Using Application Predicates

Ian Wakeman[1], Stephen Cogdon, Laurent Mathy[2], and Michael Fry[3]

[1] Dept of Informatics, University of Sussex, Brighton, UK
[2] Dept of Informatics, Lancaster University, Lancaster, UK
[3] School of IT, University of Sydney, Sydney, Australia

Abstract. In this paper, we investigate how to incorporate an application metric into the construction of a multicast tree so as to facilitate the use of range constrained multicast. We first describe the construction and delivery protocols, show through an analysis drawing on stochastic geometry that the protocol is scalable, and provide simulations showing the performance of the protocol against trees derived from reverse path forwarding construction.

1 Introduction

Multicast is widely accepted as being a useful tool in the construction of distributed applications. However, for economic reasons, and because it is believed that the amount of state needed to fully deploy native multicast is infeasible in current router designs, native multicast has been deployed only in small sections of the Internet. There has therefore been a wide set of research on application level multicast, such as HMTP [1], HostCast [2], switch-trees [3], DCMALTP [4], NICE [5], Narada [6] and TBCP [7]. These protocols aim to provide the facilities for construction of an overlay network for distributing data across the group. All of these protocols improve network performance by mapping application requirements into the delivery of messages. In this work we aim to extend this approach to design further, and investigate the benefit that can be obtained by utilising application metrics in the construction of the delivery tree, and in the propagation of messages.

Our motivating application has been to develop a protocol to allow communication between map servers for virtual worlds. Macedomia et al [8] propose that a distributed world be split across different servers so that each node serves a different geographical locality. They then propose that events be disseminated only to those nodes which need to know about them because active objects in the world managed by those nodes are able to detect the events. In any distributed world, the density of activity is uneven, and for efficiency reasons the density of nodes is uneven. We wish to design a protocol that efficiently disseminates events to those nodes that should receive them.

We assume that nodes within the application can provide a real-valued measure of the *distance* between the nodes. This measure can be anything, such as the measured latency of packet delivery, the similarity between the content in a

D. Hutchison and R.H. Katz (Eds.): IWSOS 2007, LNCS 4725, pp. 193–207, 2007.
© Springer-Verlag Berlin Heidelberg 2007

file sharing system, or a measure of the degree of trust between two nodes. We then use this measure to provide an enhanced application level multicast that can localise the delivery of messages to those nodes that meet predicates on the measure, allowing services such as locate the set of nodes which are trusted to a certain degree, or replicate content to nodes which already have similar content to provide an efficient storage redundancy. In this paper, we consider solely the *less than* ($<$) operator, and we leave the consideration of more sophisticated predicates to later work.

A massive multiplayer game (MMORPG) such as World of Warcraft is a topical example of a distributed virtual world in need of scaling. Ratnasamy et al's [9] state that "Multiple servers are each assigned a region of the virtual world and relay communication between players. Such scaling can be problematic and indeed numerous reports cite overloaded servers affecting the user experience." We believe that designs such as the ones we posit can solve such problems.

Besides distributed virtual worlds, other candidate applications that may benefit from our protocol are those with the following characteristics:

1. There are distance measures between the nodes
2. There are likely to be enough nodes or sufficient churn in group membership[1] such that full knowledge of the group is infeasible
3. The nodes want to ask queries of each other based on their distance metrics.

In the rest of this paper we first describe the protocols, then provide a mathematical analysis using techniques from stochastic geometry to show the scalability of the system. We then provide simulations which confirm our mathematical analysis, and then describe simulations showing that the protocols can provide an improved service over native shortest path first multicast for some significant set of queries.

2   Description of the Protocols

Before we develop our own protocol, it is instructive to first consider two alternative approaches. Our aim is to deliver a message to all members of a group which have an application distance from the sender less than some figure. We assume that the distances are time invariant, and require at least one exchange of messages to calculate the distance. The first approach is to calculate the pairwise distances between each sender and potential receiver, and then to distribute the message to each receiver. This would require each sender to track the membership of the group, and then to initiate a message exchange with each receiver to discover the distance between the nodes. Although the measured distances can be cached, the initial delay before sending the message to all receivers will be large.

An alternative approach is to deliver the message first, and then measure the distances to see if the message should be delivered. If we assume that the

---

[1] We do not address maintenance in this paper, but the techniques of TBCP [7] can be used to repair the tree.

Record: Node
1: $List < Address >: children$
2: $double : distanceTo$
3: $double : coi$
Procedure: probe$(cand) : Node$
4: **return** $children, distanceTo(cand), coi$
Procedure: join$(cand)$
5: $children.add(cand)$
6: **for all** $c \in children$ **do**
7:   $c.addSibling(cand)$
8: **end for**
Procedure: addSibling$(cand)$
9: $siblings.add(cand)$
Procedure: testChildren$(Node : adopter)$
10: $closest \leftarrow$ **null**
11: $minD \leftarrow$ **null**
12: **for all** $c \in adopter.children$ **do**
13:   $n \leftarrow c.probe(this)$
14:   **if** $n.d < n.coi$ **then**
15:     **if** $n.d < minD$ **then**
16:       $closest \leftarrow n$
17:       $minD \leftarrow n.d$
18:     **end if**
19:   **end if**
20: **end for**
21: **if** $closest \neq$ **null then**
22:   **if** $testChildren(closest)$ **then**
23:     **return true**
24:   **else**
25:     $becomeChild(closest)$
26:     **return true**
27:   **end if**
28: **else**
29:   **return false**
30: **end if**
Procedure: $becomeChild(p)$
31: $parent \leftarrow p$
32: $coi \leftarrow distanceTo(p)/2$
33: $p.join($**this**$)$

Fig. 1. Tree construction algorithm

message can be delivered over a source-rooted shortest path tree and that routes are symmetric, then the total cost in traffic will be the cost of the multicast tree plus the number of message exchanges to measure distance. Whilst the initial cost of dissemination of events is cheap, the cost of distance calculation is high.

Our approach has been to incorporate the calculation of distances into the construction of the delivery tree, implicitly holding the distance measurements

Procedure: propagate$(msg, radius)$
1: **if** $radius == 0$ **then**
2:    $deliverLocal(msg)$
3: **end if**
4: **for all** $c \in children$ **do**
5:    **if** $distanceTo(c) + c.coiLimit > radius$ **then**
6:       $c.propagate(msg, 0)$
7:    **end if**
8: **end for**
Procedure: checkDistance$(msg, radius, node)$
9: **if** $distanceTo(node) ¡ radius$ **then**
10:    $node.propagate(msg, 0)$
11: **else if** $distanceTo(node) - node.coi ¡ radius$ **then**
12:    $newRadius \leftarrow distanceTo(node) - radius$
13:    $node.propagate(msg, newRadius)$
14: **end if**
Procedure: checkSiblings$(msg, radius)$
15: **for all** $s \in siblings$ **do**
16:    $checkDistance(msg, distance, s)$
17: **end for**
Procedure: propagateUp$(msg, radius, distance)$
18: $checkSiblings(msg, radius + distance)$
19: **if** $distance < radius$ **then**
20:    $deliverLocal(msg)$
21: **end if**
22: **if** $parent \neq$ **null then**
23:    $rad \leftarrow radius + distance$
24:    $dist \leftarrow distance + distanceTo(parent)$
25:    $parent.propagateUp(msg, rad, dist)$
26: **end if**
Procedure: $send(msg, radius)$
27: $distance \leftarrow distanceTo(parent)$
28: $parent.propagateUp(msg, radius, distance)$
29: $checkSiblings(msg, radius)$
30: **for all** $c \in children$ **do**
31:    $checkDistance(msg, radius, c)$
32: **end for**

Fig. 2. Distance constrained propagation

in the tree topology, and using this information to localise the delivery of the message. This reduces the number of nodes needing to exchange messages to measure distances.

As in nearly all multicast protocols, our protocol can be split into a tree construction phase, shown in Figure 1, and a message propagation phase, shown in Figure 2.

In the algorithmic descriptions of the protocols in Figures 1 and 2, the $distanceTo$ method will return a real valued number representing the distance from the calling to the called node. This method is application specific, and may or may not make additional calls over the network to determine the distance. We assume that the distance comes from a metric space in that

$$A.distanceTo(A) = 0 \tag{1}$$

$$A.distanceTo(B) = B.distanceTo(A) \tag{2}$$

$$A.distanceTo(B) + B.distanceTo(C) \geq A.distanceTo(C) \tag{3}$$

We discuss how we use this assumption below.

Each node has a *circle of influence* (coi). When a node is probing to join the tree, it can only become a child of a node if the distance is less than the circle of influence. The circle of influence is a function of the distance between a node and its parents. In this paper, we assume that this is always set to $distanceTo(parent)/2$, since analysis shows that this provides good scalability. Other approaches for setting the *coi* are for future investigation.

Our tree construction is an abstracted version of the tree building control protocol (TBCP) presented in [7]. When a node wishes to join the tree, it contacts the root node, which returns a list of its children. The joining node will then query all of the children to discover in which nodes' circle of influence the requesting node lies, and which of these nodes is closest. If such a node is found, then it will attempt to join the children of the closest node. If no such node is found, then it will become a child of the current best candidate.

At the end of the tree construction phase, we will have built a hierarchical tree, in which we have the guarantee that the children of a node are all closer than its circle of influence. Each node will have a list of its siblings and its immediate children, and the relevant application distances to these nodes. We will show in Section 3 that is $O(log(n))$ in the amount of state required.

When a node N wishes to send a message, it specifies the surrounding *radius* of delivery of the message. The message shall go to at least all nodes n for which $n.distanceTo(N) < radius$. The sending node knows the distance to all of its children, siblings and parent, and the size of their coi. It can thus immediately discount those nodes whose children cannot fall within the radius of delivery. The sending node always propagates upwards to its parent unless the parent is the root node, since the parent may have siblings whose coi overlaps with the delivery radius. If a sibling falls within the radius of delivery, it is asked to propagate the message to all of its children, since the children may be within the radius of delivery. If a node falls outside of the radius of delivery, but its coi overlaps with the radius, then it is asked to propagate to those nodes which have a distance such that they would lie within the ring around the node whose inner diameter is tangent upon the radius.

Since we are using a distance metric, for two siblings A and B at distance d_{AB}, and A is sending a message at radius R for a node C at distance d_{AC}, then the children of B must be at least at a distance of $d_{AB} - coi_B$ by Equation (3). Therefore, if $R + d_{CA} < d_{AB} - coi_B$ then the children of B cannot fall within the radius. If $R + d_{CA} < d_{AB}$ but $R + d_{CA} > d_{AB} - coi_B$, then the inner circle of nodes defined by at a radius from B of $d_{AB} - R - d_{CA}$ are out of the radius also by equation (3). Similar arguments follow for the remaining recalculations of distance in Algorithm 2.

As an optimisation, when a node is asked to propagate with a radius of 0 and it has a number of children, it can recalculate the distanceTo function so as to more accurately determine which children to propagate the message to.

Each node therefore stores its parent, siblings and children. As we show in section 3, the number of children of each node tends to a constant multiplied by logarithm of the area of the circle of influence, and is independent of the total number of nodes. The amount of state at any node therefore has a finite bound, and the number of message exchanges to establish the tree is worst case logarithmic in the number of nodes.

3 Algorithm Analysis

In this section, we analyse a model of the cluster tree algorithm. The nodes wishing to join the tree are assumed to appear randomly upon a 2 dimensional plane, as part of a Poisson point process Φ with intensity λ, and the *distance* between nodes is calculated as the straight line distance between the points. We assume without loss of generality that the root of the tree is based at the origin. Each node has a circle of influence which is a function of its distance from its parent $h(r)$. When a new node appears, if it is within the circle of influence of an existing node, then it will recursively attempt to join the children of that node. If it cannot join any children, then it will become a new child of that node. This process starts at the root node.

We follow the analysis of the algorithm as a *Hard-core point process* from [10]. Each node is marked with a random number, $m(x)$, between 0 and 1. This mark can be viewed as an indication of when the node appeared in the process and thus its age. We then use a *thinning* process on the nodes to remove those younger nodes which would be subsumed in the subtree formed under a node to create a new point process Φ_h. A node is retained if there is no other point marked with a smaller random number (i.e. older) in the retained process close enough such that the first point would fall within the circle of influence of the second point. We denote the volume induced around x in which no point must lie at distance $r(x)$ by $h(.)$ as $b(r(x), h(.))$. Formally, the thinned process is given by

$$\Phi_h = \{x \in \Phi : m(x) < m(y) \; \forall y \in \Phi_h \cap b(r(x), h(.))\}$$

The intensity of Φ_h is obviously dependent upon the distance from the origin and is given by:

$$\lambda_h(r) = p_h(r)\lambda$$

An approximation of the *Palm retaining probability* $p_h(r)$ is calculated by integrating over all values of the random variable $m(x)$ and calculating the probability that there are no other points in the induced volume. We are therefore ignoring the probability that a point falls within the induced volume, but is captured by some older point.

$$p_h(r) = \int_0^1 \exp(-\lambda b(r, h(.))t)dt$$

In the simple case where $h(.)$ is a constant value H, then we have

$$p_h = \int_0^1 \exp(-\lambda\pi H^2 t)dt$$

$$= \frac{1 - \exp(-\lambda\pi H^2)}{\lambda\pi H^2}$$

The intensity of Φ_h, λ_h, is calculated from

$$\lambda_h = p_h\lambda$$

$$= \frac{1 - \exp(-\lambda\pi H^2)}{\pi H^2}$$

λ_h is interpreted as the number of nodes per unit area on the plane that are directly attached to the root node.

In general, the calculation of $b(r, h(.))$ is not a nice problem. However, it turns out that when $h(.) = r/2$, the solution is particularly neat. If we consider the diagram in Figure 3, we can show that

$$\cos(\phi) = \frac{a^2 - h^2(a) + r^2}{2ar}$$

$$\cos(\phi) = \frac{\frac{3}{4}a^2 + r^2}{2ar}$$

If we return to Cartesian Coordinates, remembering that $a^2 = x^2 + y^2$ and $\cos(\phi) = y/(x^2 + y^2)^{1/2}$ then

$$y = \frac{3/4x^2 + 3/4y^2 + r^2}{2r}$$

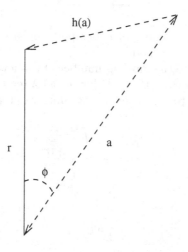

Fig. 3. Calculation of the induced area

To calculate the area as a function of r, we express the curve describing the induced area as a function of y, and then integrate with respect to y between the two roots of the polynomial.

$$x = (8/3ry - 4/3r^2 - y^2)^{1/2}$$

$$b(r, r/2) = \int_{2r/3}^{2r} (8/3ry - 4/3r^2 - y^2)^{1/2} dy$$

Completing the square, substituting $u = (y - 4r/3)$ and using standard integral tables

$$b(r, r/2) = \int_{2r/3}^{2r} (4r^2/9 - (y - 4r/3)^2)^{1/2} dy$$

$$= \int_{-2r/3}^{2r/3} (4r^2/9 - u^2)^{1/2} du$$

$$= [u(4r^2/9 - u^2)^{1/2} + 4r^2/9 \sin^{-1}(3u/2r)]_{-2r/3}^{2r/3}$$

$$= \frac{4}{9}\pi r^2$$

We can then plug this back into our equation for $p_h(r)$,

$$p_h(r) = \frac{1 - \exp(-\lambda\pi\frac{4}{9}r^2)}{\lambda\pi\frac{4}{9}r^2}$$

The instantaneous intensity of Φ_h, λ_h, is calculated from

$$\lambda_h(r) = p_h(r)\lambda$$

$$= \frac{1 - \exp(-\lambda\pi\frac{4}{9}r^2)}{\pi\frac{4}{9}r^2}$$

λ_h is interpreted as the number of nodes per unit area on the plane that are directly attached to the root node.

If we wish to calculate the limiting number of top level nodes in a particular area, we simply integrate over the area, having let λ tend to infinity. Therefore in the ring from radius d_1 to d_2, the expected number of top level nodes, $N(d_1, d_2)$ is given by

$$N(d_1, d_2) = \int_{d_1}^{d_2} \frac{2\pi r}{\frac{4}{9}\pi r^2} dr$$

$$= \frac{9}{2} \int_{d_1}^{d_2} \frac{1}{r} dr$$

$$= \frac{9}{2} [\log(r)]_{d_1}^{d_2}$$

$$= \frac{9}{2} \log(d_2/d_1)$$

For levels other than the top, the distribution of nodes is dependent not only on the radius from the parents, the number of overlapping nodes, but on the point in the process when the node appeared in the tree. For these reasons, the processes are not amenable to analysis, but we believe they can be approximated by the top level analysis.

Choosing the sphere of influence $h(.) = r/2$ gives logarithmic scaling properties in the expected number of top level nodes. However, this choice of function does suffer from a pole at the origin, and so the number of top level clusters is infinite if the radius is drawn from $(0,d)$. If there is some limit on the minimum radius, then the number of clusters will tend to some limit, calculable from above.

3.1   Expected Message Complexity

When a new node wishes to join the tree, it will query the root node for the set of children who it may need to join. For the case of the circle of influence being $r/2$, this includes any children of the root whose distance from the root, d, falls in the following range:

$$2r > d > 2r/3$$

The expected number of top level nodes the joining node will have to query can be calculated as:

$$\int_{2r/3}^{2r} 2\pi x \lambda_h dx$$

$$= \int_{2r/3}^{2r} 2\pi x \frac{1 - \exp(-\lambda \pi \frac{4}{9} x^2)}{\pi \frac{4}{9} x^2} dx$$

$$= 9/2 \int_{2r/3}^{2r} 1/x - \frac{\exp(-\lambda \pi \frac{4}{9} x^2)}{x} dx$$

$$= 9/2(\log(3r) - (Ei(-\lambda \pi \frac{16}{9} r^2) - Ei(-\lambda \pi \frac{16}{27} r^2)))$$

which in the limit as the intensity tends to infinity, shows that the number will increase as the logarithm of the distance from the root.

Exact analysis of the distribution of points below the top level is made difficult by the multiple dependencies as described above. Instead, we appeal to the following argument to characterise the number of children for any given node. As shown above, the intensity of the point distribution follows an inverse square law dependent upon the distance from the parent to a first approximation for the top level. We assume that the intensity of nodes on other levels can also be approximated by an inverse square law. We can argue that since the area of the *coi* is proportional to the square of the radius, when we calculate the expected number of nodes in the *coi* of any node, the two dependencies will cancel out, and the number of children of any node is constant to a first approximation.

The number of levels beneath any node will be dependent upon the logarithm of the distance from its parent.

4   Simulations

To demonstrate the effectiveness of the algorithm, our control will be the source-routed shortest path tree using native routing, with and without the calculation of the cost of the return paths to the source node. We argue that these are the most useful controls since they illustrate how the algorithm performs when there need be no negotiation between nodes to determine the distance between nodes,

Table 1. Transit Stub generation parameters

10 stubs per transit domain
0.1 probability of extra transit-stub edge
0.1 probability extra stub-stub edge
10 transit domains per graph fully connected to each other
20 nodes per transit domain modelled as a random graph with edge probability of 0.8
10 nodes per stub domain, modelled as random graphs with edge probability 0.5

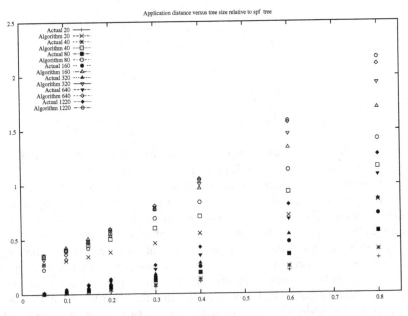

Fig. 4. Ratio of tree size for the algorithmic determination of delivery tree and the actual tree against the shortest path tree for all group members, when distance comes from position on a sphere

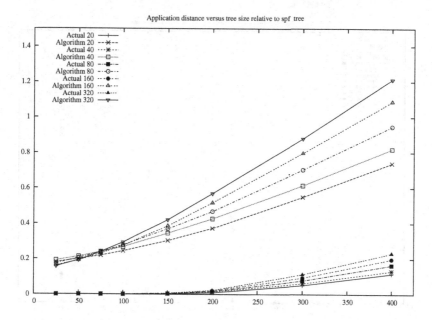

Fig. 5. Ratio of tree size for the algorithmic determination of delivery tree and the actual tree against the shortest path tree for all group members, when distance comes from weights of connecting links

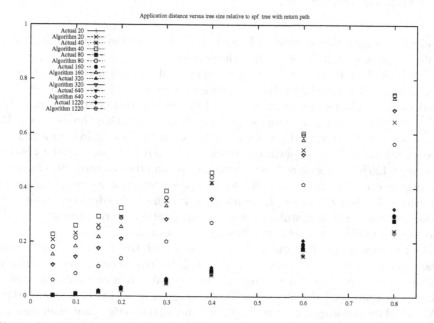

Fig. 6. Ratio of tree size for the algorithmic determination of delivery tree and the actual tree against the shortest path tree and the return path for all group members, when distance comes from position on a sphere

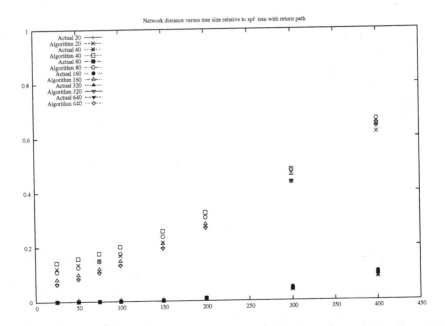

Fig. 7. Ratio of tree size for the algorithmic determination of delivery tree and the actual tree against the shortest path tree for all group members, when distance comes from weights of connecting links

and when negotiation is needed. In each case, we show the ratios of the size of delivery graphs constructed against the control.

We used four transit stub networks generated by gt-itm [11] consisting of 20200 nodes, generated from the parameters shown in Table 1. We do not claim that these topologies are representative of the current state of the Internet, but they are similar to the topologies used in other work within the area [12]. We have conducted two experiments with application metrics. In experiment 1, we have used an arbitary application metric completely unrelated to the network topology. Each node is placed on the unit sphere, and the distance between nodes is measured as the great circle diameter between their corresponding positions on the nodes. In experiment 2, we used the length of the edges connecting the nodes within the transit stub graph as the application metric, showing how the protocols can construct network efficient structures.

Our measurements first discover the shortest path tree from the source to all the other nodes within the group. The size of this tree is calculated as the sum of the weights of each unique edge in the tree. We then calculate the size of the application level delivery tree. This is composed of the intra-group edges. We calculate the length of each individual intra-group edge, and then sum the values of each intra-group edge. Note that since there is an underlying transit stub graph, some nodes will be used to carry the same traffic multiple times as the traffic traverses transit domains.

In both experiments, we vary the group and distances as shown. We then have chosen each group ten times from the possible nodes, and have chosen ten source nodes per group. We have then repeated this process for each of the four graphs. Thus each point in the graph is the mean of 400 samples. The sets of distribution radii used in the graphs are chosen so that over half the nodes in the group are covered for the largest radius.

There are two factors working against the efficiency of the cluster tree algorithm. If we use an application metric which is unrelated to the network, then the overlay tree may be very inefficient, with large degrees of network stretch. In Tan et al [12] and in [13], they show that the basic TBCP algorithm will construct trees which are up to twice as large as those produced by a shortest path spanning tree. When an arbitrary application metric is used, these inefficiencies increases with the number of members of the group. Also, the delivery algorithm is working with incomplete information, and will deliver to a number of nodes which do not need the message. As the radius of delivery increases, the number of excess nodes will also increase.

In Figure 4, we see the ratios of tree size for the arbitrary metric when there is no return path incorporated into the size. For small groups, such as 20 nodes, the algorithm is more effective since the discrepancy in tree efficiencies is low, and the algorithm has fewer false postives to discount. However, as we move to large groups, then the inefficient tree and the excess nodes make the algorithm less efficient than a basic tree delivery at a delivery radii of around 30% of nodes.

This is borne out when we use network metrics as the distance metric as in Figure 5. The underlying constructed tree is now closer to the topology of the minimum spanning tree, and the efficiencies now arise from the excess nodes. The lower lines marked as being from the *actual* nodes show the size of the subtree constructed only from the nodes actually within the delivery radius. The *algorithm* trees then become less efficient as we increase the number of excess deliveries due to incomplete information. These would be reduced on subsequent deliveries as cached distance figures become available.

In Figures 6 and 7, we show the relative ratios of tree size when a single return message to the source is required. The cost of the shortest path control is now dominated by the set of return paths. Thus the algorithm is efficient even when the overlay tree is inefficient, since it reduces the number of return messages. In figure 6, the largest groups have the best performance ratio, since the return paths grow linearly in size as the group size grows for the control, whereas the number of excess nodes in the algorithm grows sub-linearly.

5 Related Work

Besides the application layer multicast described in Sec 1, a similar approach to the design of overlays comes from the work on distributed hash tables (DHT). DHTs provide an efficient lookup mechanism for information which is necessarily distributed, providing a layer of indirection. Multicast in systems such as Scribe [14] use the DHT routing as the primary route, and then for multicast *topics* or addresses form a tree using reverse path forwarding. Utilising a cluster tree building approach over the dht may solve some of the problems discussed in [15], and is a promising level of investigation for building multicast over a CAN style DHT [16]. If we were to solve the problem of range delivery predicates within the framework of a DHT, then we would look to solutions such as Mercury [17], in which multiple range predicates are implemented using separate overlay networks for each attribute.

Our approach is similar to the formation of quadtrees and r-trees [18], where the tree structure is defined spatially by creating a hierarchy of rectangles over the space, although we we work in any metric space rather than solely in a Euclidean space. The use of such structures is natural for decomposing sensor networks as in [19].

6   Conclusions

In this paper, we investigate how to incorporate an application metric into the construction of a multicast tree so as to facilitate the use of range constrained multicast. We first describe the construction and delivery protocols, show through an analysis drawing on stochastic geometry that the protocol is scalable, and provide simulations showing the performance of the protocol against trees derived from reverse path forwarding construction.

In future work, we intend to investigate how the protocol can be made resilient to churn, and to determine its performace in real applications.

References

1. Zhang, B., Jamin, S., Zhang, L.: Host multicast: A framework for delivering multicast to end users. In: IEEE INFOCOM (2002)
2. Li, Z., Mohapatra, P.: Hostcast: A new overlay multicasting protocol. In: IEEE Int. Communications Conference (ICC), IEEE, Los Alamitos (2003)
3. Helder, D., Jamin, S.: End-host multicast communication using switchtrees protocols. In: Global and Peer-to-Peer Computing on Large Scale Distributed Systems (2002)
4. Banerjee, S., Kommareddy, C., Kar, K., Bhattacharjee, B., Khuller, S.: Construction of an efficient overlay multicast infrastructure for real-time applications. In: IEEE INFOCOM, San Francisco, USA, IEEE Computer Society Press, Los Alamitos (2003)
5. Banerjee, S., Bhattacharjee, B., Kommareddy, C.: Scalable application layer multicast. In: SIGCOMM '02: Proceedings of the 2002 conference on Applications, technologies, architectures, and protocols for computer communications, October 2002, vol. 32(4), pp. 205–217. ACM Press, New York (2002)
6. Yang-hua, C., Rao, S., Seshan, S., Zhang, H.: A case for end system multicast. Selected Areas in Communications, IEEE Journal on 20(8), 1456–1471 (2002)
7. Mathy, L., Canonico, R., Hutchison, D.: An overlay tree building protocol. In: Networked Group Communication, London, UK, pp. 76–87 (2001)

8. Macedomia, M., Zyda, M., Pratt, D., Brutzman, D., Barham, P.: Exploiting reality with multicast groups: a network architecture for large-scale virtual environments. In: Virtual Reality Annual International Symposium (VRAIS'95), p. 2 (1995)

9. Ratnasamy, S., Ermolinskiy, A., Shenker, S.: Revisiting ip multicast. In: SIG-COMM (2006)

10. Stoyan, D., Kendall, W.S., Mecke, J.: Stochastic Geometry and its Applications. Wiley, Chichester (1987)

11. Zegura, E.W., Calvert, K.L., Bhattacharjee, S.: How to model an internetwork. In: IEEE Infocom, vol. 2, pp. 594–602. IEEE, San Francisco, CA (1996)

12. Tan, S.-W., Waters, G., Crawford, J.: A performance comparison of self-organising application layer multicast overlay construction techniques. Computer Communications 29, 2322–2347 (2006)

13. Cogdon, S.: Application-level multicast for group communication. Ph.D. dissertation, University of Sussex (2003)

14. Castro, M., Druschel, P., Rowstron, A.: Scribe: A large-scale and decentralized application-level multicast infrastructure. IEEE Journal on Selected Areas in Communications 20(8) (October 2002)

15. Bharambe, A., Rao, S., Padmanabhan, V., Seshan, S., Zhang, H.: The impact of heterogeneous bandwidth constraints on dht-based multicast protocols. In: The Fourth International Workshop on Peer-to-Peer Systems (February 2005)

16. Ratnasamy, S., Francis, P., Handley, M., Karp, R., Schenker, S.: A Scalable Content-Addressable Network. In: Proc of SIGCOMM, ACM Press, New York (August 2001)

17. Bharambe, A.R., Agrawal, M., Seshan, S.: Mercury: supporting scalable multi-attribute range queries. SIGCOMM Comput. Commun. Rev. 34(4), 323–366 (2004)

18. Guttman, A.: R-trees: A dynamic index structure for spatial searching. In: Proc. ACM SIGMOD International Conference on Management of Data, pp. 45–57 (1984)

19. Soheili, A., Kalogeraki, V., Gunopulos, D.: Spatial queries in sensor networks. In: GIS '05: Proceedings of the 13th annual ACM international workshop on Geographic information systems, pp. 61–70. ACM Press, New York (2005)

Cost Aware Adaptive Load Sharing

David Breitgand[1], Rami Cohen[2], Amir Nahir[3], and Danny Raz[4]

[1] IBM Haifa Research Lab, Israel
davidbr@il.ibm.com
[2] CS Department, Technion, Haifa, Israel
ramic@cs.technion.ac.il
[3] IBM Haifa Research Lab, Israel
nahir@il.ibm.com
[4] CS Department, Technion, Haifa, Israel
danny@cs.technion.ac.il.

Abstract. We consider load sharing in distributed systems where a stream of service requests arrives at a collection of n identical servers. The goal is to provide the service with the lowest possible average waiting time. This problem has been extensively studied before, but most previous models have not incorporated the monitoring costs explicitly. This paper focuses on a rigorous study of maximizing the utility of monitoring.

We extend the Supermarket Model for dynamic load sharing by explicitly incorporating monitoring costs. These costs stem from the fact that the servers have to answer load queries, a task which consumes both CPU and communication resources. This Extended Supermarket Model (ESM) allows us to formally study the tradeoff between the usefulness of monitoring information and the cost of obtaining it. In particular, we prove that for each service request rate, there exists an optimal number of servers that should be monitored to obtain minimal average waiting time.

Based on this theoretical analysis, we develop an autonomous load sharing scheme that adapts the number of monitored servers to the current load. We evaluate the performance of this scheme using extensive simulations. It turns out that in realistic scenarios, where monitoring costs are not negligible, the self-adaptive load balancing scheme is clearly superior to any load-oblivious load sharing mechanisms.

1 Introduction

Consider a service that is being provided by a set of servers over the network. The goal of the service provider is to provide the best service (say, to minimize the service time) given the amount of available resources (*e.g.*, the number of servers). The provider can add a load sharing system (for example as suggested in RFC 2391 [1]) and improve the service time. If incoming service requests are routed to the executing servers according to the pre-computed probabilities, the load sharing scheme is static. Static load sharing schemes are simple and have

D. Hutchison and R.H. Katz (Eds.): IWSOS 2007, LNCS 4725, pp. 208–224, 2007.
© Springer-Verlag Berlin Heidelberg 2007

minimal control overhead. Unfortunately, static load sharing is poorly equipped to cope with dynamic environments, where the actual availability of each server changes over time.

To address this issue effectively, the load sharing mechanism needs to adapt to the current global state of the system. This implies that updated load information needs to be collected from the servers. Handling such load information requests requires small but nonzero resources (*e.g.*, CPU) from each server. This may reduce the actual service rate of the server, and thus, it is not easy to predict the actual amount of improvement expected from preferring a specific configuration of a dynamic load sharing scheme. For this reason it is important to identify just the right amount of resources that should be allocated to monitoring of the servers' load in order to maximize the overall system performance. Moreover, since the optimal amount of monitoring depends on external parameters (such as the arrival rate of the service requests stream), the system should self-adjust the amount of monitoring according to the current conditions.

A very efficient load sharing model was studied in [2] by Mitzenmacher. This model, termed the *supermarket model*, uses a very simple randomization strategy: when a new task arrives, $d < n$ servers are selected uniformly at random, and the task is assigned to the server with the shortest queue among these d chosen servers. For $d = 1$ this process simply assigns jobs (we refer to service requests as jobs throughout this paper) to servers uniformly at random, regardless of their load. However, for $d = 2$, the job is assigned to the least loaded server (the one with the shortest queue) among the two randomly chosen servers. It is shown in [2] that this simple process results in an exponential improvement of the expected overall time in the system compared to the (load independent) random assignment scheme. Further increasing d improves the expected service time linearly. The results of [2] suggest that even a very small amount of management information coupled with random job assignment may lead to a very efficient adaptive load balancing strategy, and as we use more information we keep on improving the scheme. However, this study assumes that the information about the local load of the servers is obtained and processed at no cost. As explained above, in many scenarios this assumption is not realistic.

In this paper we focus on developing a rigorous analytical model for adaptive load sharing where the model explicitly accounts for the inherent monitoring overhead and allows to study the effects of non-negligible monitoring cost in a quantitative way. Using this model, we address the question: what is the right amount of monitoring that is needed to maxs imize the global benefit from employing adaptive load sharing. Our approach is as follows.

We extend the aforementioned supermarket model by incorporating the management costs into it. In particular, we assume that when a server is polled about its load, it has to allocate resources in order to answer this query. The *Extended Supermarket Model (ESM)* allows us to rigorously study this intuitively obvious tradeoff between the usefulness of monitoring information and the cost of its maintenance. In other words, when a server answers a monitoring inquiry, it can spend less CPU cycles on processing the actual service requests. An important factor is the ratio between the time it takes a server to answer a load request and the mean expected service time of a job. This *load monitoring efficiency ratio*, denoted by C, reflects the amount of disturbance the monitoring task has on the actual service. The overall capacity reduction of a server is also proportional to the number of monitoring inquires per time unit, which depends linearly on d. When d increases we have more information and thus the expected average time in the system decreases thanks to better load sharing. On the other hand, each server becomes more affected by load queries and thus the service time (and therefore also the overall time in the system) increases.

The main outcome of this theoretical analysis is that for each system load and monitoring efficiency ratio C, there exists an optimal number d^* of servers that should be monitored in order to minimize the overall expected time of jobs in the system. One of the corollaries of this finding is that knowing more about the global state of the system through detailed monitoring may be not only useless, but actually harmful to the total quality of service.

The centralized load sharing device can poll d servers when it receives a service request (as described above), but it can also poll the servers load periodically. This may reduces the number of load queries but, depending on the polling rate, may affect the load balancing quality due to the staleness of the data. We study the tradeoff between the staleness and reduced overhead imposed by the periodic updates. Our results indicate that under high load conditions, using a periodic updates scheme does not help reducing the monitoring overhead considerably. The reason is as follows. Unless the periodic updates are performed frequently enough, the local information about other servers' load rapidly becomes obsolete and the quality of load sharing reduces dramatically.

Another important aspect that should be addressed is the dynamic behavior of the environment. In practical systems, many parameters such as the job requests rate (i.e., the system load), traffic load, and the availability of servers varies over time. Thus, the optimal number of servers that should be queried changes over time. An adaptive load sharing system is a system that self adapts its working point according to the current environment parameters. Such a system should be able to accurately approximate the value of the relevant parameters, and to dynamically adjust the number of the monitored servers to the optimal value. The most relevant parameter in our case is the overall system load. Note that the load results from both the mean time between arrivals and the service time. However, the actual service time is known only to the servers and not to the dispatcher. Hence estimating the current system load is not trivial.

In today's complex systems, management costs are inherent. Even though they can be reduced by configuring a system in a different way, or by using a different algorithm, they cannot be avoided completely. This paper is a first step toward better understanding the tradeoff between monitoring cost and its benefit for adaptive load sharing. To make the presentation more coherent and focused we concentrate on studying the centralized load balancing schemes. However, the reader referred to [3] for a discussion of the distributed Extended Supermarket Model.

The rest of this paper is organized as follows. In Section 2 we describe related work. In section 3 we formally define the model for the framework, explain the basic approach, and analyze the expected performance. We evaluate optimal working points in Section 3.1 and study periodic updates in section 3.2. In Section 4 we describe the adaptive scheme that adjust the amount of monitoring to the actual load and evaluate its performance. We conclude in Section 5 with a short discussion of our results.

2 Related Work

Multiple aspects of load sharing and load balancing were extensively studied in a variety of contexts over the last three decades. Due to space limitations we concentrate in this section on the prior art which is most directly relevant to this paper. The reader is encouraged to refer to [3] for a more extensive treatment of the literature.

In our study we concern ourselves with *non-preemptive* load sharing. In non-preemptive load sharing, tasks may be transferred from one server to another only upon entering the system and prior to their execution start. However, when a process starts executing, it cannot be moved to another computer.

An important example of non-preemptive load sharing was presented in [4]. In this solution, a unified cost model for heterogenous resources was presented. Using this cost model, the "marginal cost" of adding a task to a given destination is computed. A task is dispatched to a destination that minimizes the total marginal cost. In a sense, this is a variation of a greedy algorithm. However, thanks to its sophisticated cost model, this solution outperforms simple greedy algorithms. Note that the monitoring costs involved in marginal cost evaluation are not accounted explicitly.

Another approach that combines a threshold-based method with the greedy strategy in an interesting way was presented in [5]. The primary goal of this work is to achieve an autonomic adaptable strategy for peer-to-peer overlay maintenance with QoS guarantees in spite of possible server crashes. This paper, similarly to many other papers in the field, considers average time in the system as the performance parameter that should be minimized. The load balancing scheme of [5] was studied through simulations. Even though the monitoring process plays a pivotal role in [5], its cost is not explicitly accounted in these simulations.

In [6] the authors considered the case in which monitoring costs are not negligible and studied the effects of this assumption on the efficiency of adaptive load sharing. The authors compared a static load sharing scheme to a dynamic adaptive one based on load thresholds. In this scheme, if the local load of a server exceeds a pre-specified level (threshold), a new task is transferred to a server with minimal expected load. The expectations of the server loads are based on the shared data structure. The value of load threshold can be adapted dynamically to influence task redistribution policy of load balancing. The authors study their threshold based scheme under various conditions to quantitatively estimate relative importance of load threshold values and frequency of load updates and find the best empiric values for these parameters. Although an insightful study, its conclusions rely on simulations alone and, therefore, are difficult to generalize.

Monitoring costs, pertaining to network latency, delay, routes availability, *etc.*, were also considered in [7]. The authors claim that monitoring cost can be significantly reduced through cooperative measurements as opposed to uncoordinated active probing from multiple network locations. Similarly to [6], [7] resorts to simulation analysis in order to study the effects of non-negligible monitoring and provides no generic framework for a systematic study of the monitoring overhead.

In their milestone paper [8], Azar *et. al.* introduced the supermarket model (see previous section) for evenly spreading a finite number of items among n identical locations. Azar *et. al.* showed that by simply making two random choices ($d = 2$) and selecting a location with the least number of items already assigned (ties are broken arbitrarily), one reduces the maximal number of items per single location exponentially. Further increasing d results in a linear decrease of the maximal number of items per location. Azar *et. al.* studied a closed finite system in which items never leave the system and the load balancing process terminates when all items are assigned to their locations.

An application of the supermarket model to dynamic load balancing, where an infinite number of service requests arrive from a stream with a given traffic intensity and where clients leave the system once their request is serviced, was done by Mitzenmacher [2]. In this work, an infinite stochastic supermarket model is limited by a set of deterministic differential equations describing the queue length dynamics. It turns out that results similar to [8] hold, but the analysis is much more complicated.

As was already noticed in the introduction, the supermarket model of [2] does not take the cost of acquiring the local state of the servers into account. In [9], a question of how often the local state of the servers should be polled, is addressed. It turned out that obtaining a closed form solution for the expected average time in the system is difficult in this case. Therefore, [9] resorted to an extensive simulation study, showing that randomness is a powerful tool to curb *herding effect*, which becomes a dominant factor in performance degradation as the system state is acquired periodically with decreasing rate. The effect of periodic updates on the ESM performance is studied in this paper in Section 3.2.

The cost of monitoring is an integral part of the overall cost of the adaptive load sharing algorithm. Clearly, a trade-off exists between the quality of monitoring and the cost of acquiring the monitoring data. On the one hand, the more updated is the monitoring data, the higher is the total quality of load balancing [9,10,11,3]. On the other hand, since monitoring takes small, but non-zero amount of computational and communication resources, its inherent cost becomes a limiting factor for scalability. Although this trade-off has been long noticed, no formal study of it was performed until this work [3]. In contrast to other adaptive load sharing mechanisms, our solution is based on the predictive model that explicitly takes monitoring costs into account, striving to the provably optimal behavior.

3   The Extended Supermarket Model (ESM)

Following the supermarket model, we consider a system that consists of n identical servers. Each server processes its incoming service requests according to the FIFO discipline. Jobs arrive to the system in a Poisson stream of rate $\lambda \cdot n$, $0 < \lambda < 1$, service time is exponentially distributed with mean 1. In the centralized ESM model depicted in Figure 1, all clients' requests arrive at a centralized load balancing device. This device then selects $d < n$ servers uniformly at random (with replacement) and sends d inquiries about the length of the server queue to each of the selected servers. These monitoring requests have a precedence over the actual service requests, i.e., upon receiving a monitoring request, the server preempts the currently running job (if such exists) and answers the load request immediately. We assume that processing the monitoring request takes a fraction $0 < C < 1$ of the mean service time of the actual service. This factor is the load monitoring efficiency ratio that reflects the fraction of the resources (i.e., CPU) needed in order to answer a load request. When the load balancing device (dispatcher) obtains all d answers (we assume that there are no message losses), it selects the server with the minimal queue length (ties are broken arbitrarily) and forwards the job to this server. In some practical implementations the servers load monitoring is performed periodically rather than per-request. We study periodic updates in Section 3.2.

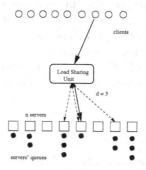

Fig. 1. The Extended Supermarket Model

Note that we do not model the time it takes the load inquiries to get back to the centralized load sharing device, or the processing time for the decision. The actual time in the system starts when the job arrives to the selected server. In this model, the system is fully described by the time it takes a server to answer a load query. This time depends on the system architecture; the load query can be handled by a separate thread or even a separate process. The load request query time is composed of receiving the load query, parsing it, getting the requested data, and sending an answer back to the dispatcher. In addition it may contain context switching time in the server's operating system. In all cases, the server must allocate resources to this task, resources that could be otherwise used to serve real client requests. The load monitoring efficiency ratio reflects the affect of this architecture with respect to the average service time of jobs.

The mean service time and thus also the mean service rate is 1. However, since the servers answer load queries immediately (thus suspending the service to regular requests), the *effective* service rate is smaller than 1. Job requests arrive at rate $\lambda \cdot n$ (where the time unit is such that the service time is one) and each job creates d load queries. Since we have n servers and they are chosen uniformly at random, each server gets a Poisson load query stream with incoming rate of $d \cdot \lambda$. Thus, handling each load query takes C time units and the effective service rate is:

$$\mu' = 1 - \lambda \cdot d \cdot C \ . \tag{1}$$

Let $\rho = \lambda/\mu'$ be the arrival rate normalized by the effective service rate; we get:

$$\rho = \frac{\lambda}{1 - \lambda \cdot d \cdot C} \ . \tag{2}$$

Clearly, ρ must be smaller than one in order to keep the system stable. In other words, if we want to have a stable state where the queues in the system have finite lengths we must have $\rho < 1$, or

$$\lambda < \frac{1}{1 + d \cdot C} \ . \tag{3}$$

Note that Equation 3 indicates that in some cases where the load is high, the monitoring process can push the system into an unstable state. This means that in some cases we would be better off without any monitoring whatsoever. In this case, a random server selection is performed.

We follow the footsteps of [2] and define $n_i(t)$ to be the number of servers with exactly i jobs in their queue at time t (This includes the job that is being served). Next we define

$$s_i(t) = \sum_{k=i}^{\infty} \frac{n_k(t)}{n} \tag{4}$$

to be the fraction of the servers with at least i jobs in the queue. When not needed we omit t; clearly $s_0 = 1$ and s_1 is the fraction of non empty servers.

It is easy to see that:

$$\sum_{i=1}^{\infty} s_i(t) = \frac{1}{n} \sum_{i=1}^{\infty} i \cdot n_i(t) \tag{5}$$

is the average queue length at time t.

For any $d > 1$, C, finite n, and λ that satisfy Equation 3, the system is in stable state, and when t is large enough there is a fixed probability to be in each of the states defined by the vector $(s_0, s_1, s_2, \ldots)$.

In such a case, a new job joins a server with a queue of size i only if all d chosen servers have queues not smaller than i, and at least one of them has a queue of size i. This happens with probability $s_i^d - s_{i+1}^d$. Similarly, the probability that a job is finished from a server with queue of size i is $s_i - s_{i+1}$. This implies that the following differential equation holds for $i \geq 1$:

$$\frac{ds_i}{dt} = \rho(s_{i-1}^d - s_i^d) - (s_i - s_{i+1}), \tag{6}$$

where $s_0 = 1$. This set of equations has a unique fix point (see (1) in [2]).

$$s_i = \rho^{\frac{d^i - 1}{d - 1}}. \tag{7}$$

In order to compute the expected time a job spends in the system, we could use the method described in Section 2.4 of [2], or use Little's Theorem [12], and divide the expected queue length by λ. Note that we need to divide by λ and not by ρ since we want to have the expected time in the system in units of the service rate and not of the effective service rate that depends on C and d. We get:

$$T_d(\lambda) = \frac{1}{\lambda} \sum_{i=1}^{\infty} \rho^{\frac{d^i - 1}{d - 1}} = \frac{1}{1 - \lambda \cdot d \cdot C} \sum_{i=1}^{\infty} \rho^{\frac{d^i - 1}{d - 1} - 1} =$$

$$T_d(\lambda) = \frac{1}{1 - \lambda \cdot d \cdot C} \sum_{i=1}^{\infty} \rho^{\frac{d^i - d}{d - 1}} \tag{8}$$

Before we proceed to the simulation analysis of $T_d(\lambda)$, we would like to develop some intuition about the function's behavior. As shown in [2], in the original supermarket model, $T_d(\lambda)$ is a monotonically decreasing function for any $0 < \lambda < 1$. However, Equation 8 suggests that as d increases, the denominator decreases, resulting in higher waiting times, and when $\lambda \cdot d \cdot C$ approaches 1, the waiting time in the system goes to infinity. Thus, we expect $T_d(\lambda)$ to decrease when d increases, but at some point as d keeps on increasing, the value of $T_d(\lambda)$ should increase as the servers put more and more resources into monitoring the load and this affects the service time.

In order to verify that the Extended Supermarket Model (ESM) indeed models correctly the behavior of server systems as described in the previous sections, we conducted an extensive set of simulation runs. This is done using an in-house event driven simulation software. The Centralized ESM is simulated for different load, d, and C values.

The left hand side of Figure 2 is obtained through simulating the Centralized ESM with 300 servers for $C = 0.003$. We observed typical service time of 300-400 milliseconds when monitoring a Web based application called PlantsBy-WebSphere on top of IBM WebSphere 6.0 application server, under normal load conditions. The load query time of 1 millisecond was observed in our preliminary experiments conducted on a realistic testbed. Due to space limitations, we do not describe these experiments here. We plotted the simulation results and the model prediction obtained from Equation 8 on the same graph. Each simulation sequence contained $1,000,000$ jobs, and each value in the graph is the average of 10 such runs, where the values of the standard deviation were well below 1%

of the obtained service times for all points except for $\lambda = 0.85$ and $d > 40$ where the standard deviation values were up to 20%.

One can see that the expected behavior indeed realized. The expected time in the system decreases when d increases, and at some point it starts increasing. When the system load is low (e.g., $\lambda = 0.55$) the value of $T_d(\lambda)$ drops from 1.34 for $d = 2$ to about 1.02 for $d = 10$, and then it increase almost linearly as d increases. This is due to the fact that when d increase from 2 to 10 the load sharing quality increases, while the effect of the cost devoted to monitoring is still small, however at this point, the return from increasing the number of peers becomes negligible while the effect of the monitoring cost keeps on increasing linearly.

When the load is higher (e.g., 85%) one can see that there is a point in which $T_d(\lambda)$ starts increasing very fast. This happens when the system approaches the instability point predicted by Equation 3. In this area the results becomes much more noisy, and the standard deviation of the simulation results increase.

The parameters chosen here reflect normal working conditions of distributed server systems. A cluster size of 300 servers may seem to be large, but this is not entirely far-fetched. Even today there exist many organizations that deploy 300 and more server farms (e.g., Yahoo, Google, etc.). The value of the load monitoring efficiency ratio (C) was chosen to be 0.003, as explained above. One can observe that the model predicts the system behavior very accurately in this range even when we used $n = 300$ servers in the simulation and the model deals with $n \to \infty$. The effect of the number of servers n is very small; all we need is a large enough ($n > 100$) number. This is due to the fact that the system load is $n \cdot \lambda$ and therefore the average load on each server depends only on the load parameter. In fact, we ran several simulations for different values of n and, similarly to the results of [2], the behavior is the same but the accuracy of the simulation improves as n increases.

The right hand side of Figure 2 depicted the simulated results and the model prediction for high load values. Again $C = 0.003$ and each point is the average of 10 runs each produced from a set of $1,000,000$ jobs. The number of peers in

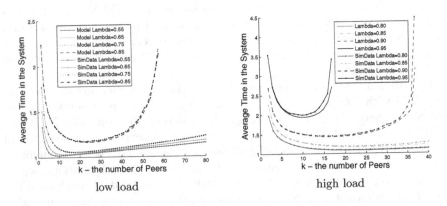

low load   high load

Fig. 2. Model Vs. simulation

this case goes from 2 to 40, and the number of servers was 500. One can observe similar behavior, and the precision of the model is quite good, except when ρ approaches one, and the system becomes less stable.

3.1   Optimal Load Balancing in ESM

In order to get the best d for a fixed set of parameters one needs to get the derivative of $T_d(\lambda)$ with respect to d. This is a rather complex expression and it has no closed formulation. Figure 3 depicts the optimal number of peers (d) as a function of the system load for different load monitoring efficiency ratio (C) values. One can see that for relatively high values of C (i.e., $C = 0.02$ where a load query takes 2% of a job service time) the optimal number of peers increases slightly as the load increases over 50% but it does not go over 10, and at around $\lambda = 0.7$ it start decreasing. This is happening since the the monitoring cost is relatively high, and cannot be justified by the improvement in performance due to better load balancing. Furthermore, when load is high, the number of peers must be small in order to keep the system stable (as indicated by Equation 3). For low values of efficiency ratio (i.e., $C = 0.001$ where a load query takes only 0.1% of a job service time) the picture is quite different. In this case the optimal d increases significantly as the load increases and reaches a value of 46 at $\lambda = 0.9$. This is due to the fact that the monitoring cost is relatively low and it is worth to improve the average time in the system by improving the load sharing quality. When the load increase over 90%, the cost effect becomes dominant and thus the optimal number of peers decreases sharply.

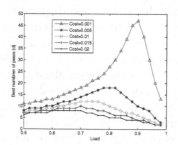

Fig. 3. Best number of peers as a function of the load

3.2   Periodical Update Model – A Simulation Study

In practical implementations, the monitoring information (in our case the load of the servers) is often acquired using periodic updates. This is done in order to conserve communication and processing resources of the network. One of the more important generic management problems is determining an optimal frequency at which the management information should be gathered.

The tradeoff here is intuitively clear. If we poll the servers' load often, then the monitoring cost is high, but the load is up to date, and therefore the load balancing decisions are improved. On the other hand, the less frequent are the updates, the lower is the cost of monitoring, but then the local information becomes stale, and the overall quality of the load balancing process decreases. We use a simulation study to investigate the effect of periodic updates in ESM, when monitoring cost is part of the model. We consider the centralized ESM, but the load sharing device uses periodic load polling instead of the per job load polling mechanism. Every q units of time, all servers are polled for their load (i.e., their queue size), this information is kept in a local database at the load sharing device. When a new job arrives, the load sharing device chooses d servers uniformly at random, and forwards the job to the server that has the smallest queue according the information in the local database. As in the centralized ESM, the load answer by each server during the load poll takes a fraction $0 < C < 1$ of the mean service time of the actual service request. However, unlike the centralized ESM case, using a larger value for d does not create more load on the servers since the load is a function of the update rate $\frac{1}{q}$, and d only affects the number of lookups in the local database.

It is important to note that the time the job is waiting for the server assignment in the centralized load sharing device is not part of the total time in the system as defined for the ESM model in Section 3. We simulated a cluster of 300 servers, processing 100,000 jobs that arrive from Poisson streams with different traffic intensities (i.e., loads). Each simulation run is repeated 10 times and the average of these runs were computed to produce the results. Similarly to Section 3, we used the load monitoring efficiency ratio $C = 0.003$.

The left hand side of Figure 4 shows dependency of the average time in the system on d (number of monitored peers) for traffic intensity $\lambda = 0.55$. Different curves on the graph correspond to different load update periods. The per-job load polling policy serves as a baseline. As evident from the figure, for this case (low traffic intensity), periodic updates do not yield substantial savings even for large values of d. Moreover, the actual mean times in the system obtained using periodic polling are marginally higher than the baseline for smaller d's. One can also observe that in the tested parameters there is no benefit to use d values that are larger than 20 since increasing d only gives marginal improvement in the average time in the system.

There are two primary reasons for this behavior. To start with, when $\rho = 0.55$, the queues are empty most of the time. Therefore, on the one hand, using CPU cycles for updating the load bulletin board delays only a small fraction of jobs. On

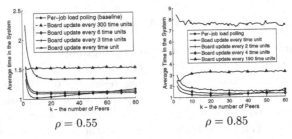

$$\rho = 0.55 \qquad\qquad \rho = 0.85$$

Fig. 4. Average Time in the System for ESM with Periodic Updates

the other hand, the value of the information gained from the load updates is low. In fact, most of the time these updates show zero length queues at many servers. The second reason has to do with the fact thet the periodic updating of the load bulletin board is done synchronously at *all* servers. Thus, the probability that jobs that considerably deviate from the mean service time would get preempted and delayed are marginally higher than those in the case when *only a fraction* of servers is polled for their queue length upon a new job arrival.

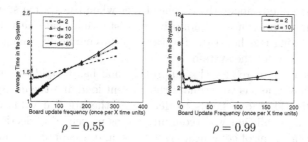

$$\rho = 0.55 \qquad\qquad\qquad \rho = 0.99$$

Fig. 5. Average Time in the System as a Function of Update Rate

One may think that when the system becomes more heavily loaded, the advantages of the periodic updates would become evident. The right hand side of Figure 4 shows the system behavior for $\rho = 0.85$. Indeed, for a large fixed value of d (60), using the appropriate update rate (once every 4 time units) periodic updates provided almost a factor of 2 improvement compared to the per-job load query case. However, for smaller d values performance of the periodic update scheme was considerably worse than that of the per-job arrival polling.

Figure 5 shows the time in the system as a function of the centralized load polling period length for $\rho = 0.55$ and $\rho = 0.99$. As one can see, when the update period increases beyond a certain value, the total quality of the load balancing decreases due to staleness of the data. We observe that the trend is similar for both traffic intensities even though the pace of the service degradation due to the management data staleness is different in each case. Not surprisingly, the higher is the traffic intensity, the faster is the load balancing degradation due to longer update periods.

As simulation results show, similarly to existence of d^*, which is the optimal number of polled peers, there exist an optimal rate of polling for each fixed d. To gain minimal time in the system, the administrator or an automated policy may vary either of these two parameters. If we compare per-job polling with optimal d values as described in Section 3.1, to the optimal rate we find here, we still see that optimally tuned per-job load queries result in better mean times in the system. Hence, periodic updates can be configured in such a way that the total quality of load balancing is not hurt, but periodic updates do not improve the total system cost effectiveness. This may sound somewhat surprising, since this counters a popular belief that periodic updates in some form or another is always more cost efficient than the per task polling.

4 Self Adaptive Heuristics

In Section 3.1 we saw that given a load and a *overhead efficiency ratio* C of the system, one can derive the optimal value of d (the number of servers that are monitored). As we described before, the overhead efficiency ratio reflects the time it takes a server to answer a load query normalized by the mean service time. This means that in practice these parameters (the load and C) may change dynamically. Thus, selecting a value of d that is optimal at one point in time may utilize the system less efficiently in other times when the system state[1] changes. In this section, we present a self adaptive heuristic that deals with this problem. In particular, using this heuristic, the value of d dynamically changes according to the current state of the system.

Clearly, in order to realize this approach, one has to address two practical difficulties. First, one has to estimate the current load and the current *overhead efficiency ratio*. Then one needs to determine the optimal value of d for the current state in an efficient way (computing d as presented in Section 3 requires a considerable amount of computing resources, and it cannot be done on-line by the dispatcher).

This last point can be solved by pre-computing the optimal value of d (off-line) for several values of load and C and storing them in a table located in the dispatcher. In this case, the dispatcher can obtain d by simple lookup at this pre-computed table. Clearly, the granularity of the table, i.e., the number of entries in the table, determines the accuracy of the estimation, but since the range of d is discrete, relative small and finite number of entries are sufficient to get an accurate estimation.

On the other hand, estimating the load and the *overhead efficiency ratio* C accurately, which is critical to the performance of the system, is more difficult. The load of the system and the *overhead efficiency ratio* C depend on the average job service time, the mean time between job arrival ($MTBA$), and the number of servers. While the number of servers may be assumed to be fixed (or, at least, known to the dispatcher), the $MTBA$, and the average job length, may change often. Therefore, the $MTBA$ and the average job length should be computed dynamically by the dispatcher.

First we deal with the estimation of the $MTBA$. The dispatcher that receives all the requests can compute the $MTBA$ of jobs arrived so far. This is done in the following way. The dispatcher divides the time axis into fixed time intervals, and it computes the $MTBA$ of each time interval separately. The $MTBA$ computed by the dispatcher in the last interval is used to estimate the current $MTBA$. The length of each time interval should be long enough to contain a large number of samples (i.e. the average number of jobs arrived at each time interval should be large enough). On the other hand, if the length of the time interval is too long, the reactions of the dispatcher to $MTBA$ changes will be too slow. A typical $MTBA$ in a server farm that contain about 100 servers could be a few milliseconds, thus a time interval of one second contains a large number of samples and it induces

---

[1] In this context the term state refers to the current values of the load and C.

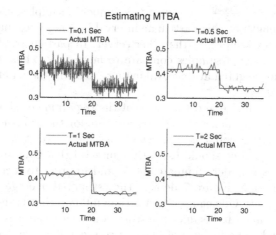

Fig. 6. Estimating the *MTBA*

a reasonable adjustment time of one second when the *MTBA* changes. Figure 6 depicts the estimated *MTBA* for different lengths of time interval. One can see that while a short interval induces fast adjustment it also creates a considerable amount of noise.

In the method described in the previous paragraph, the dispatcher estimates the *MTBA* by considering only the *MTBA* computed in the previous interval. However, one can also refer to the values of *MTBA* computed in earlier intervals by adding another parameter α that defines the importance of the last sample compared to the entire history. In this case the *MTBA* that is used in interval i is equal to

$$\alpha \cdot MTBA_{i-1} + (1 - \alpha) \cdot MTBA(i - 1). \tag{9}$$

Where $\leq \alpha \leq 1$, $MTBA_{i-1}$ denotes the *MTBA* **used** in time interval $i - 1$, and $MTBA(i - 1)$ is the *MTBA* **computed** according to interval $i - 1$ (the method described above is a special case where $\alpha = 1$). Since both parameters, α and the interval length have similar effects on the estimated *MTBA*, the value of one parameter can be eliminated by the other (i.e. a small value of α can be eliminated by selecting a short interval and vice versa).

Estimating the average length of a job is more complicated. In this case, the dispatcher that receives the requests does not know the actual service time. Only the servers that serve the requests know this information and this is only once the job processing has been completed. Thus, each server needs to compute the average job length (of jobs assigned to it), using techniques similar to those described above. When a load query arrives at a server, the server adds to its answer regarding the queue length also the average job length as computed locally during the previous interval and the number of jobs used to compute this value. Consider, for example, that the dispatcher has monitored two servers, and obtained the following job length values (computed and sent by the servers): 20 and 100 milliseconds. The dispatcher cannot conclude that the average job length, in this case, is $\frac{100+20}{2} = 60$ since the number of samples, used to compute the average job length at this particular interval in one server may be different from the number of samples in the other server. In particular, short jobs may induce a large number of samples compare to large jobs. Thus, the dispatcher must compute the weighted average, taking into account the number of samples used in each server.

The actual interval size and the value of the α parameter should be chosen in a way similar to the choice of these parameters for the $MTBA$ case. One can notice that the average job length is usually n times larger than the $MTBA$ at the dispatcher. Thus, one should use a time interval that contains a reasonable number of samples in order eliminates noise and in order to increase the accuracy of the estimation. Figure 7 depicts the estimated average job length for different interval sizes. One can see that a time interval that is shorter than 500 milliseconds may cause an insufficient estimation even when α is relative small. On the other hand, time interval of one or two seconds guarantees an accurate estimation.

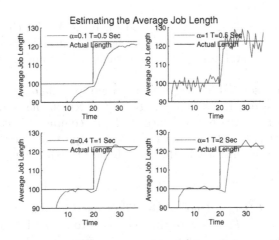

Fig. 7. Estimating the average job length

The self adaptive heuristic works as follows: The dispatcher maintains an estimated value for $MTBA$ and for the average service time. According to these values, a system parameter indicating the average reply time to a load query, and the optimal d values table, it computes the current number of servers that should be monitored. When a new job request arrives, the dispatcher queries the load of d randomly chosen servers and assigns the job to the least loaded server among them. In addition it updates its current estimation using the parameters described above. When a time interval ends, the dispatcher updates the values for the current interval using the α parameters, and using these values it finds the new optimal d value to be used.

In order to evaluate the performance of this heuristic, we simulated it, using the same configurations described in Section 3 to evaluate the performance for the fixed d cases. In particular we used $n = 300$, Average service time (job length) of 100 milliseconds, and load values between 0.8 and 0.98. The time it takes a server to answer a load query was set to be 0.3 milliseconds. As explained before this parameter depends on the system architecture and the overhead efficiency ratio is thus 0.003, similar to the one used in Section 3.

Figure 8 depicts the average response time vs. the load using the fixed values of d and with the self adaptive heuristic. For this simulation we used the values of d that are optimized with respect to a load of 0.98 and 0.8. The length of a time interval is one second and $\alpha = 0.4$ both for the (MTBA) and for the average job length. One can see that while the $d = 22$ case performs well when the load is low, when the load increases over 0.92 the system waiting time increases dramatically since the monitoring cost makes the system unstable. In a similar way if we choose $d = 5$, the system remains stable in high load but the performance when the load is 0.8 are 25% worse. The adaptive heuristic, on the other hand, performs well across all load values, and the quality of the performance is not affected by the system load.

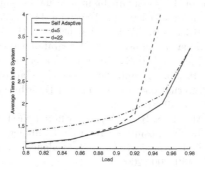

Fig. 8. Estimating system load

5   Conclusions and Future Work

The ability to quantify the benefit of a system management tool and the overhead associated with it is an important step toward developing cost effective self-enabled systems. We consider a service setting where the goal is to minimize the total average time required to provide the service to the customers. Much of the overhead associated with load balancing systems in such a setting is due to the need to monitor the load on the different servers in order to assign job requests to sub-utilized servers.

In order to understand the tradeoff between utilizing the load information and the cost of obtaining it we develop a formal model that captures both the cost of and the benefit from management processes, concentrating on load sharing among distributed servers. This model allows us to develop an adaptive scheme that adjusts the amount of the monitoring to the actual load, thus maximizing the utility of the system. The usefulness of this system in practical settings was demonstrated through extensive simulations.

This paper is just the first step in the formal study of this monitoring dilemma. One possible next step is to extend this model to a distributed setting in which load sharing is done by the servers themselves and not a separated centralized unit. Some steps in this direction are reported in [3]. Other very promising directions for further research deals with non-Poisson arrival rate, non-homogeneous servers, and the use of memory in the spirit of [13].

References

1. Srisuresh, P., Gan, D.: Load Sharing using IP Network Address Translation (LSNAT) (August 1998)
2. Mitzenmacher, M.: The power of two choices in randomized load balancing. IEEE Transactions on Parallel and Distributed Systems 12(10), 1094–1104 (2001)
3. Breitgand, D., Nahir, A., Raz, D.: To know or not to know: on the needed amount of management information. Tech. Rep. IBM-TJ-0242, IBM Research, Watson (2006)
4. Amir, Y., Awerbuch, B., Barak, A., Borgstrom, R.S., Keren, A.: An Opportunity Cost Approach for Job Assignment in a Scalable Computing Cluster. IEEE Transactions on Parallel and Distributed Systems 11(7), 760–768 (2000)
5. Adam, C., Stadler, R.: Adaptable Server Clusters with QoS Objectives. In: 9th IFIP/IEEE International Symposium on Integrated Network Management (IM 2005), Nice, France, May 2005, IEEE Computer Society Press, Los Alamitos (2005)
6. Efe, K., Groselj, B.: Minimizing control overheads in adaptive load sharing. In: Proceedings of the 9th International Conference on Distributed Computing Systems, June 1989, pp. 307–315 (1989)
7. Seshan, S., Stemm, M., Katz, R.H.: SPAND: Shared passive network performance discovery. In: USENIX Symposium on Internet Technologies and Systems (1997)
8. Azar, Y., Broder, A.Z., Karlin, A.R., Upfal, E.: Balanced allocations. SIAM Journal on Computing 29(1), 180–200 (2000)
9. Mitzenmacher, M.: How useful is old information? IEEE Transactions on Parallel and Distributed Systems 11(1), 6–20 (2000)
10. hui, C.-C., Chanson, S.T.: Improved Strategies for Dynamic Load Balancing. IEEE Concurrency 7(3), 58–67 (1999)
11. Othman, O., Balasubramanian, J., Schmidt, D.C.: Performance Evaluation of an Adaptive Middleware Load Balancing and Monitoring Service. In: 24th IEEE International Conference on Distributed Computing Systems (ICDCS), Tokyo, Japan, May 2004, IEEE Computer Society Press, Los Alamitos (2004)
12. Little, J.D.C.: A Proof for the Queueing Formula L = kW. Operations Research 9(3), 383–387 (1961)
13. Mitzenmacher, M., Prabhakar, B., Shah, D.: Load balancing with memory. In: FOCS '02: Proceedings of the 43rd Symposium on Foundations of Computer Science, Washington, DC, USA, pp. 799–808. IEEE Computer Society Press, Los Alamitos (2002)
14. Cohen, A., Rangarajan, S., Slye, H.: On the performance of TCP splicing for URL-aware redirection. In: 2nd USENIX Symposium on Internet Technologies and System, Boulder, CO, USA, October 1999 (1999)
15. IBM Developer Network. SOA and Web Services.
16. Azar, Y., Broder, A.Z., Karlin, A.R.: On-line load balancing. Theoretical Computer Science 130(1), 73–84 (1994)

Self-configuration in MANETs: Different Perspectives

Jing Wang, R. Venkatesha Prasad, and Ignas Niemegeers

EEMCS, Delft University of Technology, Mekelweg 4, 2628CD Delft, The Netherlands
{j.wang3,vprasad,ignas.n}@ewi.tudelft.nl

Abstract. The number of communication devices one uses is increasing day by day. Configuring these devices for optimal functioning, especially Mobile Ad-hoc Networks (MANETs), has not been an easy task for users. Self-configuration of the devices for optimal networking performance, being imperative, has been studied by various researchers considering the tasks at hand in various layers of the network stack. We have made a comprehensive study on self-configuration at different layers, and discussed the required interactions amongst them from different perspectives, which we think important. We break down the design complexity and identify system-level merits with the help of earlier studies. We further present our vision to provide such merits with self-configuration architecture. Our work presented here aims at understanding the potential challenges posed by the newer applications and services, and identifying research directions while providing the foundation for design of self-configurable systems.

1 Introduction

Self-Configuration is a means of organizing various devices in a network such that intended tasks requested by users are optimally performed. In this context, configuration refers to the way the network is set up, not only the configuration of individual devices. To connect a new device (such as a PDA) to the network, e.g., the appropriate network interface card needs to be configured. To be recognized by other devices a device should be properly assigned a network address. To enable the interactions with other devices, services available on devices should be advertised without human intervention. For Mobile Ad-hoc Networks (MANETs) -- which are heterogeneous, distributed and dynamic in nature – it is unrealistic to assume that users have expertise to manage such networks. A vision of autonomic communications as presented in [1] constitutes a new paradigm where network devices can configure and re-configure themselves automatically under any, predictable or unpredictable, conditions with minimal or no human intervention. Self-configuration is expected to speed up the correct responses in order to reduce the effect of network dynamics to users.

We view self-configuration in two stages. (1) *Internal* to a device, where self-configuration is an issue to optimize a set of auto-configurable parameters at different layers of the communication stack. An auto-configuration protocol defines rules for network devices executing specific configuration process to connect to other devices without fixed network infrastructure. Moreover it provides certain applications

D. Hutchison and R.H. Katz (Eds.): IWSOS 2007, LNCS 4725, pp. 225–239, 2007.

without manual input. Auto-configuration protocols are generally confined to layers. (2) *External* to a device, where interaction with external devices is also considered. Self-configuration follows the same design principle as that of network self-organization [2], but concretizes further the issues into more detailed protocols to enable interaction and coexistence among autonomous devices. That is, each self-configuration action is triggered and executed based on the awareness of situation and policies, or it is guided by sophisticated decision mechanism. In other words, the challenge comes from enabling the network to self-organize according to changing situations by re-configuring itself in a most suitable way. Thus, self-configuration is often concerned with cross-layer approaches.

Our work is motivated by the fact that most current contributions related to self-configuration issues are focusing on developing isolated layer-dependent protocols. A coherent overview is somehow missing. Further, we also believe that self-configuration can be seen from various points of view such as, at different levels of the network stack or from the user point of view. Thus it is imperative that we study the existing self-configuration techniques. Based on this overview, we bring up important requirements for self-configuration, and form the foundation by introducing self-configuration management architecture for realizing efficient ad- hoc networking. Consequently, we organize this paper based on this division. In section 2, self-configuration techniques at different levels of the network stack are enumerated. In section 3, we base our discussion on four self-configuration issues generalized from system-level. In section 4, we introduce our proposed self-configuration management architecture. In the end, we conclude in section 5.

2   Self-configuration at Different Levels

Self-configuration techniques are mostly layer specific. Often, configuration processes in the adjacent layers are closely related to each other. For simplicity, we model the OSI seven layers into three levels, with respect to connectivity, networking and services provisioning in the network, as shown in Fig.1.

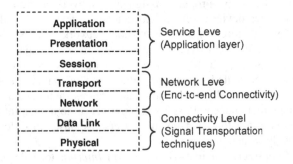

Fig. 1. Three-level model

2.1 Connectivity-Level

The connectivity level encompasses the Link and Physical layers of the OSI network stack and considers the different radio domains. Connectivity-level self-configuration is the first step for a network device to automatically set up communication links with other devices even before the formation of network. When the device is powered on, it first performs device discovery to know other devices within its transmission range - its neighbors. Device discovery process is usually specific to radio access techniques. If a device has more than one radio interface, device discovery with each interface could be performed simultaneously as long as there is no interference.

Consequently, a device could have full knowledge of its neighbors, and can potentially decide how to configure connection with them. In MANETs, such configuration benefits from *clustering*. It enables individual devices to be grouped together into virtual groups, and to form certain network topology based on specific criteria for a given scenario. Traditional metrics such as mobility and location of the devices have been exploited as cluster formation criteria [3]. Besides, learning algorithms have been explored for more sophisticated cluster formation schemes, where individual behavior is affected by either the best local or global objectives [4]. Within each cluster, different performance metrics are used for selection of cluster head, such as those based on identity (Lowest-ID) and location (Highest-Degree), or based on combined metrics of mobility, energy, degree and so on [3] [4] [5]. In this way, the network automatically builds a hierarchy and simplifies other connectivity-level configuration processes, such as channel reservation, transmission scheduling and power control etc.

2.2 Network-Level

The network level corresponds to the Network and Transport layers or the TCP/IP of the Internet architecture. At this level self-configuration is concerned with creating an end-to-end connectivity between devices, which possibly are not neighbors and in different clusters. Primary self-configuration task is for each device to obtain a valid network address in an effective way to cope with dynamics in MANETs. Depending on whether or not the addressing state is globally visible and controllable, address autoconfiguration can be identified either as stateful or stateless.

In *stateful* approaches, an allocation table is typically maintained by the cluster head, which takes the role as a dynamic host configuration protocol (DHCP) server. Several modifications have been made to take into account the dynamic network situations and to speed up the configuration process. Some examples are, refining the structure and the types of control messages to provide configuration-change-recovery mechanism [6]; combining local address selection with centralized address validation [7]. Besides, the allocation table can be split into many and may be distributed in various devices in the cluster. Different address pool splitting and updating schemes have been investigated in [8][9]. In this way, any configured device that has a part of allocation table is able to assign address to an un-configured device, thus it further enhances the robustness and scalability.

Stateless address auto-configuration protocols enable devices to configure on their own independently, and to validate their choices through agreement with other devices in the network, without necessarily having a comprehensive global view. As presented in [10], when a device joins the network, it selects an IP address by itself and performs duplicate address detection (DAD) to validate the usability of the address by broadcasting the address within the cluster. Other approaches suggest implementing on each device a distributed address generation function in such a manner to minimize the address collision as proposed in [11].

In principle, stateful approaches guarantee zero risk of address collision while stateless approaches provide more flexibility and scalability. *Hybrid* approaches are therefore proposed to combine the strength of both stateful and stateless approaches, but at the expense of complexity. For example, in [12] each device randomly selects an address from a virtual address space based on an estimation algorithm, and it further reduces the conflict probability by incorporating an address allocation table, which is maintained using cross-layer information from the traffic of routing protocol.

With valid network addresses, route discovery using ad hoc routing protocols may be used to find the routes to all the devices within or outside the cluster or even to the Internet gateway. Besides, as discussed in [7] and [8], joint consideration of addressing and routing is also able to improve the performance and simplify DAD in auto-configuration [12] [13] [14].

2.3 Service-Level

The service level corresponds to the Session and Application layers. Self-configuration at this level enables a network device to automatically provide services to- and use services from- other devices on the network. Service discovery in MANETs can use the similar schemes for infrastructure-based networks by employing a dedicated service directory. The directory is normally implemented in the cluster head, which is responsible for collecting service information and keeping it updated. When a device needs service, it simply requests the service via looking up the directory [17]. Although the service discovery process is greatly simplified by involving a centralized service directory, it is not flexible enough for devices with high mobility. Therefore many service discovery protocols proposed for MANETs operate in a completely decentralized manner, such as Bluetooth SDP, UPnP SSDP etc. Information of the available services propagates through the cluster either proactively, by service providers announcing their presence periodically, or reactively, by service providers responding to the requests from the devices that needs the service. However, as most decentralized approaches use broadcast to flood messages, they often have limited scalability.

Network resources are unevenly distributed and constrained in MANETs. Sophisticated service discovery protocols take into account additional contextual information such as power consumption [18], physical link quality [19], geographic locations of the service provider and seeker [32], etc. In order to present such contextual information along with the service, hierarchical attribute-value pair naming structures [15] and object-oriented naming structures [16] [31] are suggested as the promising candidates. They provide extensible capability to describe a service with rich meanings. Depending on context, different names can be bound to a given service at the same time, and a name can refer to different services at different times. Thus this is demanding on more flexible and sophisticated service discovery algorithms to find the optimal service for the request under given situation.

3 System-Level Self-configuration Issues

Self-configuration solutions are currently developed with a strong focus on each layer and often specific that layer. It potentially leads to incoherent or even conflicting system configuration due to the lack of system level organization. Especially with emerging MANET scenarios such as those envisioned in autonomic computing e.g., Personal Networks [20], self-configuration is crucial. This is due to the fact that these environments are heterogeneous, distributed and dynamic in nature. Regardless of network dynamics and resource diversity, the network should always be able to configure on its own to fulfill the needs of its users in a most suitable manner. In addition, federation of MANET with public or private infrastructures can be envisioned. These new complex networking scenarios and their stringent requirements are fuelling the need for sophisticated self-configuration strategy, for which we generalize several features that requires further discussions. They are user-centric, context-aware, adaptive and collaborative; and we elucidate them by firstly generalizing a system-level view with respect to their location as illustrated in Fig. 2.

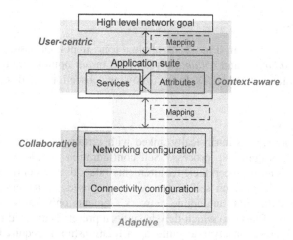

Fig. 2. A system-level view of Self-configuration issues

3.1 Three Tiers

Three tiers can be seen in the system-level view in Fig. 2. The top tier is the high-level goal of networking. This goal is purely decided by the users' requirements and is described as the concrete network configuration objective. The middle tier is application space, where an application suite containing a bundle of services is configured to represent the specific task according to the service-level abstraction. The bottom tier is the configurable network environment, where resides in all network elements to be configured to support identified services in the application space.

3.2 Two Mappings

To interpret high-level requirements into specific network and device parameter settings, two levels of mapping are necessary to connect the three tiers. During the first mapping, the high-level goal is broken down into detailed service types and further into specific service attributes. This mapping should be performed under the guidelines of the preferences of users and the contextual information, since they form the high-level goals and thus influence the service attributes. The second mapping is from the service requirements to the network setup. This mapping is bidirectional. From top-to-down, services are mapped to the specific network devices, which potentially have enough resources to provide those services with the requested quality. This is typically carried out by sophisticated service discovery protocols. On the other hand, in cases when dynamics of network results in dramatic capacity (e.g. link quality) degradation, and current service quality can not be achieved by any alternative configuration at network and connectivity levels, a bottom-to-up mapping represents the necessity to reconfigure applications to adapt to the currently available network capacity.

3.3 Four Issues

The system-level view in Fig. 2 highlights four important issues: *user-centric, context-aware, adaptive* and *collaborative*, which in our opinion are desired features of self-configuration in MANETs. In the following subsections we discuss them in the sequel.

3.3.1 User Centric

Configuring a network is intrinsically linked with certain requirements of users who influence the configuration process. Self-configuration should reduce laborious human involvement, but keep their requirements intact during configuration. In [21] the concept of "Goal-driven self-assembly" is proposed for autonomic computing systems. On initialization, autonomic network elements only know about the high-level descriptions of tasks to which they have been preset. They need to contact each other and configure themselves to build up a relationship as required to obtain the needed services. Although user-centricity has already been noticed as an important self-configuration aspect, coupling a natural user state with a technical network state, as shown in Fig. 3, is still a challenge to design and implement. Some of the challenges are discussed here.

Firstly, the network must know about the users' intentions and clarify the users' requirements in a specific situation to identify applications accordingly. However, due to implicitness and ambiguity of human expressions, getting this information is a task that requires expertise, such as advanced monitoring and recognition methods, sensing technology and even psychological models. From computer networking prospective, we steer our focus on concretizing the users' needs into a set of services which are provided by the network. For example, if a user wants to watch a movie, the network has to first identify the multimedia content, display, audio services and so on. Each service to be configured must be an instance of services needed by the

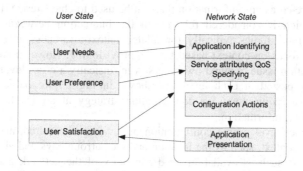

Fig. 3. User centricity in Self-configuration

application, and network formation should be configured to support these services on top. In [22], users' goal is formulated into tasks, which are explicitly encoded and represented by the quality attributes of the services used to perform those tasks.

Secondly, users' preferences must be incorporated to specify quality attributes of each service, which is the only way to impart satisfaction to a user. Preferences of users provide criteria for the system in deciding configuration strategy optimally, especially when multiple configuration possibilities are available. For example, if a user cares more about the video quality than audio quality for viewing a sport, the network should configure itself in the way to display the sport channel on a large screen with high resolution rather than to use high fidelity audio systems consuming large bandwidth. Accordingly, Quality-of-Service (QoS) requirements that network configuration has to meet can be defined based on the quality attributes of the services. The most straightforward way to incorporate user preference is to produce "if-then" policy space, but such space easily grows unboundedly due to numerous preferences and attributes. A more up-to-date method is to manipulate a utility function which is optimized when the most desirable configuration state is reached. In [22] the users' preferences are denoted as "weights" of the inputs to the function, to help in scaling the importance of the quality attributes of each service.

Thirdly, since self-configuration has to fit itself in dynamic network environment, a feedback loop is needed to evaluate the current configuration. For user-centric self-configuration, such evaluation should come from the user. As we discussed above, eliciting users' satisfaction still needs substantial work. A qualified evaluation can be a binary choice between "satisfied" and "un-satisfied", which is limited in scope. Thus quantifying "satisfaction" on a scale, similar to Mean Opinion Score (MOS) for audio system is a flexible approach.

User-centricity is one of the main concerns for the first of the two mappings illustrated in the system-level view Fig. 2. Since the users' needs and preferences are always specific to certain application, this level of mapping based on user-centricity should be distributed and can be implemented within specific applications.

3.3.2 Context-Aware

Since the long-term goal of self-configuration is to eliminate human administration, it should be aware of context as humans do. Following the definition in [23], context can be interpreted as any information that can be used to characterize the situation of an entity, which is considered relevant to the interaction between a user and a network. Current studies typically formulate context-awareness as a service-level problem, and focus on context-based service provisioning. However, with respect to self-configuration issue, context-awareness is intrinsically a cross-layer problem and is necessarily posted at each level. As shown in Fig. 2, we opine that context-awareness is a requirement for configuration strategy at all levels using all the available context information.

At the service-level proper configuration of services is required, and it should be aware of user context, such as the location and the surroundings and also be aware of network context such as availability and capability of the network and the devices. Currently, several studies have made an effort towards context-aware service configuration protocols. Basically, the context-aware manipulation of device parameters or network configuration takes place within service discovery and selection processes. In directory-based protocols, service selection can be handled by the directory servers, which make decisions during the course of service discovery process [26]. This process benefits from the well-defined context-aware naming structures, such as attribute-value pairs or name objects, by combining advanced naming resolution schemes to enable applications being aware of situations in a transparent manner [31]. If the service directory is unavailable, the service seeker has to evaluate each service on its own [17] according to selection policies, which take into account versatile network performance metrics such as time for service discovery [19], round trip latency or geographic proximity [24] between service provider and the seeker.

Network and Connectivity level self-configuration is twofold in terms of context-awareness. On one hand, it should understand application's requirements in the form of QoS, and configure network connection to support the QoS requirements with optimal or sub-optimal configuration. However, it is possible that due to serious network degradation, the QoS requirements are not likely to be guaranteed; no matter what lower level self-configuration strategy is used. In this case a notification from lower level must be sent to application that triggers new service-level self-configuration, which in turn specifies new QoS requirements for lower level self-configuration processes. On the other hand, lower level self-configuration should be aware of the available resources of network and network devices, such as mobility, density, location etc. Though some of these considerations may not have direct impact on QoS provisioning, they potentially optimize the whole system performance. For example consider energy-aware configuration, temporal over-use of energy may lead to an over-optimized solution for QoS provisioning, such as higher data rate, or shorter delay. However, from a long-term point of view, fast depletion of battery power in MANET devices increases the chance of handovers, which degrade the overall performance of a given application.

In our opinion, two considerations are important to realize context-aware self-configuration in MANETs, viz., context aggregation and context distribution. Context aggregation should define methods for individual devices to acquire user context and network context. It should comprise of interactive user interfaces, sensor input processing modules, middleware for network state querying and notification, as well as communication methods to acquire context from other devices in the network. To share one's knowledge with other devices, context distribution is needed. It is simpler in cases that certain device, e.g. the cluster head possibly configures itself as the context repository, than in cases where context is possessed by individual devices in a distributed manner. The latter case is useful to cope with mobility in MANETs but require more sophisticated schemes to propagate information across the network in a timely manner with low overhead.

3.3.3 Adaptive

Another requirement for self-configurable network is to efficiently deal with dynamic situations, i.e., the network should be aware of the situation and trigger the proper re-configuration process under certain policies or guided by some decision mechanisms. The challenge comes from enabling a network to be self-adaptive to the changing situations by re-configuring itself in a suitable way. Changes may come from the user. A trigger to move a multimedia application from PDA to a big LCD display requires higher resolution and thus more bandwidth, while a video session changing to a lower data rate codec requires less bandwidth. Changes in the network composition may result from single device independent initialization, or initialization of all the devices in the vicinity. It can also happen when devices join, when network partitions and while networks merge. As discussed in [14], these events cause change in link quality, network topology and service availability. Three levels of adaptations have been discerned for network self-configuration in [2]; here we base our discussion on a two-level adaptation strategy.

The first level is local adaptation, where adaptation is mainly achieved through device parameter reconfiguration, such as wake-up cycle, memory allocation and frame size, or via adaptive applications. The last one provides adjustable quality of an application under the same network configuration state. An example of such adaptation is the transcoding mechanism provided in MPEG-21 [25]. Local adaptation provides a relatively small scale of configuration possibility, and tries to optimize the system performance at a lower re-configuration cost.

The second level is global adaptation, which is designed to cope with severe dynamic changes that are beyond the ability of local adaptation. In this case, current network state is no longer suitable to maintain system performance requirements and a new state has to be reached through adaptive reconfiguration. At service-level an important issue is session mobility management, which incorporates fast service re-discovery and session handover to enable a multimedia session transferred from one device to another. This can be achieved directly between the communication ends, and it can also be handled by employing proxies or delegates [24][26]. At network-level address re-allocation is required in cases of terminal mobility and network mobility. For example, network merging can invoke address duplication, and network partition requires releasing some of the assigned addresses. A common solution is to periodically broadcast an identifier on the network. It enables devices to detect network merging or partition explicitly and to adapt addressing configuration accordingly. In [13] and [14], merging and partition of a network are not necessarily detected, address duplicates are tolerable, and are solved on the fly. If a device is connected to the Internet, mobility management functions such as assignment of Care-of-Address (CoA) by access router and address binding update to home agent are required.

Adaptation is important for mapping the service requirements to the network settings. To realize the mapping there is a need to develop a common messaging protocol. It should define a set of methods and mutually understandable messages for communicating requirements by applications and network configurations, which are consequently able to adapt to each other. We envision that such a messaging protocol can make use of the makeup languages such as XML. To design adaptive self-configuration protocols, an important prerequisite is to design an effective control loop which gives feedback on the fitness of current configuration. In the control loop, an essential component is the *decision making block* which is required to decide the necessity to trigger re-configuration. This means that we must take into account the adaptation cost such as service delivery delay and packet loss. As discussed in [22], adaptation is only triggered when system performance lies outside the acceptable region of performance. Besides, cognitive techniques can bring in a new dimension for decision making under dynamic environments. However defining proper learning algorithms is normally very complex in distributed network environments and is still an open issue.

3.3.4 Collaborative

MANETs are typically organized autonomously according to certain network formation criteria. In most cases, there is no central administrator or common coordinator thus it requires self-configuration to enable collaboration amongst devices and even clusters to optimize system performance. Generally, collaboration is related to specific configuration problems, and it must drive the development of more sophisticated self-configuration protocols. We believe that it is pragmatic to keep it within the protocol scope rather than go for generality.

Mutual understanding amongst devices and networks with different configuration protocols is the precondition for collaborative self-configuration strategy. Network address translation (NAT), for example, is needed for inter-networking devices with different addressing syntaxes and semantics. Similarly at service-level, a scheme for naming [27] has been proposed to enable a device to understand different naming structures.

In MANETs a device by nature autonomously configures itself to optimize its own performance in terms of achieving certain local performance goals such as those of power consumption, storage utility, etc. For example, a device may turn off the unused radio interface to achieve power optimization, or to generate a link local address rather than a global IP address for local traffic to save configuration time. However this local goal achievement is likely to conflict with global performance goals. When the device configures itself with only one of several radio interfaces on board for power saving, from global perspective concerning its capacity and mobility, it should have taken the role of a gateway to bridge communications between devices with different radio interfaces. Therefore self-configuration should be guided by a collaborative mechanism which can resolve potential contradictions between local and global needs and fulfill the performance goal that goes beyond the vision of an individual device. For example, it is especially challenging for ad hoc networks to assign roles of network infrastructure elements, such as routers, servers, etc., since no dedicated device takes such roles and devices in MANETs need to configure themselves on-the-fly [29]. It is similar to the idea of "self-assembly" in [28]; i.e., given a set of devices capable of networking, which have heterogeneous capability and functional diversity; they are able to configure themselves to form a complete system which realizes certain applications required by the user.

Besides, collaboration also means a fair use of network resources amongst devices and applications, i.e., to prevent "starving" of some applications to satisfy the needs of heavy resource-consuming ones. A design principle mentioned in [2] is to develop local behavioral rules that achieve global performance goals. Stateless address autoconfiguration protocols in [11] [12] are such examples in that each device follows the predefined local procedure for address assignment that results in network-wide non-ambiguous address allocation. Another example is a clustering algorithm based on Particle Swarm Optimization [4], a model mimicking the choreography of a bird flock or a school of fish. Further, selfish configuration behavior of individual devices can be potentially solved by game theoretical approaches [30] or by incentive schemes [33].

4 Self-configuration Management Architecture

We regard user-centric, context-aware, adaptive and collaborative configuration as the essential requirements for self-configurable networks. We noticed that previous

Table 1. Overview of self-configuration techniques

	User-centricity	Context-awareness	Adaptation	Collaboration
Service-level	*Service selection* • Utility based: [22], [28];	*Service discovery* • Context-based advertising/query [26]; • Based on Name resolution: [31]; *Service selection* • Performance metrics based: [19]; • Location based: [24] [32]; • Utility based [22]; • Based on mobile agent: [17]; *Name structure* • attribute-value pairs: INS, modified INS [15]; • Objectified name: OMG-NSS [16], ONS [31];	*Adaptable application* • MPEG-21 [25]; *Service discovery* • Routing-based: [18]; • Utility based: [22] *Session migration*: [24], [26]	*Resource allocation* • Utility based: [28]; *Naming platform* • "Ecosystem": [27]
Network-level	NA	*Context-aware addressing* • Hop distance based addressing • Network size	*IP addressing* • Based on Partition ID: [8], [11], [9]; • Based on weak	*IP translation* • Address translation: NAT • Protocol

Table 1. *(continued)*

		based variable-sized addressing [12]	DAD: [13], [14]; • Based on routing and link information: [12]; *IP Mobility* • COA-HA binding	translation: SIIT, NAT-PT *Addressing* • Decentralized address assignment: [11], [12] *Router allocation* • Game theory: [29] • Incentive based: [33];
Connectivity-level	NA	*Clustering* • Identity-based (Lowest-ID) • Location-based (Highest-Degree) • Combined metrics (mobility, energy, degree etc.): [3] [4] [5]	*Clustering* • Energy and mobility adaptive cluster head reselection [5] • Mobility based "hello period" adjustment [5]	*Clustering* • Particle swarm optimization [4] *Medium access* • Game theory: [30] *Power control* • Game theory: [30]

studies have already proposed some solutions to meet the requirements from different perspectives, and we listed them in Table 1. However, considering diversity and variety of configuration aspects, we focus on general architecture to automate self-configuration management in MANETs. We believe such architecture can provide a basis for future protocol design to meet all the requirements.

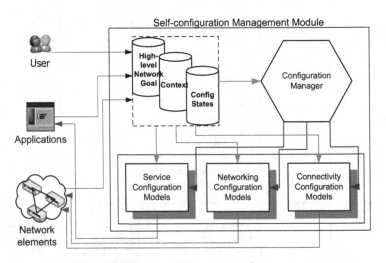

Fig. 4. Architecture of Self-configuration Management

Our proposed architecture is centered on a self-configuration management module, as shown in Fig.4. This module interacts with the three tiers discussed in the Section 3.1. The module comprises of three main components: an information abstraction, a set of level-specific configuration models and a management administrator. They potentially provide methods to fulfill the four requirements.

Information abstraction is the essential component to realize user-centricity and context-awareness. It aggregates contextual information from user, applications, network elements, and defines configuration states of both users and network. By mapping these two states as shown in Fig.3, it interprets user needs and preferences into network tasks with high-level goals. Meanwhile the contextual information can be directly used by the configuration models of different network levels to support context-aware configuration protocols.

Configuration models are grouped separately at service, networking and connectivity levels. They aim to realize specific configuration tasks within the level, and flexibly to allow sophisticated protocol design. The protocol should be designed in the way to enable harmonic collaboration amongst autonomous MANET devices and effective adaptation to changing situations, by utilizing advanced techniques and multi-dimension contextual information.

Configuration manager takes an overview of current configuration states and is responsible for coordinating configuration models beyond local levels. In some cases optimal performance can only be achieved by joint consideration and decision of all levels. For example, mobility in MANETs can be handled either at connectivity, networking or service level, thus configuration manager has to decide which one is the best solution by being aware of the nature of scenarios. The configuration manager makes it possible to achieve collaboration and adaptation from global perspective.

The entire management module can be implemented in the cluster head or the most capable device in terms of computation, storage and power resources, whilst taking into account the mobility. Resource-restricted devices only need to implement part of the module such as layer specific configuration models. In this way, centralized or hierarchical management can be organized. When centralized approach is not useful in certain scenarios, this management module can also be implemented fully or partially on each device with adequate resources to provide distributed management.

5   Conclusions

We studied the requirements of self-configuration in MANETs aiming at understanding the problems and their solutions at different layers of OSI stack. We have presented here many techniques used for self-configuration such as clustering, addressing, and service discovery. The study also considered all aspects of services and solutions currently employed for a better self-organization of networked devices. Further, from a global perspective we generalized four important issues which are challenging the self-configurable network paradigms. We proposed self-configuration management architecture as the foundation for solutions to those challenges. With this discussion we believe that a global picture of self-configuration from different perspectives is rendered. However, this is only the initial step. Further studies include implementation and generalization of protocols for self-organization that balances between local and global perspectives.

Acknowledgement

We thank IOP GenCom Future Home Network project, IST MANGET Beyond project and Freeband PNP2008 project for funding this work.

References

1. Dobson, S., et al.: Survey of autonomic communications. ACM Transactions on Autonomous and Adaptive Systems 1(2), 223–259 (2006)
2. Prehofer, C., Bettstetter, C.: Self-organization in communication networks: principles and design paradigms. IEEE Communication Magazine 43(7), 78–85 (2005)
3. Chatterjee, M., Das, S.K., Turgut, D.: WCA: A weighted clustering algorithm for mobile ad hoc networks. Journal of Clustering Computing (Special Issue on Mobile Ad Hoc Networks 5(2), 192–204 (2002)
4. Ji, C., Zhang, Y., Gao, S., Yuan, P., Li, Z.: Particle Swarm Optimization for Mobile Ad Hoc Network Clustering. Proc. of IEEE ICNSC (2004)
5. Gavalas, D., Pantziou, G., Konstantopoulos, C., Mamlis, B.: Lowest-ID with adaptive ID reassignment: A novel mobile ad-hoc networks clustering algorithm. Proc. of IEEE ISWPC (2006)
6. McAuley, A.J., Manousakis, K.: Self-configuring Networks. Proc. IEEE MILCOM (2000)
7. Sun, Y., Belding-Royer, E.: Dynamic Address Configuration in Mobile Ad hoc Networks., Technical Report 2003-11, Computer Science Department, UCSB (2003)
8. Nesargi, S., Prakash, R.: MANETconf: Configuration of hosts in a mobile ad hoc network'. Proc. IEEE INFOCOM (2002)
9. Thoppian, M.R.: A Protocol for Dynamic Configuration of Nodes in MANET, Master's thesis, Computer Science, University of Texas at Dallas, August (2002)
10. Perkins, C.E., Malinen, J.T., Wakikawa, R., Belding-Royer, E.M., Sun, Y.: IP address autoconfiguration for ad hoc networks. draft-perkin-manet-autoconf-01.txt, Work in progress (2001)
11. Zhou, H., Ni, L.M.: Prophet Address Allocation for Large Scale MANETs. Proc. IEEE INFOCOM (2003)
12. Weniger, K.: PACMAN: Passive Autoconfiguration for Mobile Ad hoc Networks. IEEE JSAC Special Issue 'Wireless Ad hoc Networks 23, 507–519 (2005)
13. Vaidya, N.H.: Weak Duplicate Address Detection in Mobile Ad Hoc Networks. In: ACM MobiHoc, ACM Press, New York (2002)
14. Sun, Y., Belding-Royer, E.: A study of dynamic addressing techniques in mobile ad hoc networks. Wireless Communications and Mobile Computing, 4(3), 315–329 (2004)
15. Chen, G., Kotz, D.: Context-sensitive resource discovery. In: Proc. IEEE PerCom, IEEE Computer Society Press, Los Alamitos (2003)
16. OMG, Naming service specification (October 2004), http://www.omg.org/docs/formal/04-10-03.pdf
17. Tyan, J., Mahmoud, Q.H.: A network layer based architecture for service discovery in mobile ad hoc networks. Proc. IEEE CCECE-CCGEI (2004)

18. Ververidis, C.N., Polyzos, G.C.: Extended ZRP: a routing layer based service discovery protocol for mobile ad hoc networks. Proc. MobiQuitous (2005)
19. Liu, J., Issarny, V.: Signal Strength based service discovery in mobile ad hoc networks. Proc. IEEE PIMRC (2005)
20. Jacobsson, M., Niemegeers, I.: Privacy and anonymity in personal networks. Proc. IEEE PerSec (2005)
21. White, S.R., Hanson, J.E., Whalley, I., Chess, D.M., Kephart, J.O.: An architectural approach to autonomic computing. Proc.IEEE.ICAC (2004)
22. Poladian, V., Sousa, J.P., Garlan, D., Shaw, M.: Dyanmic Configuration of Resource-aware services. Proc. IEEE ICSE (2004)
23. Abowd, G.D., Dey, A.K., Brown, P.J., Davies, N., Smith, M., Steggles, P.: Towards a Better Understanding of Context and Context-Awareness. Proc. ACM HUC (1999)
24. Shacham, R., Schulzerinne, H., Thakolsri, S., Kellerer, W.: The virtual device: expanding wireless communication services through service discovery and session mobility. Proc. IEEE WiMob (2005)
25. Min, O., Kim, J., Kim, M.: Design of an adaptive streaming system in ubiquitous environment. Proc. IEEE ICACT (February 2006)
26. Phan, T., Xu, K., Guy, R., Bagrodia, R.: Handoff of application session across time and space. Proc. IEEE ICC (2001)
27. Doi, Y., Wakayama, S., Ishiyama, M., Ozaki, S., Ishihara, T., Uo, Y.: Ecosystem of naming systems: discussions on a framework to induce smart space naming systems development. Proc. IEEE ARES (2006)
28. Chess, D., Segal, A., Whalley, I., White, S.: Unity: Experiences with a prototype autonomic computing system. Proc. IEEE ICAC (2004)
29. Felegyhazi, M., Buttyan, L., Hubaux, J.P.: Nash equilibriums of packet forwarding strategies in wireless ad hoc network. IEEE Transactions on Mobile Computing (May 2006)
30. MacKenzie, A.B., Wicker, S.B.: Game theory and the design of self-configuring, adaptive wireless networks. IEEE Communication Magazine 39(11), 126–131 (2001)
31. Lee, K., Lee, D., Ko., Y., Lee, J., Chung, Y.C.: An objectified naming system for providing context transparency to context-aware applications. Proc. IEEE SEUS-WCCIA (2006)
32. Lenders, V., May, M., Plattner, B.: Service discovery in mobile ad hoc networks: a field theoretic approach. Proc. IEEE WoWMoM (2005)
33. Garyfalos, A., Almeroth, K.C.: Coupons: wild scale information distribution for wireless ad hoc networks. Proc. IEEE GLOBECOM (2004)

Knowledge-Based Reasoning Through Stigmergic Linking

Kieran Greer[1], Matthias Baumgarten[1], Maurice Mulvenna[1], Chris Nugent[1], and Kevin Curran[2]

[1] School of Computing and Mathematics
Faculty of Engineering, University of Ulster
Northern Ireland, UK
[2] School of Computing and Intelligent Systems
Faculty of Engineering, University of Ulster
Northern Ireland, UK

Abstract. A knowledge network is a generic structure that organises distributed knowledge into a system that will allow it to be efficiently retrieved. The primary features of this network are its lightweight autonomous framework. The framework allows for smaller components such as pervasive sensors to operate. Stigmergy is thus the preferred method to allow the network to self-organise and maintain itself. To be able to return knowledge, the network must be able to reason over its stored information. As part of the query process, links can be stigmergically created between related sources to allow for query optimisation. This has been proven to be an effective and lightweight way to optimise. These links may also contain useful information for providing knowledge. This paper considers a number of possibilities for using these links to return knowledge through a distributed lightweight reasoning engine, thus upholding the main features of the network.

1 Introduction

CASCADAS [1] is a European Framework VI funded research project aimed at developing the next generation of autonomous network systems. The main focus of our work on this project is the lightweight organisation and request-based provision of knowledge and towards this goal the concept of a knowledge network has been developed. The network must be able to accommodate relatively small network devices, for example sensors and so a lightweight framework is required. Because of this, stigmergy is central to allowing the components to self-organise with regard to knowledge. Stigmergy (for example Ricci et al. 8) is a lightweight way of building an understanding through changes in the environment rather than through any real knowledge. One example of stigmergy is the Ant Colony Optimisation algorithm (Dorigo et al. 2). ACO works by copying the actions of ants as they try to find the optimal route from one position to another. They

---

[1] For more information on the overall project see http://www.cascadas-project.org

D. Hutchison and R.H. Katz (Eds.): IWSOS 2007, LNCS 4725, pp. 240–254, 2007.

randomly select a number of routes and leave a pheromone behind indicating the route they took. The shortest route will build up the strongest pheromone amounts and so eventually all ants will choose this route. The ants do not know what the optimal route is, but rather discover it through the experience of all the routes that they take. Their reasoning does not require any knowledge of the environment, but only to be able to read the pheromone trail.

A knowledge network (for example Baumgarten et al. 1 or Mulvenna et al. 5) is a generic structure that organises distributed knowledge of any format into a system that will allow it to be efficiently retrieved. The knowledge network acts as a middle layer that connects to a multitude of sources, organises them based on various concepts and makes this knowledge available to individual services and applications. To turn the network of information into a real knowledge network, it would be necessary to be able to derive new information from the information that it stores. It would be useful to introduce some kind of cognitive process into the network that would allow it to reason about its contents. Reasoning generally requires an understanding of knowledge, which can be a much more heavyweight process. To retrieve information the network must be queried. This paper will suggest using the query results to stigmergically organise the information in the network. The framework can then also be used to perform some basic reasoning. This means that the network can do more than simply information retrieval, while remaining lightweight. This way of using links appears quite novel, due to the lack of related work.

The rest of the paper is organised as follows: Section 2 gives a brief description of the knowledge network. Section 3 describes the linking method that has been proposed and summarises some tests. Section 4 describes some related work in the areas of linking and stigmergic reasoning. Section 5 describes the stigmergic reasoning methods proposed in this paper. Section 6 develops these further to include autonomous behaviour and Section 7 gives some conclusions on the work presented in this paper.

2   Knowledge Network

A knowledge network is a hierarchical structure with sources at leaf nodes providing information. A source can provide live data that may also be volatile (constantly changing). For example a number of sensors may return current weather information. Other sources can contain embedded data. This is data that does not change and represents a particular condition that occurred. The knowledge network architecture does not limit the type of source and so static data could also be represented by something like a Web page. One node can aggregate other nodes, acting as the organisational component of the network. The organisation of sources can take place based on the values that sources produce, their locations or their meaning. If using semantics, sources can be aggregated in a hierarchical way into other higher level concepts, building up a network of knowledge. We allow each component to store services that can be used to perform calculations on the data. These services can be dynamically loaded into the network and can be of any type, so there is really no limit to the functionality that can be provided. A node refers to any node in the network, including nodes that aggregate or reference other nodes. A source refers to a leaf node that actually references a source that provides data.

3 Stigmergic Linking

Some kind of querying process is required to query the network, to retrieve the information. As the network can be of any size, query optimisation is important. To try and optimise the network in a stigmergic manner, temporary views can be generated, or links can be created between sources, based on the experience of the querying process. Sources can be linked to each other in a stigmergic manner by linking sources that answer the same types of query. This is stigmergic because the sources to be linked are selected from the query results, or the querying experience. For a particular type of query there may be many potential sources that could answer the query. It may be useful to be able to tell the network to look at only a certain number of these sources, thus reducing the amount of search required. So far, testing has been done using a select-from-where statement, where the 'where' clause contains a number of comparisons that need to be satisfied. For example, a query might look like:

SELECT A.Value1, B.Value2 FROM A, B, C WHERE (A.Value3 EQ B.Value4) AND (B.Value2 LE C.Value1 OR B.Value3 GT C.Value2)

This statement contains information about source types ('A' for example), value types ('Value1' for example) and comparison operators ('EQ' or 'LE' for example) that are related to each other through the query. It is possible to use this information to create a linking structure that describes parts of the query. This structure can be made from sets of nested hashtables, where the keys represent the source types, value types and operators. The most successful structure that was tested contained, for the source in question, the following elements for source linking:

For the source in question store a structure with: value type - related source type - related value type - operator - related source instance

As an example, for answering the above query the engine firstly retrieves all C source instances and B source instances, where B and C represent source types. It stores the B source instances that satisfy the comparison 'B.Value3 GT C.Value2'. Then only these B source instances are used to satisfy the other query parts, thus reducing the number of sources looked at and thus the search space. This process filters through the whole query, where a full search is performed only when a new source type is encountered. The query engine would traverse the network to retrieve the C source instances, evaluate the first comparison and then use the results to limit the search for the rest of the query. The search to find the first set of sources or a new set of sources however can still potentially retrieve a large number of source instances. It is possible to use a linking structure similar to the one that links sources to create a view of the network. This view can be specific to a particular application and can store the nodes that are frequently used by that application only. When performing a search, before doing a full search of the network, the application can firstly search its view to determine if it contains any relevant nodes and then use only these nodes if they exist. This will help to reduce the search process. The view linking structure may store only source type and value type as keys, for example. On top of this, if one source is retrieved for a query part, its linking structure can be checked to see if it links to any instances of the other source type in the comparison. If links exist then only these sources need be looked at to evaluate the comparison.

The linking structure can store different levels of references, where each new level has a threshold value that must be met before a linked source is allowed to be stored there. Each source that is stored is given a weight value that can be incremented or decremented, allowing the source to be moved up or down the levels. The top level is the linked level and it is only these sources that are returned as links. This means that sources must be consistently related to each other before they are recognised as links. It is also possible to control the amount of memory used by the linking structure, by limiting the number of allowed entries, ensuring that the structure stays lightweight. It is also possible to include learning algorithms to learn certain parameters. Consider evaluating the query part 'B.Value3 GT C.Value2'. A number of B source instances returned that satisfy the query part could be linked to the C source instances through a set of nested hashtables with the keys Value2 - B - Value3 - GT - B source instances. If these instances are then consistently associated they can build up their weight values until they reach the link threshold, when they will then be returned as links, or sources to look at instead of searching the whole network. Note that the storage mechanism is quite flexible as it stores links for query parts and not whole queries, allowing the query parts to be related to different queries. Each source stores a linking structure that contains the hastables for sources linked to it. If the source is associated with a query part, then the hashtable related to that query part is retrieved. The referenced sources are then looked at.

If a linked source is used in the current query answer, then its weight value is incremented, while if it is not used its weight value is decremented. This may mean that sources are moved up or down levels in the linking structures and so memory management becomes a factor. A source can be moved down if its weight now falls below the threshold value for a level, or if a weight for another source is now greater. While this is a technical issue, it is possible to manage this. It is also possible to allow one source to borrow memory from another, so that a heavily used source could borrow memory from a lightly used one. In these tests this was done between sources of the same type only and ensures that the total amount of memory remained the same. Also implemented was an algorithm that allowed the linking structure to learn the best weight decrement value. This was the fraction of the increment amount that a source weight would then be decremented by if it was not subsequently used. The decrement was less than the increment so that a source could stay at a certain level for a period of time even if not always used. These tests allowed each branch of the linking structure to store a reserve entry, which was an extra entry that a source could be referenced by before it was moved into the linking structure itself. This was shown to produce much better results. But because it was then difficult to manage memory, it became optional. Later tests then also allowed a local view to store links that could also be used along with the global network links. The extra reserve entry became a sort of benchmark that could be compared to a configuration of limited memory in the network itself but with an additional local view. Tests showed that the two setups could produce similar performances when appropriate configurations were used. The linking structure is the most basic possible. It is simply a link between two nodes, represented by a weight. As such it is completely flexible and dynamic. Alternative approaches are certainly possible, such as building up a case of statistics. But if these were to be evaluated through more heavyweight knowledge-based algorithms they may lose some of their dynamism. It is not possible from these tests to say what approach would be best.

3.1   Example of Test Evaluations

Some tests have been carried out to determine the possible effectiveness of the linking with regard to query optimisation. Only a brief summary of some test results can be given here as this is not the main focus of this paper. We hope to publish more complete analyses in other papers. To give some idea of possible performance, a network with 300 source nodes was queried with queries skewed towards certain types of request. The work assumes that the queries would need to be skewed towards certain types for the linking to be effective. The skewing was done by placing source and value types into probability bands and then randomly generating queries by selecting source or value types from each band. The results are provided in the graphs of Fig. 1. The queries generated only used the equivalence operator in the 'where' clause comparisons. This was shown to benefit most from the links as compared to queries with all types of comparison and is equivalent to text or concept matching. The results are for queries generated from a 0.7:0.3 probability distribution, where one of 7 source types or 2 value types would be selected 70% of the time and one of the remaining 3 source types or 3 value types would be selected the remaining 30% of the time. The linking methods limited the amount of memory available by limiting the number of allowed entries. The other configuration factors were as follows: There were 10 different source types and 5 different value types. There were a total of 30 instances of each source type, where each source instance contained all 5 value types. Each value type could have a range of integer values from 1 to 10. For the linking methods that managed memory, each of the storage structures (3 levels in these tests) was allowed a total of 50 entries for each atom. Each query was made up of a maximum of 2 sources in the 'select' clause and 3 sources in the 'from' clause. If the 'select' clause had 2 sources, the 'from' clause had 2 or 3 sources. Other tests with a 0.9:0.1 distribution split showed that in some cases, queries with all comparison operators could be optimised with reasonable

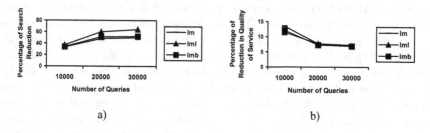

Fig. 1. Graph a) Percentage of reduction in the number of nodes searched when using one of the three linking methods described for queries with the equals operator only. Lm stands for limited memory, lml stands for limited memory with learning and lmb stands for limited memory with borrowing. Graph b) Percentage of reduction in quality of service when using one of the three linking methods described for equals only queries.

quality as well. Graph 1.a shows the amount of search reduction produced by the linking and Graph 1.b shows the related reduction in quality of service.

Limited memory means that the number of entries allowed in the linking structure itself was limited. This would also limit the search process to what links would be available. These tests however also allowed the reserve entry and so did not limit memory with regard to overall memory size. But the reserve entry would not be used in the search process, only the highest level links. Borrowing would then allow nodes to borrow memory from each other at each linking level and learning would allow the nodes to learn the best weight decrement value. As these are only initial tests, the algorithms used are relatively crude. This was the intention so as to give some baseline indications. More sophisticated algorithms could certainly be tried.

The metrics used to measure performance were as follows: To measure the node reduction a search that used linking was compared to a full search. The full search was still guided by the hierarchical structure of the network, but would look at all sources of a particular type. However, if a previous evaluation indicated that only certain sources were suitable for that query part, then only these would be looked at for the other evaluations. The linking search would consider only linked sources when links were available. If no links were available then it would look at all possible sources. If the linked search did not return a result, then a full search would be performed as well. If the full search returned an answer, then its node count would be added to the linked search count. If the linked search did return a result, then the full search node count would not be added, but a comparison would be made between the quality of answer between the full search and the linked search. All source values were represented by integers. Thus a range of 1 to 10 would mean 10 possible different values. This would be just as suitable for text matching as for numerical comparison. However, to measure quality of service a way was needed to define what the best sources were. To do this an evaluation function was used that tried to maximise the sum total for all of the sources that were asked for by a query. Thus the sources with the larger values would be defined as the better services. A full search would find the best services, but the linked search might not contain links to all of the best services. Thus the difference in the average value returned for a single source between the full and linked searches could be taken as a measure of QoS. As the search size is reduced, fewer nodes are searched and so the quality of service also reduces. But the tests showed that a reasonable QoS could still be maintained. They also indicated a substantial reduction in search size through using links. While these tests only used the equivalence operator, other tests with more skewed data showed that the links could still be effectively used with queries that had all comparison operators as well. For example, queries skewed with a 0.9:0.1 distribution could produce only a 3% worse QoS for equals only queries, but then the search reduction might be around only 35%. All of these figures are relative however. A real environment with a much greater ratio of instances to source types might expect a much greater search reduction, for example. Initial tests including views also show reasonable results for queries with all comparison operators.

3.2   Knowledge Provided by the Links

The structure that stores the links may actually contain useful information that could be used to infer new knowledge. These links may be created primarily through the experience of the users that issue requests. There is thus some cognitive process used in creating them and they will link sources that should be sensibly associated together. These may be sources that are not obviously linked by semantics or any ontology of the network that may exist. Section 5 will look at the possibility of using the linking information to derive new knowledge in a lightweight manner, but before that some related work will be discussed.

4   Related Work

Related work can be split into work that tries to use links for optimisation and work that tries to reason stigmergically. One example of using links in networks can be found in Koloniari et al. (3), where they try to cluster nodes in a peer-to-peer network based on query workloads. They measure how similar a node's content is to a type of query, which will mean that it is more likely to return an answer to that type of query. They then try to cluster nodes with similar workloads together in workload-aware overlay networks. This will maximise the number of relevant nodes that can be visited in a time period to answer a particular query, by having them just a few links apart. They describe that the mechanism for calculating the workload value is still an open issue and could be based on a node storing statistics on the queries that pass through it. They state that 60% more queries can be answered in the same time frame when nodes are aggregated this way. Another example can be found in Raschid et al. (6) or Vidal et al. (11). They apply linking to the problem of finding routes through web resources in the area of Life Sciences. In this set of resources, there are known to be different routes to different resources that may answer the same query. They create a directed acyclic graph to describe the possible routes and then adjust weights in transition matrices to produce a ranking of sources to investigate next.

There does not seem to be a lot of work focused on stigmergic reasoning. Serugendo et al. (9) describe self-organising methods for the Internet or mobile communications and includes stigmergic examples. Data mining might be a related topic as the links would generate new clusters of data from the existing information. Ramos and Abraham (7) describe a stigmergic method to self-organise by generating clusters through data mining. Two examples of directly related work seem to be Torres (10) or Ricci et al. (8). Torres (10) discusses the use of stigmergy to produce collective intelligence in robots. A global problem is broken down into simpler tasks that each robot can perform, where collectively they exhibit some form of intelligence. Ricci et al. (8) describes a process and framework for producing cognitive stigmergy. Stigmergy essentially reacts to its environment and cognitive stigmergy tries to add some cognitive or intelligent processes to this. In their paper they suggest doing this through the use of artifacts. Artifacts are first-class entities representing the environment that mediate agent interaction and enable emergent cooperation. The stigmergic process itself is not changed, but artifacts are stored in the agents or in the environment and they can trigger certain events. They can be combined and as such, can individually or collectively produce cognitive behaviour in the network as a whole. They note that stigmergic processes that might be included in artifacts include diffusion (diffuse information to nearby nodes to improve awareness), aggregation (translate a set of annotations into a single annotation for evaluation, for example), and selection and ordering (order or select annotations according to their importance to a particular artifact). These artifacts could provide the same sort of functionality as the links, but they may be able to represent anything. This paper proposes generating the links through the querying process, based on the knowledge of the users. Artifacts would appear to be more heavyweight knowledge-based entities that are then stigmergically used to provide collective intelligence. They are not necessarily organising mechanisms, but more like services. However, more heavyweight options will also be suggested in this paper, so there are clear similarities between the two approaches.

5   Stigmergic Reasoning

It may be possible to use the linking information to infer new knowledge using just simple mathematical operators like percentage, sum, average, or simple text comparison. The links are still created stigmergically as described in Section 3 and so only require simple weights and thresholds. The reasoning can then be done by recognising a simple operation and using one of the mathematical operators to calculate it. This would produce a sort of distributed lightweight reasoning engine. The reasoning in the network itself would still be limited, but may be able to answer questions like the following:

1. What is the best value of one value based on other values?
2. Is a certain value (or action) possible based on other values?

The information used to infer this is based on the querying experience and so depends on the currently existing links. As it is not built on knowledge itself, if a link is missing it may not be able to give the correct answer. Because of this it would be useful to provide a value describing the reliability of the answer. This could simply be a percentage that describes how reliable the final answer was compared to other possible answers. Consider the following types of reasoning that might be possible.

5.1 Reasoning Examples

Following are a number of examples of possible queries that a user may ask that require reasoning. The examples will also show the select-from-where statement that needs to be executed to evaluate the query. These examples are based on a weather network with other related concepts. These examples could probably be answered by different queries and are slightly contrived, but they give an idea of the sort of reasoning that is possible.

What is the best temperature for wearing t-shirts? We have a number of sources in the network relating to t-shirts. We retrieve these sources and look to see if they have any weather links. If weather links exist, then these sources are accessed and their values are retrieved. The temperature values are then averaged to produce the best value. The query for this might look like:

Select best weather.temp From weather, clothes Where clothes.item EQ t-shirt.

If there exist links between clothes and weather then you could argue that both sources must initially have been related through a common concept that allowed them to be linked as part of the where clause evaluation and so could be queried directly. However, the query is asking for an aggregated value. It is not providing a specific temperature value to check. Also, if a common concept exists associating the two source types, then there may be many variations of possible related values. The user is asking for the best value based on knowledge. This knowledge is provided by the other users of the system and they decide what the best temperature for wearing a t-shirt is.

Can I go swimming in Belfast based on the current rain and temperature weather? We retrieve the current live rain and temperature conditions from sensor sources. We then navigate to swimming sources and retrieve any weather links. The weather links stored contain embedded data that describe weather conditions during which swimming took place. If any links exist with weather conditions that match the current weather then swimming is possible. The query for this might look like:

Select exists swimming From swimming, weather, current_weather Where weather.rain EQ current_weather.rain And weather.temp EQ current_weather.temp.

The reply could be the percentage of swimming sources with linked weather sources that match and also the current weather conditions.

What clothes should I wear in Belfast when it is windy? We retrieve the Belfast source and from this the wind conditions sources. We select the wind conditions sources that indicate windy weather. From these sources we retrieve links to related clothes sources. We look at the clothes values to see what clothes people have worn. For example, if 3 sources specify coat, while 2 sources specify jumper, as coat is specified more often it can be returned as the answer. The answer can again be returned as a percentage indicating some level of confidence. The query for this might look like:

Select best clothes From clothes, city, weather Where city.name EQ Belfast AND city.weather EQ weather.conditions AND weather.wind EQ windy.

Note that these evaluations are done based purely on links and relatively simple mathematical operations or simple comparisons. This allows the reasoning engine to be distributed over the whole network and allows each individual reasoning component to be relatively lightweight. If no linked sources exist then the reply can specify this, rather than indicating an incorrect request. It is also possible to include percentage values with the reply to indicate the level of confidence. The linking structure might contain nested keys that provide more information than is required. For example, if constructed from the where clause comparisons then a structure like that given in Section 3 may exist. But linked sources can be retrieved from different branches that contain the query request, if the request is more general, and then aggregated.

The reasoning engine must be able to recognise a set of extra keywords on top of what the query engine recognises for the select-from-where statement. For example, in the above queries, the extra keywords are 'best' or 'exists'. These would trigger a different kind of query, maybe to retrieve all relevant sources and average their values. But this process should remain lightweight and not extend the querying process by too much. There is also a slight difference in the query structure, where all sources do not need to be linked together by comparisons in the where clause. If we perform some sort of aggregation for a source based on other sources, we can look directly at the existing links. For example, in the swimming query, swimming is not directly linked to the other sources by comparisons. But we can retrieve the swimming sources, look for any links through the related source type weather and then use these to compare with existing weather conditions.

5.2 More Complex Reasoning

The query engine, or simple reasoning engine, is only able to perform a limited amount of reasoning. Other types of reasoning cannot be inferred directly from available information, but require extra information in the form of rules. For example, say we have a rule that 'the brother of a mother is an uncle'. Then we have the following statements stored in the network through links:

- Susan is John's mother.
- David is Susan's brother.

A user then asks the query: 'does John have an uncle?' If the extra rule exists then we can infer that David is John's uncle, but if it does not then we cannot say this. The knowledge network, through links, could certainly store the information that Susan is John's mother and David is Susan's brother. But we need the rule that the brother of a mother is an uncle to be included into the system as an extra piece of information. The reasoning engine must then also perform a more complex piece of reasoning to come up with the answer. This may result in a system that is becoming too heavyweight for the stigmergic framework.

If we want to allow this type of query, then one possible solution is to include a centralised, more heavyweight component that can do some pre-processing. This component could be application specific, when it would be part of the client application and store sets of rules for just that particular application. As it is not distributed through the whole network the network remains lightweight. This pre-processing essentially re-writes the query to make it possible to execute it in the network. The query can be re-written as a select-from-where statement from the existing rules, for example:

Select exists C From C, A, B Where (A.brother EQ C) and (B.name EQ John and A.name EQ Susan and B.mother EQ A)

So this re-writing will produce a new query that can then be executed in the network as normal without requiring any heavyweight reasoning. We note that the re-writing could produce a query that does not necessarily require links to be retrieved. The sources could be queried directly from this example. But the query can then be executed in a distributed and lightweight manner through the network. Maybe something like ConceptNet (Liu and Singh 4) could be used to retrieve the rules from. With programs offering machine-readable interfaces, a rule-base like ConceptNet could be queried with the concepts in question and related rules retrieved. The query would then be re-written using the most appropriate rule. Section 6 will consider autonomic reasoning and introduce some other features that may be important for a lightweight network. These can be combined into an overall system that is schematically illustrated in Fig. 2. The diagram is shown now to give you a clearer view of the concepts. The extra concepts illustrated in the diagram are a global ontology that could also store the rules and a concept matcher that can be used to determine similar source types based on the matching of concept descriptions. Also, an evaluation function can be loaded into certain nodes to evaluate certain concepts for autonomic linking. These features are discussed in the next section.

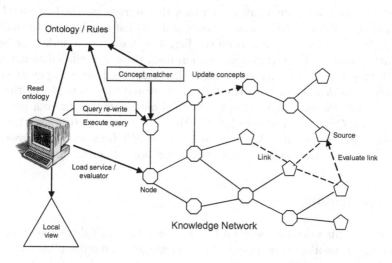

Fig. 2. Schematic view of the concepts discussed in this paper

6   Autonomic Reasoning

If we make the nodes in the network slightly more heavyweight then we can allow them to create the links autonomously thus generating their own knowledge. Each node could be loaded with an evaluation function that evaluates a certain feature. It then looks for other nodes that can provide the input to the function. The evaluation function will know what the best evaluation is and so it can retrieve values from nodes it finds and then create links to those that give it the best evaluation. The evaluation can also be distributed, where each node only evaluates a part of it, similar to evaluating one of the where clause comparisons. The links can then be used as described previously to provide reasoning. If the node continuously evaluates, it can cope with more volatile information and can adjust its links to reflect the current state of the network. This means that the knowledge is generated not just by humans executing the queries but also autonomously by the system itself and this may help to ensure that certain features are included in the network and so can be answered even if not previously queried by a user.

Depending on the evaluation process, the nodes may still be relatively lightweight and not very cognitive. It may only know about some of the sources it can link to. If a new source joins the network or an unknown source is found, then the node may not know how to use it. One solution to this is an even more heavyweight node that can match metadata in a semi-intelligent way. There is also a centralised solution to this. Say that there is an ontology built up from the network structure that defines the network concepts. This ontology can be used to build the network structure, or by a user to discover the network contents. A centralised metadata component could periodically search the ontology and retrieve sources that provide the same sort of functionality. It may be able to do this by recognising the context in which the sources are used. It could simply build up a list of similar source types and this list can be sent through the network to all nodes. If a node is on the list, or if it links to a node type on the list, then it can try to retrieve information from the other related source types to see if they will improve its evaluation. This centralised matching engine could be application dependent with the links then stored in a view rather than the network itself, making the whole process local and more specific to a particular requirement. Or it could update links in the actual network to make the whole network more intelligent, when any user can benefit from it. Fig.2 gives some idea of how the centralised and distributed components might interact.

7   Conclusions

The focus of this work has been on the development of a lightweight framework for building networks of knowledge. The lightweight framework allows sources of small sizes, for example sensors in a pervasive environment, to use the network. It does not however prevent other source types such as Web pages or databases etc. to all connect to and also be used by the network. The network should be autonomous and so should be able to maintain itself through stigmergic activities. However, stigmergy does not assume any knowledge or intelligence and to develop the next generation of network some form of cognitive process would be very desirable. This would allow the user to ask queries that require some form of reasoning and this will increase the scope of the applications that can use the network. The knowledge network can use linking of sources to optimise query performance. This linking can be local in a view or global in the network itself, when new users can then benefit from the experiences of other users. Tests have shown that the linking is indeed very useful and could justify itself purely as an optimisation technique. Tests have also shown that the linking can maintain a reasonably good quality of service with regard to the query answer. This means that any knowledge that it provides will be reasonably good. Tests with query numbers of 10000 or greater is certainly a large number. But the tests showed that performance was still improving after this number. The network would need to monitor itself before this, to determine when the links were reliable and could be made live.

The linking structures can be seen to provide information that may be useful for reasoning as well. As they are built by the cognitive processes of the users of the network, the information should be fairly sensible. With the addition of some basic mathematical operators, it is possible to build a lightweight distributed reasoning engine that can answer user's questions that are more advanced than simple information retrieval. This reasoning can also return an indication of the reliability of the information. By making the network nodes slightly more heavyweight it is possible to introduce autonomous processes that can use evaluation functions to generate these links by themselves. In this case the network can dynamically configure itself and generate knowledge for specific concepts that can provide answers even if not previously queried by any user. However, storing data will make the nodes more heavyweight and so it is possible to limit the number of allowed links. But there needs to be a balance between limiting the amount of memory and allowing enough links to be created for knowledge to be effectively inferred. A user could ask for reasoning over anything but the replies will be limited to the linked information only. For the linking to be effective it is assumed that the queries will be skewed, but it could probably be assumed that the reasoning queries would also reflect the same use of the network that the information retrieval queries did.

The stigmergic approach however still has limitations. In a typical knowledge base there will be sets of rules that can be used to derive new information from sets of facts. The network contents represent the facts, but the storage requirements for the rules may also be substantial. Such rules may not exist in the knowledge network itself, but could be included locally in a client-side application or in an application like ConceptNet. The application would store rules relating to the information that it allows a user to query. The rules might be read from an ontology. The client-side application would also need a query re-writing tool. Then, through a re-writing of the query using these rules, it would be possible to generate an alternative query to be executed that does not require the same level of reasoning. The network itself does not then need to become more heavyweight but can still provide distributed reasoning over even more complex questions. The only addition to the query language would maybe be two additional keywords. The linking mechanism is also very lightweight as two nodes only need to inform each other that they are related. These links could partially, but not completely, replace an ontology, for example. While the stigmergic linking has been shown to be effective for optimisation, the reasoning ideas suggested in this paper are still open questions. While different methods could probably be used to implement the reasoning, the simplicity of the process could make it an attractive possibility for a lightweight autonomous framework.

Acknowledgements

This work has been carried out in the project CASCADAS (IST-027807), which is supported by the European Framework VI FET Proactive Initiative IST-2004-2.3.4 programme of the European Commission.

References

Baumgarten, M., Bicocchi, N., Curran, K., Mamei, M., Mulvenna, M.D., Nugent, C., Zambonelli, F.: Towards Self-Organizing Knowledge Networks for Smart World Infrastructures. In: Invited Session on Service Development and Provisioning through Situated and Autonomic Communications at International Conference on Self-Organization and Autonomous Systems in Computing and Communications (SOAS'2006), Erfurt, Germany, (18-21 September, 2006)
Dorigo, M., Birattari, M., Stutzle, T.: Ant Colony Optimization - Artificial Ants as a Computational Intelligence Technique. IEEE Computational Intelligence Magazine (2006)

Koloniari, G., Petrakis, Y., Pitoura, E., Tsotsos, T.: Query workload-aware overlay construction using histograms. In: Proceedings of the 14th ACM International Conference on Information and Knowledge Management., pp. 640–647 (2005)

Liu, H., Singh, P.: ConceptNet: A Practical Commonsense Reasoning Toolkit. BT Technology Journal 22 (2004)

Mulvenna, M.D., Zambonelli, F., Curran, K., Nugent, C.D.: Knowledge Networks. In: Sutcliffe, G., Voronkov, A. (eds.) LPAR 2005. LNCS (LNAI), vol. 3835, pp. 99–114. Springer, Heidelberg (2005)

Raschid, L., Wu, Y., Lee, W.-J., Vidal, M.-E., Tsaparas, P., Srinivasan, P., Sehgal, A.K.: Ranking Target Objects of Navigational Queries. In: 8th ACM International Workshop on Web Information and Data Management WIDM'06, pp. 27–34 (2006)

Ramos, V., Abraham, A.: Evolving a Stigmergic Self-Organized DataMining. In: IADIS, editor, IADIS-04, International Conference on Web Based Communities (2004)

Ricci, A., Omicini, A., Viroli, M., Gardelli, L., Oliva, E.: Cognitive Stigmergy: A Framework Based on Agents and Artifacts. In: The Third International Workshop on Environments for Multiagent Systems (E4MAS) (2006)

Serugendo, G.D.M., Gleizes, M.P., Karageorgos, A.: Self-Organisation and Emergence in MAS: An Overview. Informatica 30, 45–54 (2006)

Izquierdo-Torres, E.: Collective Intelligence in Multi-Agent Robotics: Stigmergy, Self-Organization and Evolution (2004) (last accessed 7/4/07),
citeseer.ist.psu.edu/izquierdo-torres04collective.html

Vidal, M.-E., Raschid, L., Mestre, J.: Challenges in Selecting Paths for Navigational Queries: Trade-off of Benefit of Path versus Cost of Plan. In: Seventh International Workshop on the Web and Databases (WebDB 2004), 61–66 (2004)

Dynamic Ontology Mapping for Interacting Autonomous Systems

Steven Heeps[1], Joe Sventek[1], Naranker Dulay[2], Alberto Egon Schaeffer Filho[2],
Emil Lupu[2], Morris Sloman[2], and Stephen Strowes[1]

[1] Department of Computing Science, University of Glasgow
{heeps,joe,sds}@dcs.gla.ac.uk
[2] Department of Computing, Imperial College London
{n.dulay,aschaeff,e.c.lupu,m.sloman}@imperial.ac.uk

Abstract. With the emergence of mobile and ubiquitous computing environments, there is a requirement to enable collaborative applications between these environments. As many of these applications have been designed to operate in isolation, making them work together is often complicated by the semantic and ontological differences in the meta-data describing the data to be shared. Typical approaches to overcoming ontological differences require the presence of a third party administrator, an approach incompatible with autonomous systems. This paper presents an approach to automatic ontology mapping suitable for deployment in autonomous, interacting systems for a class of collaborative application. The approach facilitates the collaboration of application-level data collections by identifying areas of ontological conflict and using meta-data values associated with those collections to establish commonality. A music sharing application has been developed to facilitate the sharing of music between peers.

1 Introduction

Recent advances in ubiquitous and mobile computing have dramatically changed the role of the computer in users' lives and made mobile computing the new personal computing and communication paradigm. The overriding motivation is that computing systems should seamlessly integrate into the life of the user and interoperate with other systems to offer mobile services as and when desired.

We have previously proposed the concept of a Self-Managed Cell (SMC) as the fundamental management design pattern for autonomous systems [20]; an SMC is a policy-based architecture that provides autonomic management capabilities for ubiquitous computing environments [3,6,10,19]. In ubiquitous environments, SMCs need to collaborate without having a pre-agreed schema, and it is also desirable that there is agreement and common semantics for applications and devices. The SMC architecture currently supports integration at the system and management level where the basics for SMC interaction are handled in terms of policy, data and event exchanges [17]. Successful SMC integration at this level provides the mechanisms for services at the application level to collaborate.

D. Hutchison and R.H. Katz (Eds.): IWSOS 2007, LNCS 4725, pp. 255–263, 2007.

This paper explores the challenge of integration at the application level. Semantic differences between collaborating applications are usually managed by an administrator who maps the differences or documents a strict ontology to which systems developers and users adhere. It is likely that ontological and semantic differences between individual applications will prove a barrier to application collaboration due to the autonomous nature of the environment.

To explore application level ontology conflict and develop suitable mapping mechanisms, we have investigated the use of SMCs in the domain of peer-to-peer music sharing. The ability to see and listen to the music of others became prominent when Apple Inc. released a version of iTunes that supported the sharing of music collections on the same sub-network through the DAAP protocol [1]. This change, from music players as a single-user jukebox application to a tool for music sharing, brings with it the potential for further study, particularly with regards to the divergence of meta-data used to describe the tracks within each player. The following example highlights this problem: Bob and Alice have streaming access to each other's music collection. Bob loves "Indie" music, and searches for this in Alice's collection. Disappointingly, no matching tracks are found as Alice has not defined the genre "Indie", despite having a number of tracks that Bob would commonly classify as "Indie". There is a clear semantic difference in the way Bob and Alice define their music collections; whilst this is a standard feature of personal music collections, overcoming these differences automatically would undoubtedly enhance the users music sharing experience.

The paper is organised as follows: Section 2 describes the automatic ontology mapping mechanism; Section 3 discusses a prototype implementation in a peer-to-peer wireless music sharing environment; Section 4 presents related work, with conclusions and directions for future work presented in Section 5.

2 Automatic Ontology Mapping Mechanism

Seamless collaboration at the application level is difficult. It is unlikely that discovered services and applications will adhere to a common language or naming structure. It is likely that different devices and applications will originate from different vendors who use different semantic descriptions. Alternatively, semantics are user-defined and thus subject to great variation [18].

Ontologies are used to solve the semantic difference problem between application and application content. Ontologies capture knowledge of a given domain in a generic yet formal way, so that it can be reused and shared across applications and users. Ontologies are generally created via a man-made, time-consuming process where humans attempt to define all aspects of a system in a very explicit fashion. Frequently, different ontologies define very similar knowledge. Mapping between ontologies associates terms defined in one ontology with terms in another. Currently, such mappings are identified manually [15]. This is extremely resource intensive, not always possible and susceptible to ontology change. Automatic ontology mapping covers a large number of fields from machine learning and formal theory to database schema and linguistics. Applications also range significantly, from academic prototypes to large scale industrial applications [5]. Most systems are fairly complex, resource intensive creations and, as such, are not deployable in resource-limited, ubiquitous computing environments [11,13].

To confirm the need for ontology mapping in the music player context, we analysed the music collections of 17 users comprising 64,704 songs. There were a total of 6,040 artists and 462 distinct music genres in the libraries studied. The existence of 462 distinct genres indicates immediately that there are going to be vast ontological differences between the music of only 17 peers. Apple's iTunes, for example, only contains approximately 30 different default genres, indicating that user-defined genres are very popular. The analysis also highlighted that approximately one third of all artists had more than one genre associated with them across the libraries. Table 1 shows six popular Artists from the libraries studied and the number of unique genres with which they were associated. This was apparent for all track meta-data, such as Track Size, Length, Album, Format and Artist.

Table 1. Genres Associated with Artists

Artist	Number of Unique Genres	Genres
Miles Davis	3	Alternative and Punk, Jazz , No Genre
Mozart	3	Classical, Classicism, Concerto
Marvin Gaye	4	Dance, Electronica, RandB, No Genre
Bob Dylan	6	Folk, Pop, Rock, Soundtrack, Various, No Genre
The Beatles	7	Alternative Rock, Dance, Electronica, Pop Rock, Rock and Pop, Rock and Roll, No Genre
Oasis	8	Alternative, Alternative and Punk, Alternative Rock Brit Pop, Pop, Punk, Rock, No Genre

2.1 The Basic Mechanism

We restrict our considerations to applications that manipulate data that conform to a common schema - i.e. the application expects to access a data collection that can be modelled as a relational table; each row of the table corresponds to one object (e.g. a musical track), and each column corresponds to a metadata attribute for that type of object (e.g. Genre, Artist); finally, one, additional column containing the value of the object is included in each row (e.g. the actual encoding of a musical track).

Using the music player example, the collection of tracks used by a particular player can be represented as shown in Table 2.

Each user is associated with a "home" collection of objects; in the music sharing example, it is the collection associated with the users music player; difficulty can ensue when the application has access to one or more "foreign" collections in addition to the "home" collection. The user is most familiar with navigation through the "home" collection; in order to effectively access objects in the "foreign" collections, it is important to map the metadata values that describe the "foreign" objects into values that have meaning to the user.

Table 2. An Example Home Collection

Title	Artist	Composer	Genre	Album	Size(mb)	...	Value
Son	Jethro Tull	Ian Anderson	Rock	Benefit	2.77		mt000001.mp3
Black Hole Sun	SoundGarden	Chris Cornell	Grunge	Superunknown	5.02		mt000002.mp3
Exsultate, jubilate	Kiri Te Kanawa	Mozart	Classical		14.11		mt000003.mp3
Rusty Cage	Johnny Cash	Chris Cornell	Country	Unchained	1.31		mt000004.mp3
Hush	Tool		Metal	Opiate	1.30		mt000005.mp3
Sleeping	The Band		Country Rock	Stage Fright	3.11		mt000006.mp3
Hello	Evanescence		Gothic Rock	Fallen	3.48		mt000007.mp3
...							

In general, the metadata attributes exhibit correlated values within a collection - i.e. many objects with $attribute_i = value_i$ also have $attribute_j = value_j$. The degree of correlation between $attribute_i$ and $attribute_j$ will depend upon: the attributes chosen, the nature of the collection, and the degree of consistency in value assignment when objects are added to the collection. For example, most artists are strongly correlated with a particular genre (e.g. all tracks produced by Pearl Jam are associated with the Grunge genre), while release dates are only weakly correlated with a particular genre (e.g. Grunge is correlated with release dates 1990 and beyond, but not before). Such correlations can be asymmetric due to the fact that some attributes have broader scope than others; the correlation strength is a measure of the predictive power of one value over the value of the other (e.g. Pearl Jam strongly predicts Grunge, but Grunge predicts Pearl Jam, Soundgarden, Alice in Chains, etc.).

Consider a collection of N objects, and each object has M metadata attributes associated with it. Let us focus upon two attributes, i and j. In a particular collection, $Attr_i$ takes on values $v_{i1}...v_{in}$; similarly, $Attr_j$ takes on values $v_{j1}...v_{jn}$. We can then analyze all of the tracks in the collection to yield the following matrix (Table 3):

Table 3. Pairwise Classification of Objects in a Collection

$Attr_i / Attr_j$	V_{j1}	V_{j2}	...	V_{jn}
V_{i1}	C_{11}	C_{12}		C_{1n}
...				
V_{im}	C_{m1}	C_{m2}		C_{mn}

where C_{k1} is the number of objects in the collection that have $Attr_i = V_{ik}$ and $Attr_j = V_{jl}$. It is informative to consider two limiting cases:

1. $Attr_i$ is strongly correlated with $Attr_j$: in this case, if there are N_{ik} objects with $Attr_i = V_{ik}$, then most of those objects will have $Attr_j = V_{jl}$ for some l; note that by definition, $N_{ik} > 0$.
2. $Attr_i$ is not correlated with $Attr_j$: in this case, the N_{jk} objects with $Attr_i = V_{i,k}$ are distributed over many different values for $Attr_j$.

We can sum over the pairwise matrix in Table 3 to determine the predictive power of $Attr_i$ for $Attr_j$ as well as the predictive power of $Attr_j$ for $Attr_i$. One such formulation is as follows:

$$\text{predictive power}_{i,j} = \sum_{k=1}^{m} \frac{max_l\{c_{kl}\}}{\sum_{l=1}^{n} c_{kl}} \tag{1}$$

Obviously, the predictive $power_{j,i}$ simply requires that we swap k for l and m for n in Equation (1). Performing this analysis for all pairs of attributes yields a correlation matrix of the form shown in Table 4. The value in the i, j^{th} cell indicates how strongly correlated values of $Attr_i$ are to values of $Attr_j$; obviously, the diagonal elements have a value of 1. Armed with this correlation information for the home collection, we now describe a protocol that uses this mechanism to dynamically map objects from a foreign collection into the home object ontology.

Table 4. Predictive Power

	$Attr_1$	$Attr_2$	$Attr_3$	...	$Attr_M$
$Attr_1$	1.000	0.357	0.771		0.467
$Attr_2$	0.953	1.000	0.849		0.121
...				1.000	
$Attr_M$	0.125	0.294	0.186		1.000

2.2   The Mapping Protocol

The general protocol is as follows: if one is interested in objects in the foreign collection with $Attr_i = Value_i$, and none exist, then one searches the ith column of Table 4 from the home collection for the $Attr_j$ with the largest correlation value (excluding row i). One can then query for objects corresponding to known $Value_j$'s, and discover the $Value_i$'s that the foreign collection associates with those objects. One can then import objects with those particular $Value_i$'s, replacing the actual $Value_i$ with the value used by the home collection.

Assume that two peers are sitting on a train, each with a personal music player in the form of a PDA hosting a music streaming service; the two players have discovered each other, and the policies in the two players permit streaming of tracks from one player to the other. Once the players have bound together, the music services on each player can enter into the ontology mapping protocol. Bob's music service remotely performs a genre search on Alice's system for each value of the genre meta-data attribute defined for Bob's system; for example, suppose that one value of the genre attribute is "Grunge". Unfortunately Alice does not have any music defined as "Grunge", so the initial query returns a negative. The ontology mapping mechanism in Bob's music player selects a meta-data attribute strongly correlated with Genre, namely Artist, and queries Alice's player with a list of Artists associated with the genre "Grunge". Alice's music service then searches for those Artists in her collection, and returns the most-prevalent genre value, if any, associated with each artist in her collection. The protocol has established a Bob-specific mapping from his genre values to those used by Alice. Bob's music service can now represent tracks in Alice's system using Bob-specific genre values. Besides enabling comfortable navigation over the other individual's collection and subsequent streaming, the mapping information can also be retained for future sharing with each other, or possibly to inform future negotiations with other peers. The current protocol maps Bob's genre value to multiple genre values in Alice's collection. Another approach would be to only solicit the Alice genre value for the artist in Bob's collection with the largest number of tracks with that particular value, or the largest percentage of tracks with that particular value. The current approach maximises the number of tracks mapped to facilitate human navigation; more study is needed to determine if other approaches yield more usable results.

Table 5. Predictive Power of Music Tracks

	Genre	Artist	Name	Album	Year	BitRate	Kind
Genre	1	0.579	0.25	0.57	0.475	0.646	0.885
Artist	0.818	1	0.623	0.861	0.855	0.865	0.921
Name	0.908	0.946	1	0.912	0.905	0.939	0.941
Album	0.857	0.893	0.275	1	0.793	0.888	0.964
Year	0.283	0.259	0.139	0.256	1	0.376	0.462
BitRate	0.238	0.188	0.187	0.234	0.184	1	0.939
Kind	0.18	0.13	0.039	0.035	0.064	0.299	1

The mapping factor (attribute strongly linked to "Genre" in the preceding example) is determined through analysis of music collections. The application of Equation (1) to the meta-data from 17 unique iTunes music libraries yielded Table 5. The mapping factors for music collections indicate, for example, that there is a close relationship between Artist and Genre (0.818). In other words, if the Genre is not known then Artist is a good aspect of meta-data to map from, as is, Name and Album. Kind and Year, however, would not be suitable search attributes.

Even though our discussion is dominated by music sharing examples, other types of data collections are accessed in this way; for example, the collection of books maintained by a library. Initial results from a study of the meta-data for multiple book libraries also shows similar disparities across Subject Headings.

3   Experimental Validation

The Self-Managed Cell architecture running a music sharing service has been implemented as a test platform for our automatic ontology mapping technique. The music sharing service utilises core SMC services such as the discovery and policy service.

The SMC has been built to run on a PDA (HP iPAQ hx4700, with a 624MHz XScale PXA270 processor and 64MB RAM, running Familiar Linux 0.8.4 or Windows Mobile 5.0). The SMC is written in Java, and uses JamVM 1.4.3 [8] in a bid to cut down on memory usage. The policy service used is Ponder2 written in Java 1.4. The music player, built to run as a service on an SMC, is also written in Java 1.4. The player enables a user to search the music collection of other discovered music players and stream music found from their search via wifi to their music player. It uses the DAAP [2] which performs as an HTTP server for advertising and streaming requested songs to clients. At present the music player has been successfully tested and functions successfully under J2SE. Currently attempts are being made to run the player on a PDA under Windows Mobile 5.0 using the Mysaifu JVM [14]. The music player is approximately 4Mb in size and has a memory footprint of around 15-30mb depending on activity status i.e. idle, playing, streaming etc. The music service relies upon the mechanism documented in [17] for establishing the initial peer-to-peer binding between a pair of music players running as services on SMCs.

The ontology mapping mechanism, as used to enhance collaboration between peer music libraries, has been fully tested and evaluated. Analysis of collaborations using the 17 peers documented in Section 2 revealed significant use of the mapping system, with song returns frequently running into the hundreds where initial collaboration had revealed few or no artists. Genre-to-Artist mapping results from a peer-to-peer collaboration are shown in Table 6. Only genre searches where no song results were initially returned are shown.

Table 6. Genre-Artist Mapping

Peer 1 Genre Request	Peer 2 Returns after Mapping		
	Genres	Artists	Songs
Blues	2	337	3594
Classic Rock	2	282	2352
Electronica	1	115	587
Folk	2	282	2352
Rock/Pop	2	337	3594
Soul	1	11	109
Top 40	1	40	467

4 Related Work

Automatic Ontology mapping has seen a surge of research interest in recent years. Formal ontology mapping approaches have modelled ontologies using graphs, logic and models with mappings being developed from viewing graph, logic and model convergence [11,13]. Current software systems that automatically generate ontology mappings are ONION [13], MAFRA [4] and IFF [16]. ONION generates mappings using graph transformations. MAFRA combines different similarity measures, both lexical and structural, to establish the mappings. IFF is based on convergence between logical theories [5].

Such ontology mapping mechanisms are unlikely to be suitable for use in our ubiquitous environment. They have primarily been designed to provide automated administrative assistance when mapping well defined but conflicting ontologies in traditional conflicting environments. They require considerable user input and tend to focus on the use of a bridging ontology, a resource unlikely to be available in the ubiquitous world. Furthermore, the mapping mechanisms would likely struggle in the undefined and uncontrolled ubiquitous world. Most mechanisms are also not suitably lightweight so as to be deployable on resource limited devices.

Online music based Information Retrieval mechanisms are also gaining prominence. Last.fm [9], for example, leverages each user's musical profile to make personalised recommendations and connect users who share similar tastes. The downside of such mechanisms is the need for a common software plug-in and a network connection.

5   Conclusions and Future Work

A novel automated ontology mapping mechanism has been described that supports application-level integration within ubiquitous systems. The mechanism facilitates the successful collaboration of data collections by using meta-data contained within the collections to identify areas of commonality between them. The commonality identified is then used to automatically generate a common ontology and map between the areas of conflict. By using the meta-data information stored within music tracks, for example, we were able to successfully share music between peers despite there being no outwardly visible signs or commonality for collaboration. The techniques establish the beginnings of a common ontology and enabled a reference regarding the mapping to be held for future sharing. The system is suitably lightweight and resource efficient that it is capable of running in constrained environments such as PDAs and mobile telephones using our Self-Managed Cell architecture .

The current prototype uses exact string match during the mapping protocol. Given the anarchy that exists within some distributed collections we will investigate similarity matches between attribute values in an attempt to understand if this provides improved matching results. Likewise, future work will investigate enhancements to the quality of the mapping mechanism, particularly in relation to ranking results based on the probability a user will like them and will define how the mapping factors are regenerated over time.

Acknowledgements

The authors wish to thank the UK Engineering and Physical Sciences Research Council for their support through grants GR/S68040/01, GR/S68033/01 and GR/N15986/01.

References

1. Apple. ipod and itunes (2007), http://www.apple.com/itunes
2. Boot, C.: Digital audio access protocol (2007), http://daap.sourceforge.net/
3. Dulay, N., Heeps, S., Lupu, E., Mathur, R., Sharma, O., Sloman, M., Sventek, J.: Amuse: Autonomic management of ubiquitous e-health systems. In: Proceedings of the UK e-Science Al l Hands Meeting, UK (2005)
4. Kalfoglou, Y., Schorlemmer, M.: IF-map: an ontology mapping method based on information flow theory. Journal on Data Semantics, 98–127 (2003)
5. Kalfoglou, Y., Schorlemmer, M.: Ontology mapping: the state of the art. The Knowledge Engineering Review 18(1), 131 (2003)
6. Keoh, S.L., Twidle, K., Pryce, N., Schaeffer-Filho, A.E., Lupu, E., Dulay, N., Sloman, M., Heeps, S., Strowes, S., Sventek, J., Katsiri, E.: Forthcomming: Policy-based management for body-sensor networks. In: 4th International Work- shop on Wearable and Implantable Body Sensor Networks (2007)
7. Kong, L.C.Y., Wang, C.L., Lau, F.C.M.: Ontology mapping in per- vasive computing environment. In: International Conference on Embedded and Ubiquitous Computing, pp. 1014–1023 (2004)

8. Lougher, R.: Jamvm (2007), http://jamvm.sourceforge.net/
9. Last.fm (2007), http://www.last.fm
10. Lupu, E., Dulay, N., Sloman, M., Sventek, J., Heeps, S., Strowes, S., Twidle, K., Keoh, L., SchaefferFilho, A.E.: Amuse: autonomic management of ubiquitous systems for e-health. Special Issues of the Journal of Concurrency and Computation: Practice and Experience (2006)
11. Maedche, A., Motik, B., Silva, N., Volz, R.: A mapping framework for dis- tributed ontologies. In: 13th International Conference on Knowledge Engineering and Knowledge Management (2002)
12. Microsoft. Zune (2007), http://www.zune.net
13. Mitra, P., Wiederhold, G., Kersten, M.: A graph-oriented model for ar- ticulation of ontology interdependencies. In: 7th International Conference on Extending Database Technology (2000)
14. Mysaifu. Mysaifu (2007), http://sourceforge.jp/projects/mysaifujvm/
15. Nay, N.F., Musen, M.A.: Prompt: Algorithm and tool for automated ontology merging and alignment. AAAI (2000)
16. Romn, M., Hess, C., Cerqueira, R., Ranganathan, A., Campbell, R., Nahrst, K.: Gaia:a middleware infrastructure to enable active spaces. IEEE Pervasive Computing, 74–83 (2002)
17. Schaeffer-Filho, A., Lupu, E., Dulay, N., Keoh, S., Twidle, K., Sloman, M., Heeps, S., Strowes, S., Sventek, J.: Supporting interactions between self-managed cells. In: International Conference on Self-Adaptive and Self-Organizing Systems (2007) (submitted)
18. Heeps, S., Dulay, N., Lupu, E., Schaeffer-Filho, A.E., Sloman, M., Strowes, S., Sventek, J.: The autonomic management of ubiquitous systems meets the semantic web. In: The Second International Workshop on Semantic Web Technology For Ubiquitous and Mobile Applications (2006)
19. Strowes, S., Badr, N., Dulay, N., Heeps, S., Lupu, E., Sloman, M., Sventek, J.: An event service supporting autonomic management of ubiquitous systems for e-health. In: 5th International Workshop on Distributed Event-Based Systems (2006)
20. Sventek, J., Badr, N., Dulay, N., Heeps, S., Lupu, E., Sloman, M.: Self-managed cells and their federation. In: CAiSE Workshops vol. 2, pp. 97–107 (2005)

Trade-Off Between Performance and Energy Consumption in Wireless Sensor Networks

José-F. Martínez, Ana-B. García, Iván Corredor, Lourdes López, Vicente Hernández, and Antonio Dasilva

EUIT Telecomunicación - DIATEL
Universidad Politécnica de Madrid
Ctra. Valencia, Km. 7, 28031, Madrid, Spain
{jfmartin,abgarcia,icorred,llopez,vhernandez,
adsilva}@diatel.upm.es

Abstract. Nowadays WSNs support applications such as target tracking, environmental control or vehicles traffic monitoring. Generally, these applications have strong and strict requirements for end-to-end delaying and loosing during data transmissions. In this paper, we propose a practical scenario for application of the WSN field in order to illustrate selection of an appropriate approach for guaranteeing performance in a WSN-deployed application. The methodology we have used includes four major phases: 1) Requirements analysis of the application scenario; 2) QoS modelling in different layers of the communications protocol stack and selection of more suitable QoS protocols and mechanisms; 3) Definition of a simulation model based on an application scenario, to which we applied the protocols and mechanisms selected in phase 2; 4) Validation of decisions by means of simulation; and 5) analysis of results. This work has being partially developed in the framework of the CRISAL - M0700204174 project (partially funded by "Universidad Politécnica de Madrid" and "Comunidad de Madrid", Spain).

Keywords: Wireless Sensor Networks, QoS protocols, performance, target tracking, natural environments surveillance.

1 Introduction

Recently, we have witnessed significant progress in the field of wireless sensors. The latest stage has been characterized by improvements in sensor hardware issues (miniaturization of pieces, increased ROM and RAM capacities, more energy capacity, etc). These facts and the new field of possibilities for their application have boosted interest in Wireless Sensor Networks (WSN). WSN might be defined as follows: *Networks of tiny, small, battery-powered, resource-constrained devices equipped with a CPU, sensors and transceivers embedded in a physical environment where they operate unattendedly.* While a good deal of research and development has been carried out in architecture and protocol design, energy saving and location, only a few studies have been done on network performance in WSN (Quality of Service – QoS).

Some studies on QoS have focused on protocols and mechanisms for MAC and the network layer, and almost all these have been developed and tested through simulations.

D. Hutchison and R.H. Katz (Eds.): IWSOS 2007, LNCS 4725, pp. 264–271, 2007.

All these approaches for supporting QoS in WSN may constitute a basis for future work in this direction, and they obviously represent the starting point in our proposal. We have already conducted work on state-of-the-art QoS in WSNs [1]. This work has focused mainly on QoS-based protocols and mechanisms in MAC and network layers.

The remainder of the paper is organized as follows:

The case study is depicted in section 2. In this section the proposed application scenario is described, as are all its QoS-related characteristics. Based on these characteristics, the section concludes with a selection of the most suitable protocols for network layer available in literature on WSN. The validity of decisions on QoS protocols is verified in section 3, with the use of simulation software to perform tests. Section 4 concludes this paper with an overview of future research activities.

2 QoS in Natural Environments Surveillance (*Case Study*)

We will begin by extracting the QoS-related requirements from the real-time forest surveillance application; allow us to select the network protocols later that best suit these requirements.

2.1 Description and Analysis of Requirements of Application for Real-Time Forest Surveillance

The main objective of the application will be the early detection of forest fires to avoid ecological disasters, as well as detection and tracking of strange vehicles.

Sensor nodes collect measurement data, such as relative humidity, temperature, magnetic radiation, COx and NOx gases. Other components of the WSN supporting our application are laptops and/or PDAs (as support to firemen and safety watchmen) and a data base server. All WSN services will be accessible to remote users through web services. Figure 1 illustrates the proposed scenario.

Specifically, the application will have the following characteristics:

1) Topology and network dynamics: The WSN topology is a design parameter that should be taken into account when guaranteeing QoS.

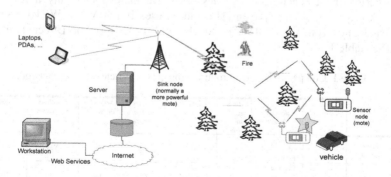

Fig. 1. Forest surveillance application scenario

2) Geographical information: Sensor nodes must obtain geographical information – i.e., coordinates – in order to locate the events within the natural reserve. For WSNs, a GPS-based approach is too expensive, thus our WSN implements a distributed location service [2].

3) Real-time requirements: Fire monitoring or target tracking reflects the physical status of dynamically changing environments such as temperatures or positions of moving targets in forest areas. This sensory data is valid only for a limited time; hence it needs to be delivered within a time deadline.

4) Unbalanced mixture traffic: Another characteristic which will considerably affect QoS decisions is *reactive-proactive* hybrid behaviour.

5) Data redundancy: High redundancy in the sensor data is a common characteristic to most WSNs. Redundancy may improve several the reliability and robustness of data delivery. However, this uses a large amount of energy. To solve this problem, we could use data fusion or data aggregation to maintain robustness while decreasing redundancy in the data. On the other hand, these mechanisms also complicate QoS design in WSNs.

6) Energy efficiency: An important challenge to this application will be energy efficiency. The need to operate over a long period of time (from 6 months to 1 year), between other factors, will require careful management of energy resources. Nowadays, achieving this energy distribution without compromising the QoS requirements is very difficult since mechanisms and protocols do not usually consider both possibilities at the same time.

7) Sensor data priority: Not all sensing data are equal; hence they have different levels of priority depending on their importance level into the application. For example, the data generated in a fire detection event will have more importance than that generated in monitoring the conditions that increase the risk of fire. QoS mechanisms will determine the data delivery priorities for the different data types existing in the WSN.

As a result, QoS support for the network will take into account almost all of the aforementioned characteristics in the application specifications.

2.1.1 Selected Network Protocols

Considering the characteristic just described, we believe that only a few of the network protocols of surveyed in [1] could be used in our WSN. We have selected two candidates from among these protocols: MMSPEED [4], and Directed Diffusion [5]. (See table 1).

Table 1. Comparative table of routing protocols in Wireless Sensor Networks

	Network topology	Data delivery model	Data aggreg ation/f usion	Traffic guarantees	Several traffic classes	Networks dynamics	Resources reservation	Scalability
Directed Diffusion	Flat	Query-driven and Event-driven	Yes	Reliability	No	Limited	Yes	Medium
MMSPEED	Flat	Event-driven and Continuous	No	Reliability and Real-time	Yes	Limited	No	High

We selected these protocols for several reasons:

MMSPEED

→MMSPEED implements localized geographic routing, which is fundamental for the network layer of our stack protocol. These mechanisms increase self-adaptability of the network to dynamic changes as well as scalability of the network. In addition, this protocol is suited for both periodic (real-time) and aperiodic traffic because routing decisions are local (i.e., no path setup and failure recovery).

→MMSPEED implements a multi-speed mechanism to assign diverse deadlines to the packets with different delay requirements. This mechanism is ideal for supporting multiple traffic types (continuous, event-driven, etc.). Its dynamic speed compensation mechanism, which is capable of immediately correcting small inaccuracies occurring in initial routing decisions, is also quite useful.

→Routing decisions in MMSPEED are also made according to the reliability level required by the packet. To route on the basis of the reliability requisite, MMSPEED has an advanced method of lending reliability to data transmissions which involves using the frame loss rate of the MAC layer to make an estimate of the reliability level of each link.

However, MMSPEED lacks a method for dealing with the data redundancy problem. In this sense, we are in the course of studying how a data aggregation mechanism (such as meta-data negotiation [3]) can be added to MMSPEED.

Directed Diffusion

→ Directed Diffusion is a data-centric and application-aware paradigm. This protocol implements a mechanism based on data aggregation to eliminate redundant data coming from different sources. This feature reduces the number of transmissions drastically, leading to two main consequences: firstly, the network saves energy and extends its lifetime, and secondly, it has higher bandwidth in the links near the sink node.

→Directed diffusion is based on a query-driven model. This means that the sink node requests data by means of broadcasting *interests*. When events begin to appear, they start to flow towards the originators of interests along multiple paths. This behaviour provides reliability and robustness to data transmissions in the network.

Although Directed Diffusion includes all these optimization mechanisms, the protocol has two shortcomings in the realm of QoS: Directed Diffusion can neither explicitly manage QoS parameters such as delay and reliability, nor differently handle more than one traffic class.

3 Simulation of Application Scenario

3.1 Simulation Model

The table 3 depicts the simplified simulation model defined for the application described in this section.

The sensor nodes are deployed around the mountain, distributed in four sectors (*North, South, West* and *East*). The sink node is placed at coordinate (0,0). (See figure 1).

Table 2. Simulation Environment Settings

Size terrain	600mx600m
Terrain morphology	A mountain of 400mx400m, centered in the terrain.
Sensor node number	176 nodes (sink included)
Radio range	80 m
Initial energy charge	1000 Joules
Bandwidth	200 Kbps
Payload	32 bytes

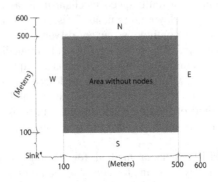

Fig. 2. WSN deployment

J-SIM is simulation software selected to implement the model. It was chosen because it is component-based, a feature that enables users to modify or improve it.

Network protocols have been configured with different parameters according to capacities.

All the parameters defined for each protocol are depicted in following sub-sections:

MMSPEED

Table 3. MMSPEED parameters

	Attaining sink probability	Max. delay (in seconds)
High priority traffic (events)	0.4	0.5
Low priority traffic (monitoring)	0.2	4

Moreover, we have defined two speed layers which have been configured with different speed levels (1000 m/s and 250 m/s, respectively).

Directed Diffusion

Directed Diffusion can be configured with multiple parameters. The most significant parameters for the simulation tests are the following: diffusion area of interests (complete area); duration of interests (all time simulation); interest refresh (every 10 seconds).

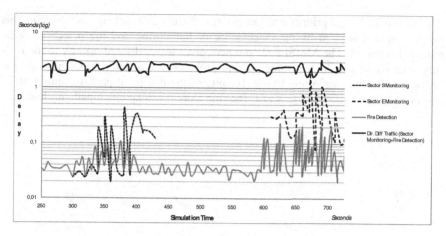

Fig. 3. Delays with MMSPEED and Directed Diffusion. Comparative graphics.

3.2 Simulation results

Deadline

-MMSPEED: The results of simulations with MMSPEED are significant in the way they show how the protocol is capable of differentiating traffic classes (see Figure 2).

When low and high-priority traffic concurs in the WSN, MMSPEED successfully supported the QoS level assigned to both traffic classes. The maximum delay configured for high-priority traffic (0.5 seconds) was never exceeded. Furthermore, the jitter (or delay fluctuation) is not excessively high, which will improve the quality of real-time data received by the application, especially if these data have been generated by the tracking of a person inside the area monitored by the WSN. In addition, low-priority traffic manages to maintain acceptable levels of delay, although the jitters are somewhat high. As it is obvious, the data proceeding from East Sector register higher delay and jitter than the data proceeding from South Sector because of this sector is placed to more distance to the sink node.

-Directed Diffusion: According to the results, it is evident that the mechanisms implemented in Directed Diffusion are insufficient to ensure the QoS level required by the WSN, specifically with regard to delays (in terms of latency and jitter). Both the delay (average value 2,5 sec) and jitter (very fluctuating reaching times) values are not suitable to provide a good QoS level to the real time traffic. These bad results are due to the data aggregation mechanism that implements Directed Diffusion needs a long processing time in intermediate nodes of the WSN.

Reliability

-MMSPEED: When MMSPEED initiates a packet flow to the sink following a period of inactivity, it is common for intermediate nodes to have incoherent routing information. Until MMSPEED recovers operational status, tenths of seconds to one second may elapse, during which a few packets might be discarded. In other cases (see figure 3), MMSPEED shows a great robustness due to its multi-path mechanism. The figure 3 shows a congestion period (3,5 seconds) during which are discarded a few packets both high priority and low priority: 30 and 270 packets, respectively. This congestion is generated by the coincidence of monitoring and fire detection traffic during this period. The MMSPEED protocol gives lower discard priority to the real-time data packets as higher attaining sink probability was assigned to the real time traffic than to the monitoring traffic. In spite of the discarded packets, almost all of low priority, no notification gets lost since the aforementioned multipath mechanism.

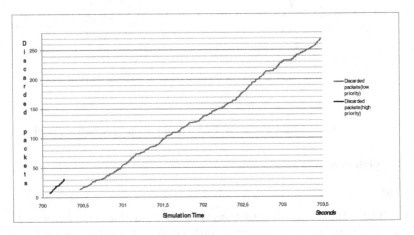

Fig 4. Discarded packets with MMSPEED during a congestion period

-Directed Diffusion: The simulation tests with Directed Diffusion were satisfactory in terms of reliability. Although Directed Diffusion does not implement an explicit mechanism to provide reliability, it achieves an acceptable reliability level by means of a multi-path routing that selects the best paths towards the sink.

Energy consumption

During the first 12 hours of simulated time, the consumption of energy of the eight nodes closest to the sink was recorded. The results can be seen in the table 4.

Table 4. Energy consumption with each protocol

	Average energy consumption	Lifetime in a real WSN
MMSPEED	3.5225 Joules/hour	9 months
Directed Diffusion	0.9575 Joules/hour	3 years

The first column shows the average levels of energy consumption in the simulation period. The second column shows the lifetime of a real WSN, assuming that sensor nodes use AA alkaline batteries.

Directed Diffusion showed the best rate of energy consumption (3 years approx.). These good results have been obtained through use of the data aggregation mechanism implemented by Directed Diffusion. This mechanism significantly reduces the number of data transmissions, and therefore helps saves a great deal of energy.

MMSPEED achieves an acceptable lifetime (9 months) and a good energy/delay balance. However, this lifetime could be increased if MMSPEED implemented a mechanism for reduction of redundant data (e.g. meta-data negotiation or data aggregation).

Taking into account all the simulation results, we conclude that the most suitable protocol for improving performance in our WSN is MMSPEED. However, this protocol should be improved with several add-on features, which should be the subject of future research.

4 Conclusions and Future Study

In this paper we have presented a study of network layer protocols that have been defined to provide QoS in wireless sensor networks. We have focused on the basic mechanisms used in these protocols for guaranteeing performance parameters to applications.

Taking this study as a basis, we have also selected a forest surveillance application in order to show how appropriate protocols for QoS could be selected by defining the performance requirements of the application and the classification criteria for protocol study.

This research has also shown what we consider to be shortcomings in the protocols. For instance, the MMSPEED protocol lacks a data aggregation or an even more preferable meta-data negotiation system. Other aspects that could be considered in more detail in MMSPEED are the energy-delay trade-off.

We are presently working on defining and subsequent deploying a WSN scenario in which a surveillance application will be run. For future research, and after the functional aspects of the application are working, we plan to include performance monitoring in the system. This will allow us to perform empirical studies of the influence of the parameters we have considered on the quality offered to the application.

References

1. Martínez, J.F., García, A.B., Corredor, I., López, L., Hernández, V., Dasilva, A.: QoS in Wireless Sensor Network: Survey and Approach. ACM DL (May 2007)
2. He, T., Huang, C., Blum, B., Stankovic, J., Abdelzaher, T.: Range-Free Localization Schemes for Large Scale Sensor Networks. Proc. Mobicom Conf. (2003)
3. Kulik, J., Heinzelman, W.R., Balakrishnan, H.: Negotiation-based protocols for disseminating information in wireless sensor networks. Wireless Networks 8, 169–185 (2002)
4. Felemban, E., Lee, C.-G., Ekici, E.: MMSPEED: multipath Multi-SPEED protocol for QoS guarantee of reliability and. Timeliness in wireless sensor networks 5(6), 738–754 (2006)
5. Intanagonwiwat, C., et al.: Directed diffusion: A scalable and robust communication paradigm for sensor networks. In: Intanagonwiwat, C., et al. (eds.) The Proc. of MobiCom'00, Boston, MA (August 2000)
6. Sobeih, A., Hou, J.C., Kung, L.-C.: J-Sim: a simulation and emulation environment for wireless sensor networks. Wireless Communications, IEEE 13(4), 104–119 (2006)

Automated Trust Negotiation in Autonomic Environments

Andreas Klenk[1], Frank Petri[1], Benoit Radier[2], Mikael Salaun[2], and Georg Carle[1]

[1] Wilhelm-Schickard-Institut,
Sand 13, 72076 Tübingen, Germany
[2] France Télécom R&D,
avenue Pierre Marzin 2, 22307 Lannion, France

Abstract. Autonomic computing environments rely on devices that are able to make intelligent decisions without human supervision. Automated Trust Negotiation supports the cooperation of devices with no prior trust relationship. They can reach an agreement by iteratively exchanging credentials during a negotiation process. These credentials can serve as authorization tokens or may carry information that becomes a parameter of the further service usage. A careful negotiation strategy helps in protecting sensitive credentials that must only be available to authorized entities. We introduce the *VersaTrust* framework that supports a stateless negotiation protocol to reach comprehensive agreements. We argue how this approach applies to autonomic environments and demonstrate its scalability.

Keywords: attribute-based access control, stateless automated trust negotiation.

1 Introduction

The growing complexity of the information technologies infrastructure leads to an increase in administrative effort to ensure availability and security of the systems. There is a lot of manual configuration associated with implementing administrative decisions. Autonomic computing research aims for facilitating administration of complex infrastructures by introducing self-management capabilities [7] into networks and devices. The coordination of autonomic entities is challenging if these entities are part of different administrative domains without unbounded mutual trust. In such scenarios, constraints of future interactions between the devices need to be considered [4] depending on the trust between the entities. The Global Grid Forum recognized the need for an automated establishment of agreements between web services with its work on the *WS-AgreementNegotiation* specification draft [5]. However, the draft neglects the protection of sensitive information during the negotiation and requires session state at the participating hosts.

The research on *Automated Trust Negotiation (ATN)* [14] [12] [2] deals with automatically establishing mutual trust between strangers by an iterative

D. Hutchison and R.H. Katz (Eds.): IWSOS 2007, LNCS 4725, pp. 272–279, 2007.

credential exchange. *Automated Trust Negotiation* systems use a policy driven iterative negotiation process to reach an agreement between two parties that need not have a prior trust relationship. The main focus is on the protection of sensitive information (credentials and policies) and the definition of policy languages for the negotiation process. However, ATN does not help to supervise or enforce the agreement. Other techniques must complement the ATN to check if the other party adheres to its promises.

In this paper, we explore the use of Automated Trust Negotiation for autonomic systems. We show how to reach an agreement via an automated exchange of policies and credentials.

1. We introduce the *VersaTrust* framework for stateless trust negotiation, explain how policies control the negotiation process and evaluate the feasibility and the performance of the implementation.
2. We argue how to represent the final agreement and the complete negotiation in one single document. That allows to demonstrate all conditions under which the negotiation succeeded, at a later point in time, say if the terms of the agreement are under dispute. This feature is a clear advantage over current ATN implementations which can only state the results of the negotiation but lack a method to prove the interrelation of the received credentials.

In Sec. 2 we survey related work. In Sec. 3 we introduce the stateless trust negotiation and show experimental results of the implementation in Sec. 4.

2   Related Work

Winsborough and Li came from the idea of credentials as tokens for authorization and introduced the idea of Automated Trust Negotiation for establishing trust between strangers in [14]. They discussed the *parsimonious strategy* to disclose only the minimal amount of credentials necessary for the successful termination of the negotiation. Sometimes the negotiation process itself discloses private information by referring to the existence of sensitive credentials during the negotiation process. The authors enhanced their negotiation with *Ack* policies to address these privacy concerns in [8].

IBM specified the *Trust Policy Language* for a role based access control scheme that uses credentials to determine which roles a principal can obtain. *Trust-Builder*[12] uses this language to implement a trust negotiation system that incorporates trust reputation measures.

PeerTrust [10] is an ATN system that can handle X.509 certificates and import RDF for its policies. Yamaki et al introduce user preferences into the trust negotiation by assigning a cost metric to the release of a credential [15]. The authors in [16] use a locally trusted third party to break cyclic dependencies between credentials that can occur during a negotiation. Frikken et al. [6] proposed a protocol that can reach a decision if the negotiation fails or succeeds without actually revealing hidden credentials. This method is appropriate if the information of the credentials is of no importance for the further service usage.

Within the scope of multi-agent systems, a large body of work exists on the negotiations between distributed agents to reach some specific goals [3]. Negotiations in multi-agent lack the capabilities of ATN systems for the protection of sensitive information and are not specifically fit to deal with credentials. ATN systems are comparable lightweight, because they reach a binary decision, (e.g. access granted/access denied), in contrast to multi-agent systems which negotiate about complex tasks, for instance, the market price of goods [9].

The *Trust-X* of Bertino, Ferrari and Squicciarini [1][2] is a recent ATN framework that had a strong influence on our work. This framework uses XML for its Trust Negotiation Language, disclosure policies and credentials. It uses DTD to specify credential types. It supports different negotiation strategies and optimization mechanisms. An important difference is that the *Trust-X* transmits individual disclosure policies and credentials during each round and relies on local state during the negotiation. Hence, it is not obvious how to proof the interrelation of the credentials retrospectively. *VersaTrust* in contrast can represent all conditions under which promises were made, that led to a specific agreement, within one single digitally signed document. Another difference is that *VersaTrust* allows for an easy recovery from system failure during the negotiation due to the stateless realization of its negotiation process.

3   Mutual Agreement with Automated Trust Negotiation

Automated Trust Negotiation governs the access to resources by attribute based authorization. Authorization can use properties connected to a subject in contrast to solely the identity. This functionality can be useful for the self-management in environments where autonomic devices without prior trust relationship join the network and establish trust at the time they interact with other services. Another scenario is the collaboration of autonomic services across administrative domains without the need for manual configuration. An important property of ATN is the disclosure of only the minimal set of credentials and the protection of sensitive information within credentials. It is even possible to authorize a resource access without revealing the actual identity of the requester.

3.1   Credentials and Disclosure Policies

ATN systems use digital credentials usually signed by a trustworthy third party. *VersaTrust* utilizes currently a XML data structure for the credentials; for real world use other credential formats, for instance, X.509, or SAML are preferable. We denote the credential set of the party that initiates the request by C_L and the credential set of the the remote party by C_R.

Disclosure policies define logical conditions that must be met before a resource can be accessed or a credential can be released. Propositional formulas help to express the conditions of the disclosure policies [13][17] using the logical symbols $\wedge$, $\vee$, $\leftarrow$ and parentheses. The formula $O \leftarrow F_O(R_1, R_2, R_3..., R_k)$ governs the access to an object O. The propositional variable R_i is true if the associated credential $C_i \in \mathcal{C}_R$ can be offered by the other party and if conditions regarding the attributes of the credential C_i are satisfied. The expression $C_j \leftarrow F_{C_j}$ states that the disclosure of credential $C_j \in \mathcal{C}_L$ is regulated by the formula F_{C_j}. Credentials without protection requirements are called *unprotected* and are by default $C_k \leftarrow true$. The implementation uses XML for the disclosure policies and the negotiation state. The formula $R_x \wedge (\bigvee_{0<y<n} R_y)$ is equivalent to the XML representation of a node R_x having a number n of children R_y.

3.2 Iterative Negotiation Process

The objective of Automated Trust Negotiation is to find a *safe disclosure sequence* of credentials $(C_1, C_2, ..., C_n)$ in a way that all preconditions attached to the release of credentials are met before releasing them. This strategy is known as *parsimonious strategy* [13]. Before a negotiating party is willing to release a credential it must check that $C_i \leftarrow F_{C_i}(\bigcup_{k>i} C_k) = true, C_i \in \mathcal{C}_L, C_k \in \mathcal{C}_R$.

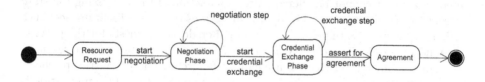

Fig. 1. State diagram of the negotiation process

The iterative exchange of *Negotiation State* messages during the automated trust negotiation contains all information about a particular negotiation process and can be evaluated without the need for session state. This is in contrast to related ATN systems which work on a tree data structure in local memory and exchange only incremental messages. The negotiation process itself is a transition of four states as depicted in the state transition diagram in Figure 1:

- **Resource Request**: The service requests access to the resource. As the resource is protected by a disclosure policy, a trust negotiation is initiated.
- **Negotiation Phase**: The objective of this phase is to find the *safe disclosure sequence* by evaluating requested credentials and their local disclosure policies.
- **Credential Exchange Phase**: This phase starts after at least one *safe disclosure sequence* was identified. The credentials that were requested most recently in the negotiation are now transmitted first. The credential exchange happens iteratively in reverse order until all credentials are disclosed.
- **Agreement**: After all required credentials were successfully exchanged, the trust negotiation terminates with a positive outcome. The objective of the negotiation is reached, for example, access to the storage service is permitted.

The *Negotiation Phase* is critical for the discovery of a safe disclosure sequence. The algorithm that processes a received *Negotiation State* is depicted

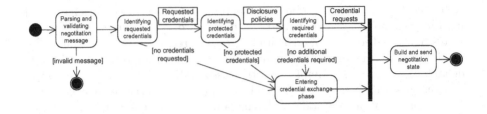

Fig. 2. Activity diagram of a processing step during the Negotiation Phase

in Figure 2. The first task is to assure syntactical and logical correctness and discard invalid messages. The next activity is to identify the requests R_i of the remote party for credentials. If a credential C_i is protected by a *Disclosure Policy* P_i, the algorithm extends the tree structure appending P_i to all leafs containing C_i in the path from root to lead. The algorithm marks leafs as *failed* that contain credentials that cannot be offered. After completion of the processing the state is sent to the other party. This algorithm iterates till a safe disclosure sequence is found, that means there are no additional credential requests for the path.

A negotiation fails during *Negotiation Phase* if the parties cannot reach an agreement. However, if there is a technical failure, or one party tries to cheat, the negotiation process can also fail at another point in time. One precaution is to exchange credentials in reverse order during the *Credential Exchange Phase*, processing the *safe disclosure sequence* in the tree from the corresponding leaf to the root. That implies that all required credentials are present and the conditions on the values of the credentials are met.

3.3 Security Aspects of the Negotiation

Security is especially challenging in trust negotiations, due to the large potential negative impact and the legal dimension of the negotiation. Both parties can protect the integrity and confidentiality against a malicious third party by using asymmetric cryptography and digital signatures with cryptographic protocols, like TLS/SSL or WS-Security.

It is more difficult to protect the negotiation against manipulations of the other negotiating party. The *VersaTrust* relies solely on the received *Negotiation State*. We are currently investigating a strategy to apply digital signatures to the *Negotiation State* to detect manipulations.

4   Experimental Results

A short overall negotiation time is important for fast service access. The outcome of one negotiation can serve as authorization for a long lasting service usage, and thereby reduce the number of required negotiations. The time for an Automated Trust Negotiation results from the iterative exchange of the negotiation messages.

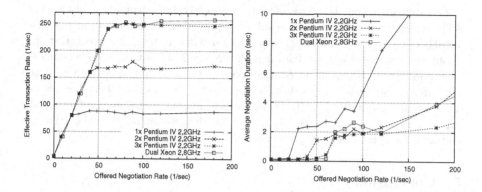

Fig. 3. Scalability under varying Load Conditions: (a) Effective Transaction Rate (b) Average Negotiation Duration

As ATN is a young direction little experience exists on the characteristics of real negotiations. We used the reference example as one test case for our measurements. It allows for a negotiation consisting of 4 transactions: 2 for the *negotiation phase* and 2 for the *credential exchange phase*. It performs additionally a constraint check on an attribute of the credential. In the first experiment, one server (2,8 GHz Dual Xeon, 2x1024KiB L2 cache) was put under stress by 5 clients (2,2 GHz Pentium IV); all running with a standard configuration of Fedora Core 4, being connected in a local area network with RTTs below 0.1 ms. Both, server and client were multi threaded to support parallel processing of requests. The clients started trust negotiations at a defined rate; each experiment lasted for 600 seconds.

The left-hand figure 3(a) shows the effective transaction rate for different negotiation rates. The Xeon server scales for up to 60 complete negotiations per second in this experiment, totaling to 240 transactions per second. Another important metric is the total negotiation time - that is the time between the construction of the request till the receipt and interpretation of the last negotiation message at the requester. Figure 3(b) shows that the average negotiation of a single server stays below 0.3 seconds for the whole negotiation till it gets into overload beyond 60 requests per second, after that point the server starts queuing.

Another experiment concerns the scalability of the system. How does the system scale with off-the-shelfe standard hardware? We used *haproxy*[1] for load balancing of up to three Pentium IV machines (see Figure 3(a)) One system can handle 80 concurrent transactions per second, two 160 and three 240, demonstrating the linear scalability of *VersaTrust*. The results in figure 3(b) show that despite the additional latency by the load balancer, the negotiation duration stays beyond 0.3 seconds besides overload conditions.

---

[1] The Reliable, High Performance TCP/HTTP Load Balancer,
http://haproxy.1wt.eu/

It is difficult to put these results into perspective; performance evaluations of ATN systems are rare. Certain results are published about a system that uses TrustBuilder in [11]. One single negotiation without integrity protection and about the release of one credential took already 7 second, and 0.5 seconds for each additional credential on comparable hardware. The comparison with the measures of our system is not fair, because we do not use X.509 certificates but much smaller proprietary XML certificates without cryptographic protection. We expect a performance decrease in our system when we introduce real certificates and cryptographic integrity protection of the negotiation.

5 Conclusion

This paper presented and studied a new Automated Trust Negotiation framework for attribute based resource access, called *VersaTrust*. Our approach reaches binding agreements by using a policy driven and privacy preserving negotiation. We introduced a novel stateless trust negotiation algorithm that operates on messages that encompass the complete negotiation state. The agreements in *VersaTrust* demonstrate the relationship between the credentials. Measurements of our prototype showed the scalability. Future work includes support of the security strategy and of other credential formats. We are hopeful that automated trust negotiation can become an important technology for the self-management of autonomic networks.

References

1. Bertino, E., Ferrari, E., Squicciarini, A.C.: Trust Negotiations: Concepts, Systems, and Languages. Computing in Science and Engineering 06(4), 27–34 (2004)
2. Bertino, E., Ferrari, E., Squicciarini, A.C.: Trust-X: A Peer-to-Peer Framework for Trust Establishment. IEEE Transactions on Knowledge and Data Engineering 16(7), 827–842 (2004)
3. Bui, H., Venkatesh, S., Kieronska, D.: An architecture for negotiating agents that learn (1995)
4. Chess, D.M., Palmer, C., White, S.R.: Security in an autonomic computing environment. IBM Syst. J. 42(1), 107–118 (2003)
5. Andrieux, A., et al.: Web Services Agreement Negotiation Specification (WS-AgreementNegotiation). Technical report, Global Grid Forum (2007)
6. Frikken, K.B., Li, J., Atallah, M.J.: Trust Negotiation with Hidden Credentials, Hidden Policies, and Policy Cycles. In: Proceedings of the Network and Distributed System Security Symposium, NDSS 2006, San Diego, California, USA. The Internet Society (2006)
7. Ganek, A.G., Corbi, T.A.: The dawning of the autonomic computing era. IBM Syst. J. 42(1), 5–18 (2003)
8. Li, N., Winsborough, W.: Towards Practical Automated Trust Negotiation. In: POLICY'02. POLICY '02: Proceedings of the 3rd International Workshop on Policies for Distributed Systems and Networks, p. 92. IEEE Computer Society Press, Washington (2002)

9. Lopes, F., Mamede, N., Novais, A.Q., Coelho, H.: A negotiation model for autonomous computational agents: Formal description and empirical evaluation (2002)
10. Nejdl, W., Olmedilla, D., Winslett, M.: PeerTrust: automated trust negotiation for peers on the semantic web (2003)
11. Olson, L., Winslett, M., Tonti, G., Seeley, N., Uszok, A., Bradshaw, J.: Trust Negotiation as an Authorization Service for Web Services. In: ICDEW'06. ICDEW '06: Proceedings of the 22nd International Conference on Data Engineering Workshops, IEEE Computer Society Press, Los Alamitos (2006)
12. Smith, B., Seamons, K.E., Jones, M.D.: Responding to Policies at Runtime in TrustBuilder. In: POLICY, pp. 149–158 (2004)
13. Winsborough, W., Seamons, K., Jones, V.: Automated Trust Negotiation. Technical report, North Carolina State University at Raleigh, Raleigh, NC, USA (2000)
14. Winsborough, W.H., Li, N.: Protecting sensitive attributes in automated trust negotiation. In: WPES '02. Proceedings of the 2002 ACM workshop on Privacy in the Electronic Society, pp. 41–51. ACM Press, New York (2002)
15. Yamaki, H., Fujii, M., Nakatsuka, K., Ishida, T.: A Dynamic Programming Approach to Automated Trust Negotiation for Multiagent Systems. rrs, 0:55–66 (2005)
16. Ye, S., Makedon, F., Ford, J.: Collaborative Automated Trust Negotiation in Peer-to-Peer Systems. In: P2P'04. P2P '04: Proceedings of the Fourth International Conference on Peer-to-Peer Computing, pp. 108–115. IEEE Computer Society Press, Washington, DC (2004)
17. Yu, T., Winslett, M., Seamons, K.E.: Interoperable strategies in automated trust negotiation. In: CCS '01. Proceedings of the 8th ACM conference on Computer and Communications Security, pp. 146–155. ACM Press, New York (2001)

Collaborative Anomaly-Based Attack Detection

Thomas Gamer[1], Michael Scharf[1], and Marcus Schöller[2]

[1] Institut für Telematik, Universität Karlsruhe (TH), Germany
[2] Computing Department, Lancaster University, UK

Abstract. Today networks suffer from various challenges like distributed denial of service attacks or worms. Multiple different anomaly-based detection systems try to detect and counter such challenges. Anomaly-based systems, however, often show high false negative rates. One reason for this is that detection systems work as single instances that base their decisions on local knowledge only.

In this paper we propose a collaboration of neighboring detection systems that enables receiving systems to search specifically for that attack which might have been missed by using local knowledge only. Once such attack information is received a decision process has to determine if a search for this attack should be started. The design of our system is based on several principles which guide this decision process. Finally, the attack information will be forwarded to the next neighbors increasing the area of collaborating systems.

1 Introduction

Today, the Internet is used by companies frequently since it simplifies daily work, speeds up communication, and saves money. But the more popular the Internet gets the more it suffers from various challenges that appear with increasing frequency. Challenges currently threatening networks include attacks like denial-of-service (DDoS) attacks [1] and worm propagations [2] besides others. DDoS attacks, for example, aim to overload a victim's resources like link capacity or memory by flooding the system with more traffic than it can process. The attack traffic is generated by many slave systems called zombies which an attacker has compromised prior to the attack. The attacker only has to coordinate all these slave systems to start the attack nearly at the same time. A DDoS attack is a distributed attack where zombies are located in various domains of the Internet. Every zombie generates only a small bandwidth attack flow to prevent detection of the zombies. This traffic runs on different routes through the Internet to the victim aggregating at intermediate systems the nearer it gets to the victim (see figure 1). Keeping the zombie systems undetected enables the reuse of them for a later attack.

Current efforts aim at detecting attack flows and blocking them to prevent them from reaching the victim. The detection systems are usually deployed in the access networks. Because of the small bandwidth at the zombies' detection systems close to them can hardly detect the attack flows and therefore are not able to block them in most cases. A detection close to the victim can still protect the victim's system against an attack but only if the detection system itself is not overwhelmed by the attack. There are two possibilities to bridge the gap between the point in the network where you want to block attack traffic and the place where you can detect it: on the one hand attack specific

D. Hutchison and R.H. Katz (Eds.): IWSOS 2007, LNCS 4725, pp. 280–287, 2007.

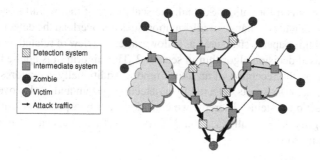

Fig. 1. Traffic aggregation during a DDoS attack

information can be exchanged in order to enable the systems close to the zombies to detect the attack traffic [3,4,5]. On the other hand, detection systems can be pushed deeper into the network if they pay special attention to the resource constraints there. The deeper the detection system is placed in the network the more attack traffic can have aggregated. A detection of this aggregated traffic is easier than a detection of the low bandwidth traffic close to a zombie.

In [6] we proposed an anomaly-based detection system that can be deployed within the network in order to detect adverse events as early as possible. This ensures a fast reaction and therefore, an effective protection of the victim. Furthermore, the network itself and its resources can be better protected by such a detection system since detection takes place on routers within the network instead of at the victim's edge. We identified two disadvantages of such an approach: unfavorable aggregation of attack traffic and upstream activated countermeasures. If attack flows aggregate only close to the victim our approach suffers from the same effects as deployment of detection systems close to the edge of the network. An early detection is unlikely. Furthermore, if one of our systems detects an attack flow and starts blocking it detection systems downstream will experience attack traffic with a smaller bandwidth. Again, this decreases the likelihood of detection. As a consequence anomaly-based detection systems that decide on the existence of an adverse event based on local knowledge only show false negative errors, i.e., some adverse events are not detected. These detection problems could be solved or at least diminished if the knowledge of multiple detection systems can be shared and thereby detection systems collaborate in a self-organized manner. Therefore, we propose to combine in-network deployment of detection systems with information exchange in order to build an effective system that detects and prohibits adverse events.

By combining local knowledge and remote information we built a system that organizes itself and that enables each node to autonomously decide if the suspicious traffic is an adverse event or just a legal traffic anomaly. Thereby, a coordinated collaboration of independent systems is achieved. Furthermore, the detection of adverse events is improved by the fact that a false negative of one detection system is compensated by exchanging information between neighborhood detection systems. Thus, the probability of detecting adverse events increases. But such an information exchange comes with risks, too. First, an attacker can try to launch an attack on the detection system itself by injecting bogus information. Second, the scalability of the overall system must be guaranteed so that the exchange of information does not overload the detection systems.

The rest of this paper is structured as follows: after a review of related work we detail the metric-based decision algorithm in section 2. This algorithm is applied for reasonably reacting on the reception of an attack report. Additionally, we address the issues of describing adverse events detected by a detection system and of discovering neighborhood detection systems. Thereby, we consider security aspects like authentication, trust, or integrity, too. Finally, section 3 gives a short summary about this paper and mentions future work.

1.1 Related Work

Today there is a great research effort in intrusion detection systems. Common intrusion detection techniques can be divided into misuse and anomaly detection [7]. Misuse detection, e.g. snort [8], relies on signatures that define byte patterns of known attack packets. They provide a low false positive rate but are not able to detect previously unknown or protocol-conform attacks. Anomaly detection systems like NSOM [9] or NETAD [10], on the other side, monitor network traffic and search for anomalous behavior, e.g. by applying neural networks or threshold-based mechanisms. Thus, they are able to detect previously unknown attacks at the expense of a higher false positive and false negative rate. These and other similar intrusion detection systems [11,12,13] use local knowledge to form a local opinion on an observed traffic flow. So they consider a detection system just as a single instance. and do not use the possibility of exchanging information with other detection systems.

There also exist other approaches like Emerald [14], INDRA [15] or CITRA [16] that deal with the coordination of distributed detection systems and use its advantages. The INDRA-Project proposes the cooperation of different detection systems through a subscription-based group communication with a peer-to-peer architecture. CITRA-devices report detected attacks to a central Discovery Coordinator that coordinates countermeasures based on a complete view on the network. Such frameworks and infrastructures are able to detect domain-wide threats and thus, could improve their detection reliability. But they have to deal with higher communication efforts and close trust relationships between the involved entities in order to prevent an attacker from abusing the detection system. Additionally, the communication of these approaches often is based on a central control entity.

At last, there exist several other approaches to cope with adverse events like DDoS attacks. Traceback techniques like Itrace [3] or SPIE [4] allow to identify the origin of an attack even in case that spoofed source addresses are used by zombies. The Pushback [5] mechanism examines congestion situations as an indication for a DDoS attack and reacts to it with a request to preceding routers to initiate rate limiters. This could defuse the congestion situation at the victim.

Our approach has the advantage that it can detect and react to adverse events during the build-up of the event. Furthermore, different detection systems could inform each other about their recognitions and thus, enhance their detection reliability without need for a central control entity. The fact that detection systems need not trust each other absolutely reduces the requirements for security aspects.

2 Design

The self-organizing extension for our detection system can be separated into three parts: first of all, a neighbor discovery has to be performed. Afterwards, a system communicates its information to a neighborhood detection system. The receiving detection system then autonomously has to decide on how to react to this information. Therefore, we propose a metric-based decision algorithm. Third, the adverse event detected by the sending system has to be described in a comprehensible way, so that all detection systems involved in the coordination are able to process a message correctly.

2.1 Neighbor Discovery

In section 1 we mentioned that a self-organizing collaboration of detection systems is necessary to improve an anomaly-based detection of adverse events. Such a collaboration is based on an exchange of information between neighborhood detection systems. Therefore, a detection system must be able to discover neighborhood detection systems in order to communicate the locally gathered information. Having discovered a neighborhood detection system a communication channel can be established that must have certain characteristics specified by the sender, e.g. reliable message transfer. Before exchanging the available information a security association – authentication of communicating systems or data integrity – should be established in order to prevent an exchange of wrong or forged information and DoS attacks against the detection system itself.

We believe that detection systems are sparsely distributed in the Internet and thus, two detection systems are rarely connected directly with each other. Furthermore, a detection system in our opinion has no detailed knowledge about the whole topology and the distribution of all detection systems in the Internet since only a minimal amount of long-lived state information should be maintained. Additionally, a neighbor discovery mechanism has to regard dynamics of the Internet, e.g. changing of routes or dynamically activated detection systems which cause new neighbor relationships. Thus, we need a mechanism that is able to locate neighborhood detection systems on demand. In order to discover neighborhood detection systems multiple methods are possible: expanding ring search, path-coupled mechanisms, overlay networks, and others.

With an expanding ring search [17], for example, the metric that defines the notion *neighborhood detection system* in most cases is a maximal hop count. A problem of most expanding ring search mechanisms currently deployed is that no security is provided. Another approach that provides discovery of neighborhood detection systems are path-coupled mechanisms as provided by the signaling framework NSIS [18]. This framework additionally enables a sender to specify communication requirements, e.g. reliable and confidential message exchange.

2.2 Metric-Based Decision

In our previous work as well as in many related approaches an anomaly-based detection system represents an independent and autonomous network device that detects adverse events based on local measurement. We propose to reduce the false negative rate of such an anomaly-based detection system by a self-organizing exchange of information about adverse events already detected elsewhere in the network. The system we have developed is based on the following design principles:

Verify received information. *No system should commit itself to a tight trust relationship with other detection systems but rather rely on its own recognitions.*

This ensures robustness against bogus information as well as message injection attacks due to local verification of received information. Furthermore, a close trust relationship between detection systems would constrain flexibility in dynamic environments and cause an increased overhead.

Consider current workload. *The available resources of the detection system limit the number of parallel executed detection threads.*

If the detection system is heavily loaded, i.e. it already does multiple fine-grained detections in parallel, it should reject an additional parameterized detection to prevent overload situations.

Rate report granularity. *The more fine-grained the data in the attack report is the fewer resources must be spent on its verification.*

If only few characteristics of the adverse event are known, i.e. the anomaly description is rather coarse-grained, the receiver must apply some stages of refinement. Thus, the verification of received information may waste valuable free resources.

Count duplicates. *The more systems report an ongoing attack the higher is the likelihood to detect the attack locally, too.*

If the same adverse event is reported by different detection systems the importance of the information increases dependent on the number of duplicate messages.

Check significance. *The traffic volume of the detected attack must be compared to the overall traffic on the detection system.*

If the sender of a message, compared to the receiver, only scans a rather small amount of traffic for anomalies the adverse event he reports may be negligible for the receiver. The message of a detection system that is located at the edge of the network and processes an average traffic amount of 100 Mbit/s, for example, is nonsignificant for an in-network detection system that processes 5 Gbit/s. But the information about a detected attack of this in-network system is of great significance to the edge system. If, however, multiple detection systems at the edge report the same attack, the significance of this information increases due to the principle *count duplicates.*

Measure distance. *The farther apart two neighborhood detection systems are the more likely the attack will be detected at the system receiving an attack report.*

The distance between two neighborhood detection systems in regard to IP hops recommends a parameterized detection since the probability that attack traffic aggregates between these system is the higher the longer the distance between the communicating detection systems is. This also increases the probability that a receiving detection system is able to detect the reported adverse event even if preceding systems apply countermeasures since attack traffic may be present again due to aggregation in intermediate systems.

Use verification failure history. *An attack report that has been checked unsuccessfully by the predecessors is not likely to be detected.*

If the message has passed several detection systems that scanned for the adverse event described without detecting it the message becomes less important. Additionally, the message is discarded after a maximum number of failed verifications in order to keep communication localized.

Use verification history. *If an attack report has been forwarded unchecked by the predecessors the distance to the detection instance which verified the adverse event last must be regarded.*

This ensures that communication of information about an adverse event does not run endlessly without being verified by a neighborhood detection system. Therefore, importance of a message grows the more often a verification is refused by a receiving detection system. In combination with the parameter verification failure history this parameter guarantees that a verification is done once in a while and thus, communication ends after a certain time if the adverse event reported cannot be verified at multiple systems.

Having considered all aspects described previously the detection system receiving an attack report should react in the following way:

– The system has to check if it has already detected the reported adverse event on its own. If so it can silently discard the message because it has informed its neighborhood systems before.
– If the system decides to start an anomaly detection that is parameterized by the message content and the verification succeeds it should start appropriate countermeasures. Furthermore, it communicates its own recognitions to a neighborhood detection system. If the verification, however, fails it updates the received message with its local knowledge, e.g. the verification failure history, and then forwards the message. Additionally, the content of the message is stored for a certain time in order to detect duplicate messages in the future, i.e., a soft-state approach is applied.
– If the system decides not to start an anomaly detection and the adverse event reported is not yet known to the receiving system it updates the received message with its local knowledge, e.g. the verification history, and then forwards the message. The content of the message is stored in order to detect duplicate messages.
– If the system receives a duplicate message the action depends on former decisions. If the message reports an adverse event the receiving system did not verify before it must reconsider its decision. In case that the parameters now recommend a verification and this verification is successful the system communicates its new knowledge to a neighborhood system. Otherwise – if the verification fails or if a verification was already done earlier – the message is discarded without doing something since it was forwarded before and no new knowledge is available.

In order to achieve the behavior described above a receiver needs not only the description of the adverse event but some additional information: current workload, report granularity and the number of duplicate messages can be obtained based on local knowledge. The distance between neighborhood detection systems has to be obtained by neighbor discovery (see section 2.1). Other parameters like significance of the sending detection system, verification history and verification failure history must be transmitted in addition to the description of the adverse event.

2.3 Description of Adverse Events

Having received an attack report the detection system may decide, based on the metric-based decision described in the previous section, to start a fine-grained detection itself if it is able to interpret the message correctly. At this point it must be considered that detection systems of different domains may scan for different protocol anomalies, i.e., not all systems necessarily must know the same anomalies due to different local policies or knowledge. Thus, a message possibly contains information the receiver cannot understand. Therefore, a generic and extensible message format, e.g. based on a type-length-value (TLV) structure, must be used for description of detected anomalies. In this case different data records are differentiated by the type field. The length field indicates the byte length of the following data field and thereby, enables a system to skip unknown data records. Thus, a receiving system extracts only that information it is able to understand and ignores unknown data records. Another approach that enables a flexible and extensible description of anomalies is the usage of a structured description language like XML [19].

3 Summary and Outlook

In this paper we presented a collaboration of anomaly-based detection systems which use local knowledge and measurements and combine them with remote information received by their neighbors. After establishing communication channels with detection systems in the neighborhood attack information can be freely exchanged. The reaction to such a message is determined by a metric-based decision process. The design of this decision process is guided by a set of principles. Depending on available resources, trust, and history of the message the system starts a search for the described attack locally to verify the message, forwards the message unprocessed or drops it silently.

We additionally implemented the collaborative attack detection proposed in this paper. Therefore, we extended our detection system [6] with a decision and a communication engine. Neighbor detection is implemented as an external process in order to allow for transparent addition of new neighbor detection and security mechanisms. Some first evaluations in different usage scenarios show how the decision process derives a reaction to an attack information received from a neighborhood detection system. But these evaluations representing small simulated environments only describe the microscopic behavior of such a collaboration of detection systems. Thus, future work has to address the question how the collaboration behaves in a macroscopic scenario. We plan to implement our anomaly-based detection system as well as the extension we proposed in a network simulator which allows a simulation of more and larger networks. This macroscopic analysis enables an examination how each single design principle influences the metric-based decision algorithm and thus, enables the choice of an optimal decision function. Additionally, it will show which effect the collaboration has on the false negative rate. Finally, different neighbor discovery mechanisms have to be evaluated and some work has to be done regarding the description of adverse events that are only known domain-wide and have to be communicated to a detection system outside the domain of the sending system.

References

1. Hussain, A., Heidemann, J., Papadopoulos, C.: A framework for classifying denial of service attacks-extended. Technical report, USC/Information Sciences Institute (2003)
2. Shannon, C., Moore, D.: The spread of the witty worm. IEEE Security and Privacy 2(4), 46–50 (2004)
3. Bellovin, S., Leech, M., Taylor, T.: Icmp traceback messages. Internet draft, Internet Engineering Task Force, Work in Progress (2003)
4. Snoeren, A.C.: Hash-based IP traceback. In: SIGCOMM, pp. 3–14 (2001)
5. Mahajan, R., Bellovin, S.M., Floyd, S., Ioannidis, J., Paxson, V., Shenker, S.: Controlling high bandwidth aggregates in the network. SIGCOMM Computer Communication Review 32(3), 62–73 (2002)
6. Gamer, T.: A system for in-network anomaly detection. In: Kommunikation in Verteilten Systemen, February 2007, pp. 275–282. Springer, Heidelberg (2007)
7. Kumar, S.: Classification and Detection of Computer Intrusions. PhD thesis, Purdue University (1995)
8. Roesch, M.: Snort, intrusion detection system (1999), http://www.snort.org
9. Labib, K., Vemuri, V.R.: NSOM: A tool to detect denial of service attacks using self-organizing maps (2004)
10. Mahoney, M.V.: Network traffic anomaly detection based on packet bytes. In: Proceedings of the ACM symposium on Applied computing (SAC), pp. 346–350. ACM Press, New York (2003)
11. Lakhina, A., Crovella, M., Diot, C.: Diagnosing network-wide traffic anomalies. In: Proceedings of the 2004 conference on Applications, technologies, architectures, and protocols for computer communications (SIGCOMM), pp. 219–230 (2004)
12. Paxson, V.: Bro: a system for detecting network intruders in real-time. Compututer Networks 31(23-24), 2435–2463 (1999)
13. Wang, K., Stolfo, S.J.: Anomalous payload-based network intrusion detection. In: Jonsson, E., Valdes, A., Almgren, M. (eds.) RAID 2004. LNCS, vol. 3224, pp. 203–222. Springer, Heidelberg (2004)
14. Porras, P.A., Neumann, P.G.: EMERALD: Event monitoring enabling responses to anomalous live disturbances. In: Proc. 20th NIST-NCSC National Information Systems Security Conference, October 1997, pp. 353–365 (1997)
15. Janakiraman, R., Waldvogel, M., Zhang, Q.: Indra: A peer-to-peer approach to network intrusion detection and prevention. In: Proceedings of 12th IEEE Workshops on Enabling Technologies, Infrastructure for Collaborative Enterprises (WETICE), June 2003, pp. 226–231. IEEE Computer Society Press, Los Alamitos (2003)
16. Schnackenberg, D., Holliday, H., Smith, R., Djahandari, K., Sterne, D.: Cooperative intrusion traceback and response architecture (CITRA). In: Proceedings of the DARPA Information Survivability Conference and Exposition (DISCEX), June 2001, pp. 56–68 (2001)
17. Boggs, D.R.: Internet Broadcasting. PhD thesis, Stanford University (1982)
18. Hancock, R., Karagiannis, G., Loughney, J., den Bosch, S.V.: Next steps in signaling (NSIS): Framework. RFC 4080, Internet Engineering Task Force (2005)
19. Bray, T., Paoli, J., Sperberg-McQueen, C.M., Maler, E., Yergeau, F., Cowan, J.: Xml 1.1, 2nd edn. W3C recommendation, W3C (2006)

Modeling and Management of Service Level Agreements for Digital Video Broadcasting(DVB) Services

Thapelo Tlhong and Jeff S. Reeve

Electronics and Computer Science
University of Southampton
SO17 1BJ United Kingdom

Abstract. This paper describes a metamodeling strategy of Service Level Agreements for Digital Video Broadcasting services based on Service Level Agreement Language(SLAng). The purpose of the paper is to provide a detailed analysis of SLAs in this domain and provide a motivation for modeling and automating their management. We also discuss why precise and machine readable SLAs can improve the levels of automation in SLA Management thereby reducing potential violations. The meta-modeling approach based on the Model Driven Architecture(MDA) described in this paper also simplifies the integration of a SLA Management systems with other infrastructure that delivers the service to the client.

1 Introduction

Service Level Agreements(SLAs) have traditionally been considered as a legal binding between a service provider and a customer. However, the advent of Service Oriented Architectures(SOA) and service based business models has seen the IT industry move away from considering SLAs only as a legal document but instead as means of enforcing and managing user requirements and expectations. As proposed in [1] a SLA can be used as a basis for the specification and development of a contracted service. Hence, it is necessary to integrate the specification and management of SLAs with other systems that are involved in the development, provision, maintenance and management of the service. A potential benefit of this integration is the ability to monitor the real time conformance of the service performance and related metrics to the SLA requirements. This can enable service reconfiguration and adjustment to minimise SLA violations. Traditionally SLAs are defined in legalese whilst the actual service being constrained by the SLA is specified, designed and implemeted in a technical context [2]. This paper presents an analysis of requirements for modeling SLAs in a Digital Video Broadcasting domain and a modeling strategy based on SLAng [3].

D. Hutchison and R.H. Katz (Eds.): IWSOS 2007, LNCS 4725, pp. 288–294, 2007.
© Springer-Verlag Berlin Heidelberg 2007

2   Related Work

2.1   SLAng

Service Level Agreement Language(SLAng) [4] forms the basis of the our work on SLA. However, the original SLAng was developed for Web (application) Services. The language is based on OMG's Meta Object Facility(MOF)[1] and Object Constraint Language(OCL)[2]. SLAng's dependence on MDA standards(XMI, UML, MOF and OCL) means that the extensibility mechanisms offered by such standards can also be utilized in extending SLAng. Moreover, the separation of concepts between the domain of the language and the language itself used in specifying the language enables SLAng to be extended to different domains for which it was not originally defined. This makes it an appropriate candidate for adaptating existing specifications for DVB services.

2.2   X-Contract

In [5] a contract representation scheme is proposed based on the Finite State Machine(FSM) semantics. In this strategy, a contract is interpreted as a set of rights and obligations which are divided into sets R and O of Rights and Obligations respectively for each signatory party in the contract. According to [6] a right is defined as *an action that a signing entity can do if it wishes to* and an obligation is *a duty that an entity is expected to perform*. Consequently failure to perform an obligation may result in a breach of the contract or violation. The designers of the language propose a state machine-like approach implementation of the contracts(SLAs) written in the language, by having one state machine for the R and another for O per party. The violations of or conformance to the SLA can then be inferred from any given state in the FSM. This approach might be sufficient for small scale SLAs but for large systems of SLAs it will lead to complicated state machines that are not maintainable.

2.3   RBSLA

The Rule Based Service Level Agreement Language is a rule based modeling and implementation of SLAs [2] based on Predicate Logic and Event calculus. The language extends RuleML with concepts that enable SLA specification to be written as accurately as possible. These concepts include Event, Condition and Action model, Procedural Attachments, External Data Integration and Typed Logic, which enable assigning logical terms a type. The formal approach taken in defining RBSLA provides accurate type and constraints checking as well as allowing formal analysis of SLA written in the language. The dependence of the language on the use of rules for the specification of SLAs restricts the deployment of such SLAs to rule based systems which is in contradiction implementation independence we propose in Section 4.1.

---

[1] http://www.omg.org/cgi-bin/doc?formal/2006-01-01
[2] http://www.omg.org/docs/ptc/03-10-14.pdf

3   DVB over Terrestrial Services

DVB[3] is a standard based on the MPEG-2 video compression scheme for the provision of digital video services [7]. Since DVB has several flavours and their associated standards - Satellite(DVB-S), Terrestrial(DVB-T), Cable(DVB-C) and Handheld(DVB-H), the scope of this paper is on Digital Terrestrial Television (DVB-T) services. Traditionally the broadcaster or content producer was also the transmission provider. However, this operational model has changed over recent years. This has lead to a fragmentation of the broadcast transmission chain as shown in Figure 1. Compared to traditional analogue video broadcasting,

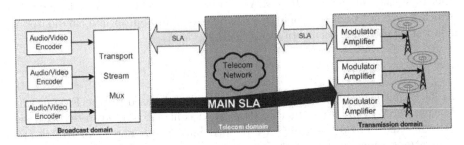

Fig. 1. DTT transmission chain showing inter-domain and intra-domain SLA interfaces

DVB offers the ability to deliver high quality video and audio, subtitles, audio descriptions, wide screen signalling, MHEG data services, Conditional Access Control and Schedule Management in the form of Electronic Programme Guide(EPG). The multipartite service composition means that the potential for errors occuring in the service increases since error probability varies linearly with the number of service components. Moreover, some broadcast services are time-shared or time-exclusive, which means that multiple services can share a single service slot within the service multiplexer and the switching between the services is performed on time based schedules. As the signal traverses the chain from one domain to the other it is necessary to have well defined metrics and parameters that describe the service at each domain interface. This requires well specified and precise SLAs that explicitly state the roles and responsibilities of each party at each service interface point.

4   DVB SLAs

A DVB Service traverses at least three domains - Broadcast, Telecom and Transmission, before reception by the viewer(Figure 1). The interfaces between these domains represent inter-domain SLAs and vertically within each domain there is also intra-domain SLAs. In the context of this paper the Broadcast(DVB) SLA refers to a bilateral agreement between a transmission service provider and

---

[3] http://www.dvb.org

a broadcaster that controls the management, performance, quality and monitoring of the service (denoted MAIN-SLA in Figure 1). We informally define such SLA as a set of contract clauses that specify and constrain the behaviour of entities along the transmission chain to ensure reception of a decodable and viewable service.

Several languages and Information Models for SLAs have been proposed. Instead of proposing a new model, we chose to build on an existing model(SLAng) and use such to evaluate the specification and management of DVB SLAs. The choice of the relevant strategy is based on the following requirements that we impose on the existing strategies.

4.1   Requirements for SLA Modeling and Specification

Integration - SLA specification language should ease the integration of the SLA management framework with other subsystems that cooperate to provide the service to the customer.

Machine Readability- is the basis for automating the management of service level agreements and enabling exchange of SLA specific information with other systems involved in the provision of the service e.g. Fault Management and Correlation systems.

Automatability - is meant to ensure a more proactive monitoring of the runtime system that provides the service to the SLA requirements. Automatability is necessary for a system of SLAs to self manage and adapt itself to minimise violations.

Implementation Independence - It is necessary to describe SLAs in a language that is technology neutral and platform independent to cohere with the Integration requirement described above.

Reusability - is necessary to allow SLAs for different domains(e.g. Web Services, Broadcast, Telecoms) to be expressed in the specified language without requiring extensive modification to the language.

Extensibility - an SLA language/specification should provide extension mechanisms that allows language users to extend it for other peculiar service domains and service provision scenarios for which it was not originally defined. This requirement is an extension to the Reusability requirement discussed above.

Other requirements for SLA specifications are described in detail in [3] and these include Precision, Expressiveness, Understandability and Analysability.

5   DVB SLA Model

This section describes the adaptations and extensions we made to SLAng to make it appropriate for modeling SLAs for DVB services. As per the SLAng paradigm, we have two sets of models, the Domain and the Service models as described in the next section.

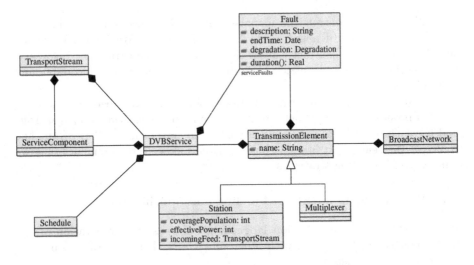

Fig. 2. Domain model for which Transmission service SLAs apply

5.1   Broadcast Domain Model

A broadcast SLA imposes constraints on the behaviour of entities and services within the broadcast domain, as such it is imperative to first model the domain on which the SLA constraints apply. The structure and relationship between the entities within the domain are shown in Figure 2. A DVB service is contained within an MPEG2 Transport Stream which also contains service components and transport level tables that enable the services in the stream to be decoded. The service is carried over a transmission network by different types of transmission elements. The domain model provides the context for the constraints and enables us to attach semantic definitions to the SLA elements in Figure 3. For instance, a *StationClause* and *StationDefinition* are used to express constraints in OCL on the behaviour of Station objects.

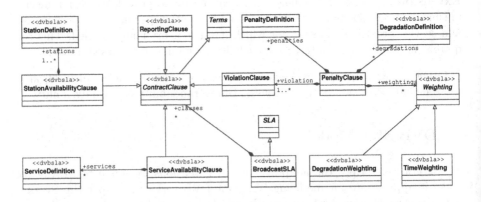

Fig. 3. Service(SLA) model for a broadcast SLA

In the original SLAng model, parties to the SLA are defined as entities that are signatory to the contract. However, we found this approach to be limiting in terms of modeling other relationships that exist in the the SLA. Instead of this approach, we model a SLA Party as an individual or enterprise whose "behaviour" can affect the conformance of the service to the SLA. Roles are then attached e.g. ProviderRole, MonitoringRole, ClientRole to these parties as required. This approach enables us to model complex interactions of parties involved in the SLA and also enables Accountability for observed service behaviour to be assigned to the appropriate party based on their role.

```
abstract  class  services :: BroadcastService  extends
          :: slang :: ServiceDefinition
{
          serviceFaults  :  Fault[0,*] opposite sLA unique
          serviceFaults  :  Fault[0,*] unique
          components  :  ServiceComponent[1,*]
          status  :  :: types :: ServiceStatus

          invariant {
                    serviceFaults ->forAll(f1  :  Fault,  f2  :  Fault |
                            f1.degradation  =  f2.degradation
                            and f1.startTime  =  f2.startTime
                            and f1.endTime  =  f2.endTime
                            implies f1  =  f2
                    )
          }
}
```

Listing 1.1. Sample EMOF/OCL Specification for a BroadcastService

5.2 Constraints Checking

The textual syntax of the SLA is in EMOF and OCL. Listing 1.1 shows a sample specification for *DVBService* and an invariant that specifies that any accountable fault affecting the service must not be duplicated. This is to restrain from applying penalties for a single fault occurence multiple times. The specification is converted into a JMI [4] repository using the UCLMDA Tools [8]. Constraints checking is performed by creating instances of the SLA from the repository and populating the instance fields with SLA requirements e.g. availability values for a given time period. Each entity in the specification has an embedded constraint checking. For example, an instance of the *BroadcastService* shown in Listing 1.1 with have an methods, *verifyConstraints()* and *verifyConstraintsDeep()* that when called will verify that there are no duplicates for all faults affecting this service.

---

[4] http://java.sun.com/products/jmi/index.jsp

6 Conclusion and Further Work

In this paper we have presented the an detailed analysis of service level agreements in the broadcast domain and also the motivation for focusing on broadcast services, one of which is the lack of research or publications concerning management of SLAs in this domain. We also discussed the challenges posed by the transition from analogue to digital video broadcasting. The metamodeling approach for specifying broadcast SLAs adopted in this paper based on SLAng language provides a basis for developing an SLA Management framework that we can use to study the effects and level of automation SLA management that can be achieved based on this approach. Further work will focus on the integration of a machine readable broadcast SLA and predictive modeling techniques to determine how previous SLA impact information can be used to predict the potential of SLA violation based on classification of service impact. This can enable the system to reconfigured adaptively against potential violations instead of reconfiguring the system only based on the current status of different SLAs.

References

[1] Bournan, J.: Specification of service level agreements clarifying concepts on the basis of practical research. Software Technology and Engineering Practice, 1999, pp. 169–178 (1999)

[2] Paschke, A.: A declarative rule-based service level agreement language based on ruleml. In: International Conference on Intelligent Agents, Web Technology and Internet Commerce, University of Munich, November 2005, vol. 2, pp. 308–314 (2005)

[3] James, S.: Slang - language for service level agreements. Ph.D. dissertation, Department of Computer Science, Univerity College London, Malet Place, London WC1E 6BT, UK (2006)

[4] Lamanna, D.D., Skene, J., Emmerich, W.: Slang: a language for defining service level agreements. In: FTDCS 2003. Proceedings the Ninth IEEE Workshop on Future Trends of Distributed Computing Systems, pp. 100–106. IEEE Computer Society Press, Los Alamitos (2003)

[5] Molina-Jimenez, C., Shrivastava, S., Solaiman, Warne, J.: Contract representation for run-time monitoring and enforcement. In: Proceedings IEEE International Conference on E-Commerce. CEC 2003, pp. 103 – 110 (2003),
http://dx.doi.org/10.1109/COEC.2003.1210239

[6] Molina-Jimenez, C., Shrivastava, S., Solaiman, E., Warne, J.: Run-time monitoring and enforcement of electronic contracts. Electronic Commerce Research and Applications 3(2), 108–125 (2004),
http://dx.doi.org/10.1016/j.elerap.2004.02.003

[7] Dubery, P., Wilson, D.: Policing slas for digital video. Cable and Satellite International, Tech. Rep. (May 2003) [Online]. Available:
http://www.cable-satellite.com/features/may_jun%2003/csi%2029_30.pdf

[8] Skene, J.: The uclmda tools, robust implementation of omg standards for research and development. Open Source Project

Author Index

Lecture Notes in Computer Science

Sublibrary 5: Computer Communication Networks and Telecommunications

For information about Vols. 1– 4725
please contact your bookseller or Springer

Vol. 4003: Y. Koucheryavy, J. Harju, V.B. Iversen (Eds.), Next Generation Teletraffic and Wired/Wireless Advanced Networking. XVI, 582 pages. 2006.

Vol. 3996: A. Keller, J.-P. Martin-Flatin (Eds.), Self-Managed Networks, Systems, and Services. X, 185 pages. 2006.

Vol. 3976: F. Boavida, T. Plagemann, B. Stiller, C. Westphal, E. Monteiro (Eds.), NETWORKING 2006. Networking Technologies, Services, and Protocols; Performance of Computer and Communication Networks; Mobile and Wireless Communications Systems. XXVI, 1276 pages. 2006.

Vol. 3970: T. Braun, G. Carle, S. Fahmy, Y. Koucheryavy (Eds.), Wired/Wireless Internet Communications. XIV, 350 pages. 2006.

Vol. 3964: M.Ü. Uyar, A.Y. Duale, M.A. Fecko (Eds.), Testing of Communicating Systems. XI, 373 pages. 2006.

Vol. 3961: I. Chong, K. Kawahara (Eds.), Information Networking. XV, 998 pages. 2006.

Vol. 3912: G.J. Minden, K.L. Calvert, M. Solarski, M. Yamamoto (Eds.), Active Networks. VIII, 217 pages. 2007.

Vol. 3883: M. Cesana, L. Fratta (Eds.), Wireless Systems and Network Architectures in Next Generation Internet. IX, 281 pages. 2006.

Vol. 3868: K. Römer, H. Karl, F. Mattern (Eds.), Wireless Sensor Networks. XI, 342 pages. 2006.

Vol. 3854: I. Stavrakakis, M. Smirnov (Eds.), Autonomic Communication. XIII, 303 pages. 2006.

Vol. 3813: R. Molva, G. Tsudik, D. Westhoff (Eds.), Security and Privacy in Ad-hoc and Sensor Networks. VIII, 219 pages. 2005.

Vol. 3462: R. Boutaba, K.C. Almeroth, R. Puigjaner, S. Shen, J.P. Black (Eds.), NETWORKING 2005. XXX, 1483 pages. 2005.